D1766016

Digital Power Electronics and Applications

Digital Power Electronics and Applications

Fang Lin Luo
Hong Ye
Muhammad Rashid

ELSEVIER

AMSTERDAM • BOSTON • HEIDELBERG • LONDON • NEW YORK • OXFORD
PARIS • SAN DIEGO • SAN FRANCISCO • SINGAPORE • SYDNEY • TOKYO

This book is printed on acid-free paper

Elsevier Academic Press
525 B Street, Suite 1900, San Diego, California 92101-4495, USA
http://www.elsevier.com

Elsevier Academic Press
84 Theobald's Road, London WC1X 8RR, UK
http://www.elsevier.com

Library of Congress Control Number: 2005929576

British Library Cataloguing-in-Publication Data
A catalogue record for this book is available from the British Library.

ISBN-13: 978-0-12-088-757-6
ISBN-10: 0-1208-8757-6

Typeset by Charon Tec Pvt. Ltd, Chennai, India
www.charontec.com
Printed and bound in United States
06 07 08 9 8 7 6 5 4 3 2

Contents

Preface

The purpose of this book is to provide a theory of Digital Power Electronics and its applications. It is well organized in 400 pages and over 300 diagrams. Traditionally, Power Electronics is analyzed by the analog control theory. For over a century, people have enjoyed analog control in Power Electronics, and good results in the analog control and its applications in Power Electronics mislead people into an incorrect conclusion that Power Electronics **must** be in analog control scheme. The mature control results allowed people to think that Power Electronics is a sunset knowledge. We would like to change these incorrect conclusions, and confer new life onto the traditional Power Electronics. In this book the authors initially introduce the digital control theory applied to Power Electronics, which is completely different from the traditional control scheme.

Power Electronics supplies electrical energy from its source to its users. It is of vital importance to all of industry as well as the general public – just as the air that we breathe and water that we drink are taken for granted, until they are no longer available, so it is with Power Electronics. Therefore, we have to carefully investigate Power Electronics. Energy conversion technique is the main focus of Power Electronics. DC and AC motor drive systems convert the electrical energy to mechanical energy and vice versa. The corresponding equipment that drives DC and AC motors can be divided into four groups:

- AC/DC rectifiers;
- DC/AC inverters;
- DC/DC converters;
- AC/AC (AC/DC/AC) converters.

All of the above equipment are called power supplies. They are switching circuits working in a discrete state. High-frequency switch-on and switch-off semiconductor devices allow switching circuits to have the advantage of high power rate and efficiency, low cost, small size and high power density. The size of a flat-transformer working in 250 kHz is much less than 1% of the volume of a normal transformer working in 50 Hz with the same power rating. Switching circuits perform in switching-on and switching-off states periodically. The switching period, T, is the sampling interval ($T = 1/f$), where f is the switching frequency. Switching circuits, including all converters, transfer energy from a source to the end-users in discontinuous manner; i.e. the energy is not continuously flowing from a source to load. The energy is pumped by energy-quantization via certain energy-storage elements to load in a sampling interval.

In order to apply digital control theory to Power Electronics, the authors define new parameters such as the energy factor (EF), pumping energy (PE), stored energy (SE), time constant, τ, and damping time constant, τ_d. These parameters are totally different from the traditional parameters such as the power factor (PF), power transfer efficiency (η), ripple factor (RF) and total harmonic distortion (THD). Using the new parameters we successfully describe the characteristics of the converters' systems. Correspondingly, new mathematical modeling has been defined:

- A zero-order-hold (ZOH) is used to simulate all AC/DC rectifiers.
- A first-order-hold (FOH) is used to simulate all DC/AC inverters.
- A second-order-hold (SOH) is used to simulate all DC/DC converters.
- A first-order-hold (FOH) is used to simulate all AC/AC (AC/DC/AC) converters.

The authors had successfully applied the digital control theory in the AC/DC rectifiers in 1980s. The ZOH was discussed in digitally-controlled AC/DC current sources. Afterwards, the FOH was discussed in digitally-controlled DC/AC inverters and AC/AC converters. Finally, the SOH has been discussed in digitally-controlled DC/DC converters. The energy storage in power DC/DC converters have carefully been paid attention to and the system's characteristics have been discussed, including the fundamental features: system stability, unit-step response and impulse-response for disturbance.

These research results are available not only for all types of the converters, but for other branches in Power Electronics as well. We describe the digital control scheme in all types of the converters in this book, and some applications in other branches such as power factor correction (PFC) and power system synchronous static compensation (STATCOM). Digital Power Electronics is a fresh theory and novel research method.

We hope that our book attracts considerable attention from experts, engineers and university professors and students working in Power Electronics. This new control scheme could be described as fresh blood injected into the traditional Power Electronics field, and hopefully may generate new development. Therefore, this book is useful for both engineering students and research workers.

Fang Lin Luo
Hong Ye
Muhammad Rashid

Autobiography

Dr. Fang Lin Luo (IEEE M'84, SM'95) received a Bachelor of Science Degree, First Class with Honours, in Radio-Electronic Physics at the Sichuan University, Chengdu, Sichuan, China and his Ph.D. Degree in Electrical Engineering and Computer Science (EE & CS) at Cambridge University, England, UK in 1986.

Dr. Luo was with the Chinese Automation Research Institute of Metallurgy (CARIM), Beijing, China as a Senior Engineer after his graduation from Sichuan University. In 1981 and 1982, he was with the Enterprises Saunier Duval, Paris, France as a Project Engineer. Later, he worked with Hocking NDT Ltd, Allen-Bradley IAP Ltd and Simplatroll Ltd in England as a Senior Engineer, after he received his Ph.D. Degree from Cambridge University. He is with the School of Electrical and Electronic Engineering, Nanyang Technological University (NTU), Singapore, and is a Senior Member of IEEE.

Dr Luo has published seven teaching textbooks and 218 technical papers in IEEE-Transactions, IEE-Proceedings and other international journals, and various international conferences. His present research interest is in the Digital Power Electronics and DC and AC motor drives with computerized artificial intelligent control (AIC) and digital signal processing (DSP), and DC/AC inverters, AC/DC rectifiers, AC/AC and DC/DC converters.

Dr. Luo was the Chief Editor of the international journal, *Power Supply Technologies and Applications*. He is currently the Associate Editor of the *IEEE Transactions* on both *Power Electronics* and *Industrial Electronics*.

Dr. Hong Ye (IEEE S'00-M'03) received a Bachelor Degree (the first class with honors) in 1995 and a Master Engineering Degree from Xi'an Jiaotong University, China in 1999. She completed her Ph.D. degree in Nanyang Technological University (NTU), Singapore.

From 1995 to 1997, Dr. Ye was with the R&D Institute, XIYI Company Ltd, China, as a Research Engineer. She joined the NTU in 2003.

Dr. Ye is an IEEE Member and has authored seven teaching textbooks and written more than 48 technical papers published in *IEEE-Transactions*, *IEE-Proceedings*, other international journals and various international conferences. Her research interests are in the areas of DC/DC converters, signal processing, operations research and structural biology.

 Muhammad H. Rashid is employed by the University of Florida as Professor of Electrical and Computer Engineering, and Director of the UF/UWF Joint Program in Electrical and Computer Engineering. Dr. Rashid received a B.Sc. Degree in Electrical Engineering from the Bangladesh University of Engineering and Technology, and M.Sc. and Ph.D. Degrees from the University of Birmingham in the UK. Previously, he worked as Professor of Electrical Engineering and was the Chair of the Engineering Department at Indiana University, Purdue University at Fort Wayne. He has also worked as Visiting Assistant Professor of Electrical Engineering at the University of Connecticut, Associate Professor of Electrical Engineering at Concordia University (Montreal, Canada), Professor of Electrical Engineering at Purdue University Calumet, Visiting Professor of Electrical Engineering at King Fahd University of Petroleum and Minerals (Saudi Arabia), as a Design and Development Engineer with Brush Electrical Machines Ltd (UK), a Research Engineer with Lucas Group Research Centre (UK), and as a Lecturer and Head of Control Engineering Department at the Higher Institute of Electronics (Malta).

Dr. Rashid is actively involved in teaching, researching, and lecturing in power electronics. He has published 14 books and more than 100 technical papers. He received the 2002 IEEE Educational Activity Award (EAB) Meritorious Achievement Award in Continuing Education with the following citation "*for contributions to the design and delivery of continuing education in power electronics and computer-aided-simulation*". From 1995 to 2002, Dr. Rashid was an ABET Program Evaluator for Electrical Engineering and he is currently an Engineering Evaluator for the Southern Association of Colleges and Schools (SACS, USA). He has been elected as an IEEE-Industry Applications Society (IAS) as a Distinguished Lecturer. He is the Editor-in-Chief of the *Power Electronics and Applications Series* with CRC Press.

Our acknowledgment goes to the executive editor for this book.

Chapter 1

Introduction

Power electronics and conversion technology are exciting and challenging professions for anyone who has a genuine interest in, and aptitude for, applied science and mathematics. Actually, the existing knowledge in power electronics is not completed. All switching power circuits including the power DC/DC converters and switched DC/AC pulse-width-modulation (PWM) inverters (DC: direct current; AC: alternative current) perform in high-frequency switching state. Traditional knowledge did not fully consider the pumping–filtering process, resonant process and voltage-lift operation. Therefore, the existing knowledge cannot well describe the characteristics of switching power circuits including the power DC/DC converters. To reveal the disadvantages of the existing knowledge, we have to review the traditional analog Power Electronics in this Chapter.

1.1 HISTORICAL REVIEW

Power Electronics and conversion technology are concerned to systems that produce, transmit, control and measure electric power and energy. To describe the characteristics of power systems, various measuring parameters so-called the factors are applied. These important concepts are the power factor (PF), power-transfer efficiency (η), ripple factor (RF) and total harmonic distortion (THD). For long-time education and engineering practice, we know that the traditional power systems have been successfully described by these parameters.

These important concepts will be introduced in the following sections.

1.1.1 WORK, ENERGY AND HEAT

Work, W, and energy, E, are measured by the unit "**joule**". We usually call the kinetic energy "*work*", and the stored or static energy potential "*energy*". Work and energy can be transferred to *heat*, which is measured by "**calorie**". Here is the relationship (**Joule–Lenz law**):

$$1 \text{ joule} = 0.24 \text{ calorie}$$

or

$$1 \text{ calorie} = 4.18 \text{ joules}$$

In this mechanism, there is a relationship between power, P, and work, W, and/or energy, E:

$$W = \int P \, dt \quad E = \int P \, dt$$

and

$$P = \frac{d}{dt} W \quad P = \frac{d}{dt} E$$

Power P is measured by the unit "**watt**", and

$$1 \text{ joule} = 1 \text{ watt} \times 1 \text{ second}$$

or

$$1 \text{ watt} = 1 \text{ joule}/1 \text{ second}$$

1.1.2 DC AND AC EQUIPMENT

Power supplies are sorted into two main groups: DC and AC. Corresponding equipment are sorted into DC and AC kinds as well, e.g. DC generators, AC generators, DC motors, AC motors, etc.

DC Power Supply

A DC power supply has parameters: voltage (amplitude) V_{dc} and ripple factor (RF). A DC power supply can be a battery, DC generator or DC/DC converter.

AC Power Supply

An AC power supply has parameters: voltage (amplitude, root-mean-square (rms or RMS) value and average value), frequency (f or ω), phase angle (ϕ or θ) and total harmonic distortion (THD). An AC power supply can be an AC generator, transformer or DC/AC inverter. An AC voltage can be presented as follows:

$$v(t) = V_p \sin(\omega t - \theta) = \sqrt{2} \, V_{rms} \sin(\omega t - \theta) \tag{1.1}$$

where $v(t)$ is the measured AC instantaneous voltage; V_p, the peak value of the voltage; V_{rms}, the rms value of the voltage; ω, the angular frequency, $\omega = 2\pi f$; f, the supply frequency, e.g. $f = 50 \text{ Hz}$ and θ, the delayed phase angle.

1.1.3 LOADS

Power supply source transfers energy to load. If the characteristics of a load can be described by a linear differential equation, we call the load a linear load. Otherwise, we call the load a non-linear load (i.e. the diodes, relays and hysteresis-elements that cannot be described by a linear differential equation). Typical linear loads are sorted into two categories: passive and dynamic loads.

Linear Passive Loads

Linear passive loads are resistance (R), inductance (L) and capacitance (C). All these components satisfy linear differential equations. If the circuit current is I as shown in Figure 1.1, from Ohm's law we have:

$$V_R = RI \tag{1.2}$$

$$V_L = L\frac{dI}{dt} \tag{1.3}$$

$$V_C = \frac{1}{C}\int I\, dt \tag{1.4}$$

$$V = V_R + V_L + V_C = RI + L\frac{dI}{dt} + \frac{1}{C}\int I\, dt \tag{1.5}$$

Equations (1.2)–(1.5) are all linear differential equations.

Linear Dynamic Loads

Linear dynamic loads are DC and AC back electromagnetic force (EMF). All these components satisfy differential equation operation.

The back EMF of a DC motor is DC back EMF with DC voltage that is proportional to the field flux and armature running speed:

$$EMF = k\Phi\omega \tag{1.6}$$

where k is the DC machine constant; Φ, the field flux and ω, the machine running speed in rad/s.

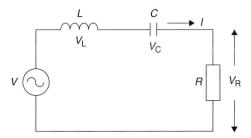

Figure 1.1   An L–R–C circuit.

The back EMF of an AC motor is AC back EMF with AC voltage that is proportional to the field flux and rotor running speed.

1.1.4 IMPEDANCE

If an R–L–C circuit supplied by a voltage source with mono-frequency ($\omega = 2\pi f$) sinusoidal waveform as shown in Figure 1.1, we can simplify the differential equation (1.5) into an algebraic equation using the concept "impedance", Z:

$$V = ZI \tag{1.7}$$

We define impedance Z as follows:

$$Z = R + j\omega L - j\frac{1}{\omega C} = R + jX = |Z|\angle\theta \tag{1.8}$$

where

$$X = \omega L - \frac{1}{\omega C}$$

$$|Z| = \sqrt{R^2 + \left(\omega L - \frac{1}{\omega C}\right)^2} \tag{1.9}$$

$$\theta = \tan^{-1}\left(\frac{\omega L - \frac{1}{\omega C}}{R}\right) \tag{1.10}$$

in which θ is the conjugation phase angle. The real part of an impedance Z is defined as resistance R, and the imaginary part of an impedance Z is defined as reactance X. The reactance has two components: the positive part is called *inductive reactance* $j\omega L$ and the negative part is called *capacitive reactance* $-j/\omega C$. The power delivery has been completed only across resistance. The reactance can only store energy and shift phase angle. No power is consumed on reactance, which produces reactive power and spoils power delivery.

From Ohm's law, we can get the vector current (**I**) from vector voltage (**V**) and impedance (**Z**):

$$I = \frac{V}{Z} = \frac{V}{R + j\omega L - j\frac{1}{\omega C}} = |I|\angle\theta \tag{1.11}$$

Most industrial application equipment are of inductive load. For example, an R–L circuit is supplied by a sinusoidal voltage V, and it is shown in Figure 1.2. The impedance Z obtained is:

$$Z = R + j\omega L = R + jX = |Z|\angle\theta \tag{1.12}$$

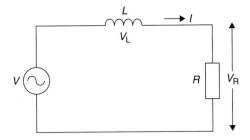

Figure 1.2   An L–R circuit.

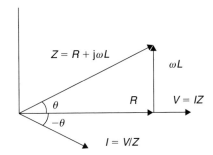

Figure 1.3   The vector diagram of an L–R circuit.

with

$$|Z| = \sqrt{R^2 + (\omega L)^2} \quad \text{and} \quad \theta = \tan^{-1}\left(\frac{\omega L}{R}\right)$$

The conjugation angle (θ) is a positive value. The corresponding vector diagram is shown in Figure 1.3.

We also get the current as follows:

$$I = \frac{V}{Z} = \frac{V}{R + j\omega L} = |I| \angle -\theta \tag{1.13}$$

Select the supply voltage **V** as reference vector with phase angle zero. The current vector is delayed than the voltage by the conjugation angle θ. The corresponding vector diagram is also shown in Figure 1.3. The voltage and current waveforms are shown in Figure 1.4.

1.1.5 POWERS

There are various powers such as apparent power (or complex power), S, power (or real power), P, and reactive power, Q.

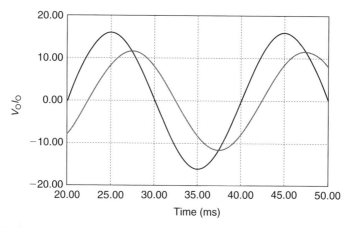

Figure 1.4   The corresponding voltage and current waveforms.

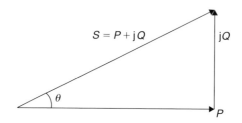

Figure 1.5   The power vector diagram of an L–R circuit.

Apparent Power *S*

We define the apparent power *S* as follows:

$$S = VI^* = P + jQ \tag{1.14}$$

Power *P*

Power or real power *P* is the real part of the apparent power *S*:

$$P = S\cos\theta = I^2 R \tag{1.15}$$

Reactive Power *Q*

Reactive power *Q* is the imaginary part of the apparent power *S*:

$$Q = S\sin\theta = I^2 X \tag{1.16}$$

Referring to the R–L circuit in Figure 1.2, we can show the corresponding power vectors in Figure 1.5.

1.2 TRADITIONAL PARAMETERS

Traditional parameters used in power electronics are the power factor (*PF*), power-transfer efficiency (η), total harmonic distortion (*THD*) and ripple factor (*RF*). Using these parameters has successfully described the characteristics of power (generation, transmission, distribution, protection and harmonic analysis) systems and most drive (AC and DC motor drives) systems.

1.2.1 POWER FACTOR (*PF*)

Power factor is defined by the ratio of real power P over the apparent power S:

$$PF = \frac{P}{S} = \cos\theta = \frac{I^2 R}{VI^*} = \frac{IR}{V} \tag{1.17}$$

Figure 1.5 is used to illustrate the power factor (*PF*).

1.2.2 POWER-TRANSFER EFFICIENCY (η)

Power-transfer efficiency (η) is defined by the ratio of output power P_O over the input power P_{in}:

$$\eta = \frac{P_O}{P_{in}} \tag{1.18}$$

The output power P_O is received by the load, end user. The input power P_{in} is usually generated by the power supply source. Both the input power P_{in} and output power P_O are real power.

1.2.3 TOTAL HARMONIC DISTORTION (*THD*)

A periodical AC waveform usually possesses various order harmonics. Since the instantaneous value is periodically repeating in fundamental frequency f (or $\omega = 2\pi f$), the corresponding spectrum in the frequency domain consists of discrete peaks at the frequencies nf (or $n\omega = 2n\pi f$), where $n = 1, 2, 3, \ldots \infty$. The first-order component ($n = 1$) corresponds to the fundamental component V_1. The total harmonic distortion (*THD*) is defined by the ratio of the sum of all higher-order harmonics over the fundamental harmonic V_1:

$$THD = \frac{\sqrt{\sum_{n=2}^{\infty} V_n^2}}{V_1} \tag{1.19}$$

where all V_n ($n = 1, 2, 3, \ldots \infty$) are the corresponding rms values.

1.2.4 RIPPLE FACTOR (*RF*)

A DC waveform usually possesses DC component V_{dc} and various high-order harmonics. These harmonics make the variation (ripple) of the DC waveform. Since the instantaneous value is periodically repeating in fundamental frequency f (or $\omega = 2\pi f$), the corresponding spectrum in the frequency domain consists of discrete peaks at the frequencies nf (or $n\omega = 2n\pi f$), where $n = 0, 1, 2, 3, \ldots \infty$. The zeroth-order component ($n = 0$) corresponds to the DC component V_{dc}. The ripple factor (*RF*) is defined by the ratio of the sum of all higher-order harmonics over the DC component V_{dc}:

$$RF = \frac{\sqrt{\sum_{n=1}^{\infty} V_n^2}}{V_{dc}} \qquad (1.20)$$

where all V_n ($n = 1, 2, 3, \ldots \infty$) are the corresponding rms values.

1.2.5 APPLICATION EXAMPLES

In order to describe the fundamental parameters better, we provide some examples as the application of these parameters in this section.

Power and Efficiency (η)

A pure resistive load R supplied by a DC voltage source V with internal resistance R_O is shown in Figure 1.6. The current I is obtained by the calculation expression:

$$I = \frac{V}{R + R_O} \qquad (1.21)$$

The output voltage V_O is:

$$V_O = \frac{R}{R + R_O} V \qquad (1.22)$$

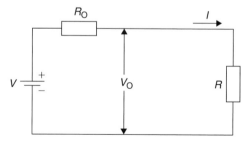

Figure 1.6   A pure resistive load supplied by a DC source with internal resistance.

The output power P_O is:

$$P_O = I^2 R = \frac{R}{(R + R_O)^2} V^2 \tag{1.23}$$

The power-transfer efficiency (η) is:

$$\eta = \frac{P_O}{P_{\text{in}}} = \frac{I^2 R}{IV} = \frac{R}{R + R_O} \tag{1.24}$$

In order to obtain maximum output power, we can determine the condition by differentiating Equation (1.23):

$$\frac{\mathrm{d}}{\mathrm{d}R} P_O = \frac{\mathrm{d}}{\mathrm{d}R} \left[\frac{R}{(R + R_O)^2} V^2 \right] = 0 \tag{1.25}$$

$$\frac{1}{(R + R_O)^2} - \frac{2R}{(R + R_O)^3} = 0$$

Hence,

$$R = R_O \tag{1.26}$$

When $R = R_O$, we obtain the maximum output power:

$$P_{O\text{-max}} = \frac{V^2}{4R_O} \tag{1.27}$$

and the corresponding efficiency:

$$\eta = \frac{R}{R + R_O} |_{R=R_O} = 0.5 \tag{1.28}$$

This example shows that the power and efficiency are different concepts. When load R is equal to the internal resistance R_O, maximum output power is obtained with the efficiency $\eta = 50\%$. Vice versa, if we would like to obtain maximum efficiency $\eta = 1$ or 100%, it requires load R is equal to infinite (if the internal resistance R_O cannot be equal to zero). It causes the output power, which is equal to zero. The interesting relation is listed below:

Maximum output power $\eta = 50\%$

Output power $= 0$ $\eta = 100\%$

The second case corresponds to the open circuit. Although the theoretical calculation illustrates the efficiency $\eta = 1$ or 100%, no power is delivered from source to load.

Another situation is $R = 0$ that causes the output current is its maximum value $I_{\text{max}} = V/R_O$ as (1.21) and:

Output power $= 0$ $\eta = 0\%$

An R–L Circuit Calculation

Figure 1.7 shows a single-phase sinusoidal power supply source with the internal resistance $R_O = 0.2\,\Omega$, supplying an R–L circuit with $R = 1\,\Omega$ and $L = 3\,mH$. The source voltage is a sinusoidal waveform with the voltage 16 V (rms voltage) and frequency $f = 50\,Hz$:

$$V = 16\sqrt{2}\sin 100\pi t \text{ V} \tag{1.29}$$

The internal impedance is:

$$Z_O = R_O = 0.2\,\Omega \tag{1.30}$$

The impedance of load is:

$$Z = 1 + j100\pi \times 3m = 1 + j0.94 = 1.3724\angle 43.23°\,\Omega \tag{1.31}$$

$$Z + Z_O = 1.2 + j0.94 = 1.524\angle 38.073°\,\Omega \tag{1.32}$$

The current is:

$$I = \frac{V}{Z + Z_O} = 10.5\sqrt{2}\sin(100\pi t - 38.073°)\,\text{A} \tag{1.33}$$

The output voltage across the R–L circuit is:

$$V_O = ZI = 14.4\sqrt{2}\sin(100\pi t - 5.16°)\,\text{A} \tag{1.34}$$

The apparent power S across the load is:

$$S = V_O I^* = 14.4 \times 10.5 = 151.3 \text{ VA} \tag{1.35}$$

The real output power P_O across the load is:

$$P_O = P_R = I^2 R = 10.5^2 \times 1 = 110.25 \text{ W} \tag{1.36}$$

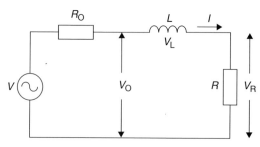

Figure 1.7   An R–L circuit supplied by an AC source with internal resistance.

The real input power P_{in} is:

$$P_{in} = I^2(R + R_O) = 10.5^2 \times 1.2 = 132.3 \text{ W} \tag{1.37}$$

Therefore, the power factor PF of the load is:

$$PF = \frac{P_O}{S} = \cos\theta = \cos 43.23 = 0.73 \text{ (lagging)} \tag{1.38}$$

The corresponding reactive power Q is:

$$Q = S \times \sin\theta = 151.3 \times \sin 43.23 = 103.63 \text{ VAR} \tag{1.39}$$

Thus, the power-transfer efficiency (η) is:

$$\eta = \frac{P_R}{P_{in}} = \frac{110.25}{132.3} = 0.833 \tag{1.40}$$

Other way to calculate the efficiency (η) is:

$$\eta = \frac{R}{R + R_O} = \frac{1}{1.2} = 0.833$$

To obtain the maximum output power we have to choose same condition as in Equation (1.26),

$$R = R_O = 0.2 \ \Omega \tag{1.41}$$

The maximum output power P_O is:

$$P_{\text{O-max}} = \frac{V^2}{4R_O} = \frac{16^2}{4 \times 0.2} = 320 \text{ W} \tag{1.42}$$

with the efficiency (η) is:

$$\eta = 0.5 \tag{1.43}$$

A Three-Phase Circuit Calculation

Figure 1.8 shows a balanced three-phase sinusoidal power supply source supplying a full-wave diode-bridge rectifier to an R–L load. Each single-phase source is a sinusoidal voltage source with the internal impedance 10 kΩ plus 10 mH. The load is an R–L circuit with $R = 240 \ \Omega$ and $L = 50$ mH. The source phase voltage has the amplitude 16 V (its rms value is $16/\sqrt{2} = 11.3$ V) and frequency $f = 50$ Hz. It is presented as:

$$V = 16 \sin 100\pi t \text{ V} \tag{1.44}$$

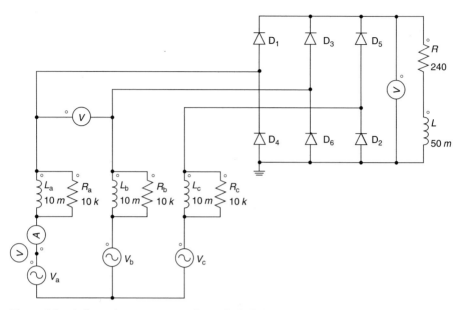

Figure 1.8   A three-phase source supplies a diode full-wave rectifier to an R–L load.

The internal impedance is:

$$Z_O = \frac{10,000}{j100\pi \times 10\,m} = \frac{10,000 \times j3.1416}{10,000 + j3.1416} \approx j3.1416\,\Omega \tag{1.45}$$

The impedance of the load is:

$$Z = 240 + j100\pi \times 50\,m = 240 + j15.708 = 240.5\angle 3.74°\,\Omega \tag{1.46}$$

The bridge input AC line-to-line voltage is measured and shown in Figure 1.9. It can be seen that the input AC line voltage is distorted. After the fast Fourier transform (FFT) analysis, the corresponding spectrums can be obtained as shown in Figure 1.10 for the bridge input AC line voltage waveforms.

The input line–line voltage fundamental value and the harmonic peak voltages for *THD* calculation are listed in Table 1.1.

Using formula (1.19) to calculate the *THD*, we have,

$$THD = \frac{\sqrt{\sum_{n=2}^{\infty} v_{AB\text{-}n}^2}}{v_{AB\text{-}1}} = \frac{\sqrt{0.737^2 + 0.464^2 + 0.566^2 + 0.422^2 + \cdots}}{27.62} \times 100\% = 4.86\%$$

$$\tag{1.47}$$

We measured the output DC voltage in Figure 1.11. It can be seen that the DC voltage has ripple. After FFT analysis, we obtain the corresponding spectrums as shown in Figure 1.12 for the output DC voltage waveforms.

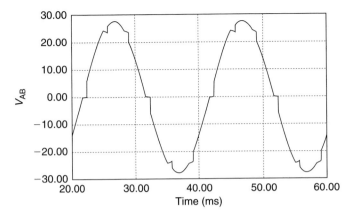

Figure 1.9 The input line AC voltage waveform.

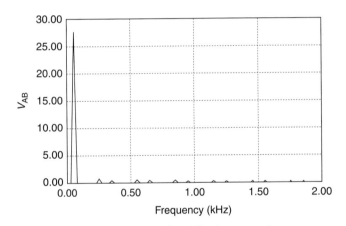

Figure 1.10 The FFT spectrum of the input line AC voltage waveform.

Table 1.1

The harmonic peak voltages of the distorted the input line–line voltage

Order no.	Fundamental	5	7	11	13	17	19
Volts	27.62	0.737	0.464	0.566	0.422	0.426	0.34
Order no.	23	25	29	31	35	37	*THD*
Volts	0.297	0.245	0.196	0.164	0.143	0.119	4.86%

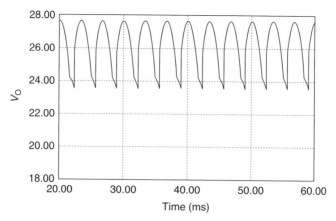

Figure 1.11   The output DC voltage waveform.

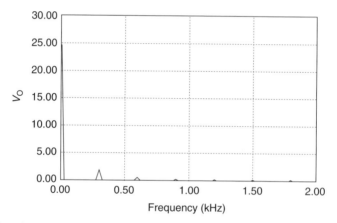

Figure 1.12   The FFT spectrum of the output DC voltage waveform.

The output DC load voltage and the harmonic peak voltages for *RF* calculation are listed in Table 1.2.

Using formula (1.20) to calculate the *RF*, we have,

$$RF = \frac{\sqrt{\sum_{n=1}^{\infty} v_{O\text{-}n}^2}}{v_{O\text{-}dc}} = \frac{\sqrt{1.841^2 + 0.5^2 + 0.212^2 + 0.156^2 + \cdots}}{26.15 \times \sqrt{2}} \times 100\% = 5.24\%$$

$$(1.48)$$

From input phase voltage and current, the partial power factor (PF_p) is obtained,

$$PF_p = \cos\theta = 0.9926 \tag{1.49}$$

Table 1.2

The harmonic peak voltages of the DC output voltage with ripple

Order no.	DC	6	12	18	24	30	36	RF
Volts	26.15	1.841	0.500	0.212	0.156	0.151	0.134	5.24%

Table 1.3

The harmonic peak voltages of the input phase current

Order no.	Fundamental	5	7	11	13	17	19
Amperes	0.12024	2.7001e–2	1.2176e–2	9.3972e–3	5.9472e–3	4.5805e–3	3.2942e–3
Order no.	23	25	29	31	35	37	Total *PF*
Amperes	2.3524e–3	1.8161e–3	1.2234e–3	9.7928e–4	7.3822e–4	5.9850e–4	0.959

Table 1.4

The harmonic peak voltages of the output DC current

Order no.	DC (0)	6	12	18	24	30	36	η
Amperes	0.109	7.14e–3	1.64e–3	5.72e–4	3.49e–4	2.85e–4	2.19e–4	0.993

The input phase current peak value and the higher-order harmonic current peak values are listed in Table 1.3.

$$I_{\text{a-1}} = \frac{0.12024}{\sqrt{2}} = 0.085 \text{ A} \quad I_{\text{a-rms}} = \sqrt{\sum_{n=0}^{\infty} i_n^2} = 0.088 \text{ A}$$

Total power factor

$$PF_{\text{total}} = \frac{I_{\text{a-1}}}{I_{\text{a-rms}}} \cos \theta = \frac{0.085}{0.088} \times 0.9926 = 0.959$$

The average DC output load current and the higher-order harmonic current peak values are listed in Table 1.4.

$$V_{\text{O-rms}} = \sqrt{\sum_{n=0}^{\infty} v_n^2} = 26.186 \text{ V} \quad I_{\text{O-rms}} = \sqrt{\sum_{n=0}^{\infty} i_n^2} = 0.1096 \text{ A}$$

The efficiency (η) is:

$$\eta = \frac{P_{\text{dc}}}{P_{\text{ac}}} = \frac{V_{\text{O-dc}} I_{\text{O-dc}}}{V_{\text{O-rms}} I_{\text{O-rms}}} \times 100\% = \frac{26.15 \times 0.10896}{26.186 \times 0.1096} \times 100\% = 99.28\% \quad (1.50)$$

From this example, we fully demonstrated the four important parameters: power factor (*PF*), power-transfer efficiency (*η*), total harmonic distortion (*THD*) and ripple factor (*RF*). Usually, these four parameters are enough to describe the characteristics of a power supply system.

1.3 MULTIPLE-QUADRANT OPERATIONS AND CHOPPERS

Multiple-quadrant operation is required in industrial applications. For example, a DC motor can perform forward running or reverse running. The motor armature voltage and armature current are both positive during forward starting process. We usually call it the forward motoring operation or "Quadrant I" operation. The motor armature voltage is still positive and its armature current is negative during forward braking process. This state is called the forward regenerative braking operation or "Quadrant II" operation.

Analogously, the motor armature voltage and current are both negative during reverse starting process. We usually call it the reverse motoring operation or "Quadrant III" operation. The motor armature voltage is still negative and its armature current is positive during reverse braking process. This state is called the reverse regenerative braking operation or "Quadrant IV" operation.

Referring to the DC motor operation states, we can define the multiple-quadrant operation as below:

Quadrant I operation: Forward motoring; voltage and current are positive;
Quadrant II operation: Forward regenerative braking; voltage is positive and current is negative;
Quadrant III operation: Reverse motoring; voltage and current are negative;
Quadrant IV operation: Reverse regenerative braking; voltage is negative and current is positive.

The operation status is shown in the Figure 1.13. Choppers can convert a fixed DC voltage into various other voltages. The corresponding chopper is usually called which quadrant operation chopper, e.g. the first-quadrant chopper or "A"-type chopper. In the

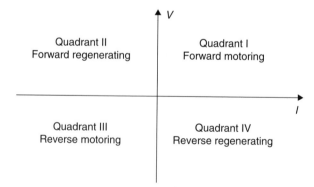

Figure 1.13   The four-quadrant operation.

following description we use the symbols V_{in} for fixed voltage, V_p for chopped voltage and V_O for output voltage.

1.3.1 THE FIRST-QUADRANT CHOPPER

The first-quadrant chopper is also called "A"-type chopper and its circuit diagram is shown in Figure 1.14(a) and the corresponding waveforms are shown in Figure 1.14(b). The switch S can be some semiconductor devices such as BJT, integrated gate bipolar transistors (IGBT) and power MOS field effected transistors (MOSFET). Assuming all parts are ideal components, the output voltage is calculated by the formula:

$$V_O = \frac{t_{on}}{T} V_{in} = kV_{in} \tag{1.51}$$

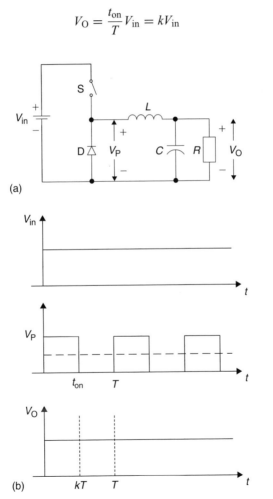

Figure 1.14   The first-quadrant chopper. (a) Circuit diagram and (b) voltage waveforms.

where T is the repeating period $(T = 1/f)$, in which f is the chopping frequency; t_{on} is the switch-on time and k is the conduction duty cycle $(k = t_{on}/T)$.

1.3.2 THE SECOND-QUADRANT CHOPPER

The second-quadrant chopper is also called "B"-type chopper and its circuit diagram is shown in Figure 1.15(a) and the corresponding waveforms are shown in Figure 1.15(b). The output voltage can be calculated by the formula:

$$V_O = \frac{t_{off}}{T} V_{in} = (1 - k)V_{in} \tag{1.52}$$

where T is the repeating period $(T = 1/f)$, in which f is the chopping frequency; t_{off} is the switch-off time $(t_{off} = T - t_{on})$ and k is the conduction duty cycle $(k = t_{on}/T)$.

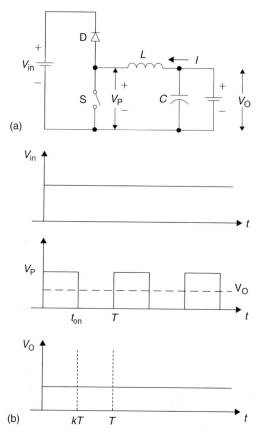

Figure 1.15   The second-quadrant chopper. (a) Circuit diagram and (b) voltage waveforms.

1.3.3 THE THIRD-QUADRANT CHOPPER

The third-quadrant chopper is shown in Figure 1.16(a) and the corresponding wave-forms are shown in Figure 1.16(b). All voltage polarities are defined in the figure. The output voltage (absolute value) can be calculated by the formula:

$$V_O = \frac{t_{on}}{T} V_{in} = k V_{in} \tag{1.53}$$

where t_{on} is the switch-on time and k is the conduction duty cycle ($k = t_{on}/T$).

1.3.4 THE FOURTH-QUADRANT CHOPPER

The fourth-quadrant chopper is shown in Figure 1.17(a) and the corresponding wave-forms are shown in Figure 1.17(b). All voltage polarities are defined in the figure.

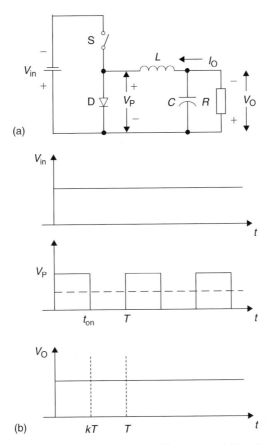

Figure 1.16 The third-quadrant chopper. (a) Circuit diagram and (b) voltage waveforms.

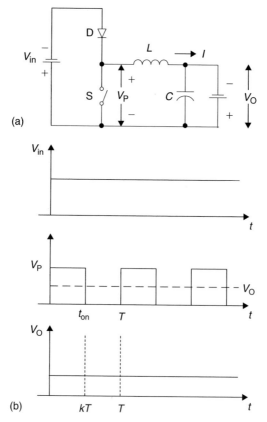

Figure 1.17   The fourth-quadrant chopper. (a) Circuit diagram and (b) voltage waveforms.

The output voltage (absolute value) can be calculated by the formula:

$$V_O = \frac{t_{\text{off}}}{T} V_{\text{in}} = (1 - k)V_{\text{in}} \qquad (1.54)$$

where t_{off} is the switch-off time ($t_{\text{off}} = T - t_{\text{on}}$) and k is the conduction duty cycle ($k = t_{\text{on}}/T$).

1.3.5 THE FIRST–SECOND-QUADRANT CHOPPER

The first–second-quadrant chopper is shown in Figure 1.18. Dual-quadrant operation is usually requested in the system with two voltage sources V_1 and V_2. Assume the condition $V_1 > V_2$, the inductor L is the ideal component. During Quadrant I operation, S_1 and D_2 work, and S_2 and D_1 are idle. Vice versa, during Quadrant II operation, S_2 and D_1 work, and S_1 and D_2 are idle. The relation between the two voltage sources can

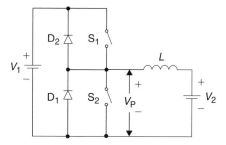

Figure 1.18   The first–second quadrant chopper.

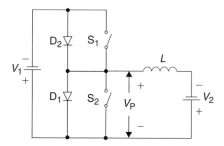

Figure 1.19   The third–fourth quadrant chopper.

be calculated by the formula:

$$V_2 = \begin{cases} kV_1 & \text{Quadrant I operation} \\ (1-k)V_1 & \text{Quadrant II operation} \end{cases} \tag{1.55}$$

where k is the conduction duty cycle ($k = t_{on}/T$).

1.3.6 THE THIRD–FOURTH-QUADRANT CHOPPER

The third–fourth-quadrant chopper is shown in Figure 1.19. Dual-quadrant operation is usually requested in the system with two voltage sources V_1 and V_2. Both the voltage polarities are defined in the figure, we just concentrate on their absolute values in analysis and calculation. Assume the condition $V_1 > V_2$, the inductor L is the ideal component. During Quadrant III operation, S_1 and D_2 work, and S_2 and D_1 are idle. Vice versa, during Quadrant IV operation, S_2 and D_1 work, and S_1 and D_2 are idle. The relation between the two voltage sources can be calculated by the formula:

$$V_2 = \begin{cases} kV_1 & \text{Quadrant III operation} \\ (1-k)V_1 & \text{Quadrant IV operation} \end{cases} \tag{1.56}$$

where k is the conduction duty cycle ($k = t_{on}/T$).

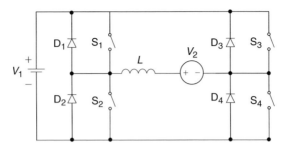

Figure 1.20   The four-quadrant chopper.

Table 1.5

The switches' and diodes' status for four-quadrant operation

Switch or diode	Quadrant I	Quadrant II	Quadrant III	Quadrant IV
S_1	Works	Idle	Idle	Works
D_1	Idle	Works	Works	Idle
S_2	Idle	Works	Works	Idle
D_2	Works	Idle	Idle	Works
S_3	Idle	Idle	On	Idle
D_3	Idle	Idle	Idle	On
S_4	On	Idle	Idle	Idle
D_4	Idle	On	Idle	Idle
Output	V_2+, I_2+	V_2+, I_2-	V_2-, I_2-	V_2-, I_2+

1.3.7 THE FOUR-QUADRANT CHOPPER

The four-quadrant chopper is shown in Figure 1.20. The input voltage is positive, output voltage can be either positive or negative. The status of switches and diodes for the operation are given in Table 1.5. The output voltage can be calculated by the formula:

$$V_2 = \begin{cases} kV_1 & \text{Quadrant I operation} \\ (1-k)V_1 & \text{Quadrant II operation} \\ -kV_1 & \text{Quadrant III operation} \\ -(1-k)V_1 & \text{Quadrant IV operation} \end{cases} \quad (1.57)$$

1.4 DIGITAL POWER ELECTRONICS: PUMP CIRCUITS AND CONVERSION TECHNOLOGY

Besides choppers there are more and more switching circuits applied in industrial applications. These switching circuits work in discrete-time state. Since high-frequency switching circuits can transfer the energy in high power density and high efficiency, they have been applied on more and more branches of power electronics. The energy

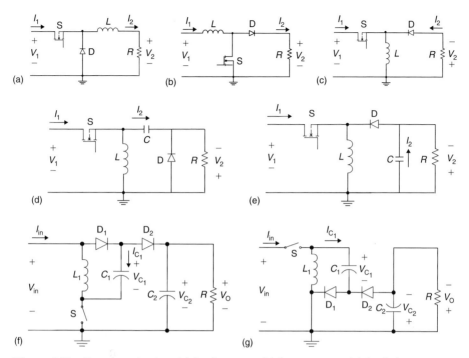

Figure 1.21 Pumping circuits: (a) buck pump, (b) boost pump, (c) buck–boost pump, (d) positive Luo-pump, (e) negative Luo-pump, (f) positive super Luo-pump and (g) negative super Luo-pump.

and power delivery from source to the users are not in continuous mode. Therefore, digital control theory has to be applied in this area.

All conversion technologies (such as pumping circuits, AC/DC rectifiers, DC/AC inverters, DC/DC converters and AC/AC (and/or AC/DC/AC) converters) are theoretically based on the switching circuit. It is urgent to investigate the digital power electronics rather than the traditional analog control applied in analog power electronics. The following typical circuits are examples of switching circuits working in the discrete-time mode.

1.4.1 FUNDAMENTAL PUMP CIRCUITS

All power DC/DC converters have pumping circuit. Pumping circuits are typical switching circuits to convert the energy from an energy source to energy-storage components in discrete state. Each pump has a switch S and an energy-storage component that can be an inductor L. The switch S turns on once in a period $T = 1/f$, where f is the switching frequency. Therefore, the energy transferred in a period is a certain value that can be called energy quantum. Figure 1.21 shows seven (buck, boost, buck–boost, positive Luo, negative Luo, positive super Luo and negative super Luo) pumping circuits, which are used in the corresponding DC/DC converters.

All pumping circuits are switching circuits that convert the energy from source to load or certain energy-storage component in discrete state. Each pumping circuit has at least one switch and one energy-store element, for example an inductor. The switch is controlled by a PWM signal with the period T ($T = 1/f$, where f is the switching frequency) and the conduction duty cycle k. The energy was absorbed from the energy source to the inductor during switching-on period kT. The energy stored in the inductor will be delivered to next stage during switching-off period $(1 - k)T$. Therefore, the energy from the source to users is transferred in discrete-time mode.

1.4.2 AC/DC RECTIFIERS

All AC/DC controlled rectifiers are switching circuits. Figure 1.22 shows few rectifier circuits (namely single-phase half-wave, single-phase full-wave, three-phase half-wave, and three-phase half-wave controlled rectifier), which are used in the corresponding AC/DC converters.

All AC/DC rectifier circuits are switching circuits that convert the energy from an AC source to load in discrete state. Each AC/DC controlled rectifier has at least one switch. For example, a half-wave controlled thyristor (silicon controlled rectifier, SCR) rectifier has one SCR switch. The switch is controlled by a firing pulse signal with the repeating period T ($T = 1/f$, where f is the switching frequency for the single-phase rectifiers) and the conduction period. The energy was delivered from the energy source to the load during switching-on period. The energy is blocked during switching-off period. Therefore, the energy from the source to loads is transferred in discrete-time mode.

1.4.3 DC/AC PWM INVERTERS

All DC/AC inverters are switching circuits. Figure 1.23 shows three (single-phase, three-phase, three-level three-phase) DC/AC PWM inverter circuits, which are used in the corresponding DC/AC inverters.

All DC/AC PWM inverter circuits are switching circuits that convert the energy from a DC source to load in discrete state. Each DC/AC inverter has multiple switches. The switches are controlled by PWM signals with the repeating period T ($T = 1/f$, where f is the switching frequency for the single-phase rectifiers) and the modulation ratio m. The energy was delivered from the energy source to the load during switching-on period. The energy is blocked during switching-off period. Therefore, the energy from the source to loads is transferred in discrete-time mode.

1.4.4 DC/DC CONVERTERS

All DC/DC converters are switching circuits. Figure 1.24 shows seven (buck, boost, buck–boost, positive output Luo, negative output Luo, positive output super-lift Luo and negative output super-lift Luo converters) DC/DC converter circuits.

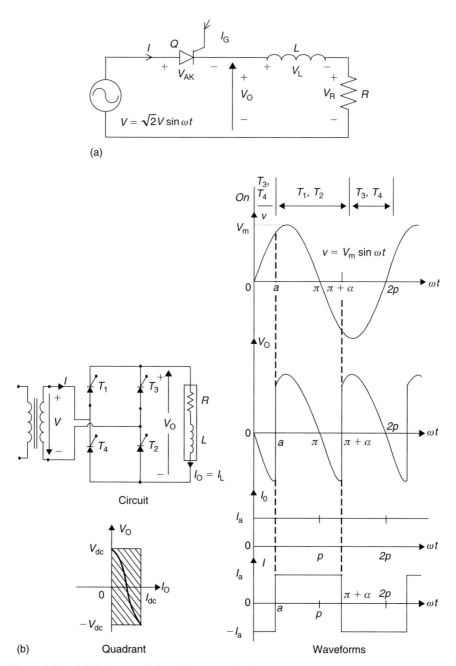

Figure 1.22 AC/DC controlled rectifiers: (a) Single-phase half-wave controlled rectifier and (b) single-phase full-wave controlled rectifier.

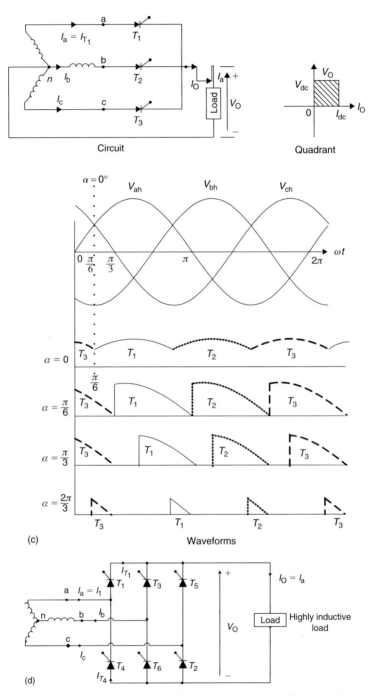

Figure 1.22 (*contd.*) (c) Three-phase half-wave controlled rectifier and (d) three-phase half-wave controlled rectifier.

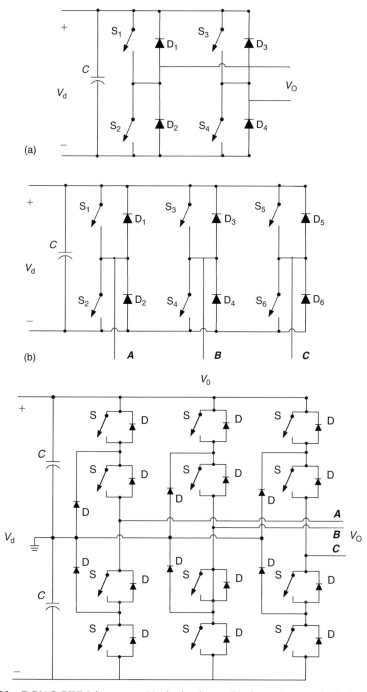

Figure 1.23 DC/AC PWM inverters: (a) single-phase, (b) three-phase and (c) three-level three-phase.

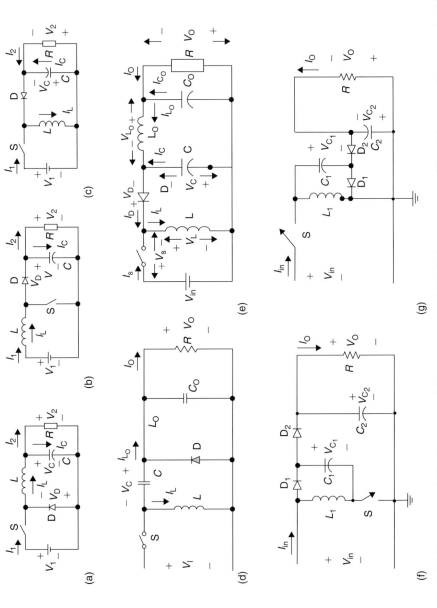

Figure 1.24 DC/DC converters: (a) buck converter, (b) boost converter, (c) buck–boost converter, (d) positive output Luo-converter, (e) negative output Luo-converter, (f) positive output super-lift Luo-converter and (g) negative output super-lift Luo-converter.

All DC/DC converters' circuits are switching circuits that convert the energy from a DC source to load in discrete state. Each power DC/DC converter has at least one pumping circuit and filter. The switch is controlled by a PWM signal with the repeating period T ($T = 1/f$, f is the switching frequency) and the conduction duty cycle k. The energy was delivered from the energy source to the load via the pumping circuit during switching-on period kT. The energy is blocked during switching-off period $(1 - k)T$. Therefore, the energy from the source to loads is transferred in discrete-time mode.

1.4.5 AC/AC CONVERTERS

All AC/AC converters are switching circuits. Figure 1.25 shows three (single-phase amplitude regulation, single-phase and three-phase) AC/AC converter circuits.

All AC/AC converter circuits are switching circuits that convert the energy from an AC source to load in discrete state. Each AC/AC converter has multiple switches. The

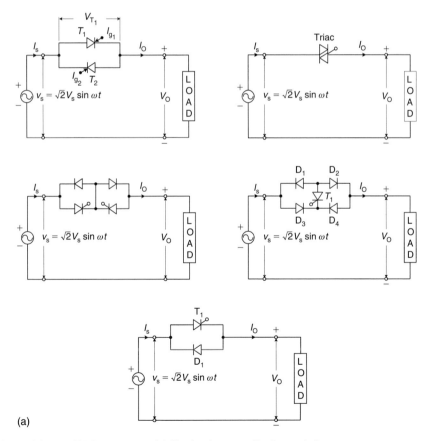

(a)

Figure 1.25 AC/AC converters. (a) Single-phase amplitude regulation.

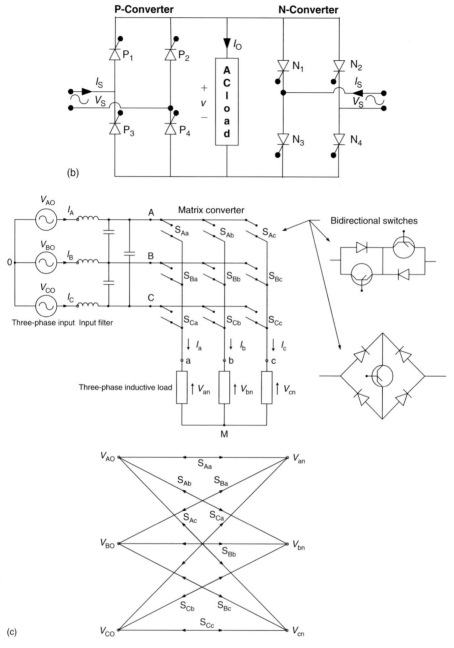

Figure 1.25 (*contd.*) (b) Single phase AC/AC cyclo converter and (c) three-phase AC/AC matrix converter.

switches are controlled by PWM signals with the repeating period T ($T = 1/f$, where f is the switching frequency for the single-phase rectifiers) and the modulation factor. The energy was delivered from the energy AC source to the load during switching-on period. The energy is blocked during switching-off period. Therefore, the energy from the source to loads is transferred in discrete-time mode.

1.5 SHORTAGE OF ANALOG POWER ELECTRONICS AND CONVERSION TECHNOLOGY

Analog power electronics use the traditional parameters: power factor (PF), efficiency (η), total harmonic distortion (THD) and ripple factor (RF) to describe the characteristics of a power system or drive system. It is successfully applied for more than a century. Unfortunately, all these factors are not available to be used to describe the characteristics of switching circuits: power DC/DC converters and other high-frequency switching circuits.

Power DC/DC converters have been usually equipped by a DC power supply source, pump circuit, filter and load. The load can be of any type, but most investigations are concerned to resistive load R and back EMF or battery. It means that the input and output voltages are nearly pure DC voltages with very small ripple, e.g. output voltage variation ratio is usually less than 1%. In this case, the corresponding RF is less than 0.001, which is always ignored.

Since all powers are real power without reactive power jQ, we cannot use power factor (PF) to describe the energy-transferring process.

As only DC components exists without harmonics in input and output voltage, THD is not available to be used to describe the energy-transferring process and waveform distortion.

To simplify the research and analysis, we usually assume the condition without power losses during power-transferring process to investigate power DC/DC converters. Consequently, the efficiency $\eta = 1$ or 100% for most of description of power DC/DC investigation. Otherwise, efficiency (η) must be considered for special investigations regarding the power losses.

In general conditions, all four factors are not available to apply in the analysis of power DC/DC converters. This situation lets the designers of power DC/DC converters confusing for very long time. People would like to find other new parameters to describe the characteristics of power DC/DC converters.

There is no correct theory and the corresponding parameters to be used for all switching circuits till 2004. Dr. Fang Lin Luo and Dr. Hong Ye firstly created new theory and parameters to describe the characteristics of all switching circuits in 2004.

Energy storage in power DC/DC converters has been paid attention long time ago. Unfortunately, there is no clear concept to describe the phenomena and reveal the relationship between the stored energy and the characteristics of power DC/DC converters. We have theoretically defined a new concept, energy factor (EF), and researched the relations between EF and the mathematical modeling of power DC/DC converters.

EF is a new concept in power electronics and conversion technology, which thoroughly differs from the traditional concepts such as power factor (*PF*), power-transfer efficiency (*η*), total harmonic distortion (*THD*) and ripple factor (*RF*). *EF* and the subsequential other parameters can illustrate the system stability, reference response and interference recovery. This investigation is very helpful for system design and DC/DC converters characteristics foreseeing.

1.6 POWER SEMICONDUCTOR DEVICES APPLIED IN DIGITAL POWER ELECTRONICS

High-frequency switching equipment can convert high power, and its power density is proportional to the applying frequency. For example, the volume of a 1-kW transformer working in 50 Hz has the size 4 in. × 3 in. × 2.5 in. = 30 in.^{3} The volume of a 2.2-kW flat-transformer working in 50 kHz has the size 1.5 in. × 0.3 in. × 0.2 in. = 0.09 in.^{3} The difference between them is about 1000 times.

To be required by the industrial applications, power semiconductor devices applied in digital power electronics have been improved in recent decades. Their power, voltage and current rates increase in many times, the applying frequency is greatly enlarged. For example, the working frequency of an IGBT increases from 50 to 200 kHz, and the working frequency of a MOSFET increases from 5 to 20 MHz.

The power semiconductor devices usually applied in industrial applications are as follows:

- diodes;
- SCRs (thyristors);
- GTOs (gate turn-off thyristors);
- BTs (power bipolar transistors);
- IGBTs (insulated gate bipolar transistors);
- MOSFETs (power MOS field effected transistors);
- MSCs (MOS controlled thyristors).

All devices except diode are working in switching state. Therefore, the circuits consists them to be called switching circuits and work in discrete state.

FURTHER READING

1. Luo F. L. and Ye H., *Advanced DC/DC Converters*, CRC Press LLC, Boca Raton, Florida, USA, 2004. **ISBN: 0-8493-1956-0**.
2. Luo F. L., Ye H. and Rashid M. H., DC/DC conversion techniques and nine series luo-converters. In *Power Electronics Handbook*, Rashid M. H. and Luo F. L. *et al.* (Eds), Academic Press, San Diego, USA, 2001, pp. 335–406.
3. Mohan N., Undeland T. M. and Robbins W. P., *Power Electronics: Converters, Applications and Design*, 3rd edn., John Wiley & Sons, New York, USA, 2003.

4. Rashid, M. H., *Power Electronics: Circuits, Devices and Applications*, 2nd edn., Prentice-Hall, USA, 1993.
5. Nilsson J. W. and Riedel S. A., *Electric Circuits*, 5th edn. Addison-Wesley Publishing Company, Inc., New York, USA, 1996.
6. Irwin J. D. and Wu C. H., *Basic Engineering Circuit Analysis*, 6th edn., John Willey & Sons, Inc., New York, USA, 1999.
7. Carlson A. B., *Circuits*, Brooks/Cole Thomson Learning, New York, USA, 2000.
8. Johnson D. E., Hilburn J. L., Johnson J. R. and Scott P. D., *Basic Electric Circuit Analysis*, 5th edn., John Willey & Sons, Inc. New York, USA, 1999.
9. Grainger J. J. and Stevenson Jr. W. D., *Power System Analysis*, McGraw-Hill International Editions, New York, USA, 1994.
10. Machowski J., Bialek J. W. and Bumby J. R., *Power System Dynamics and Stability*, John Wiley & Sons, New York, USA, 1997.
11. Luo F. L. and Ye H., *Energy Factor and Mathematical Modelling for Power DC/DC Converters*, IEE-Proceedings on EPA, vol. 152, No. 2, 2005, pp. 233–248.
12. Luo F. L. and Ye H., *Mathematical Modeling for Power DC/DC Converters*, Proceedings of the IEEE International Conference POWERCON'2004, Singapore, 21–24/11/2004, pp. 323–328.
13. Padiyar K. R., *Power System Dynamics, Stability and Control*, John Wiley & Sons, New York, USA, 1996.

Chapter 2

Energy Factor (*EF*) and Sub-sequential Parameters

Switching power circuits, such as power DC/DC converters, power pulse-width-modulation (PWM) DC/AC inverters, soft-switching converters, resonant rectifiers and soft-switching AC/AC matrix converters, have pumping–filtering process, resonant process and/or voltage-lift operation. These circuits consist of several energy-storage elements. They are likely an energy container to store certain energy during performance. The stored energy will vary if the working condition changes. For example, once the power supply is on, the output voltage starts from zero since the container is not filled. The transient process from one steady state to another depends on the pumping energy and stored energy. Same reason affects the interference discovery process since the stored energy, similar to inertia, affects the impulse response.

All switching power circuits work under the switching condition with high frequency f. It is thoroughly different from traditional continuous work condition. The obvious technical feature is that all parameters perform in a period $T = 1/f$, then gradually change period-by-period. The switching period T is the clue to investigate all switching power circuits. Catching the clue, we can define many brand new concepts (parameters) to describe the characteristics of switching power circuits. These new factors fill in the blanks of the knowledge in power electronics and conversion technology. We will carefully discuss the new concepts and their applications in this chapter.

2.1 INTRODUCTION

From the introduction in previous chapter, we have got the impression of the four important factors: power factor (*PF*), power transfer efficiency (η), total harmonic distortion (*THD*) and ripple factor (*RF*) that well describe the characteristics of power systems. Unfortunately, all these factors are not available to be used to describe the characteristics of power DC/DC converters and other high-frequency switching circuits.

Power DC/DC converters have usually equipped by a DC power supply source, pump circuit, filter and load. The load can be of any type, but most of the investigations are concerned with resistive load, R, and back electromagnetic force (EMF) or battery. It means that the input and output voltages are nearly pure DC voltages with very small ripple (e.g. output voltage variation ratio is usually less than 1%). In this case, the corresponding RF is less than 0.001, which is always ignored.

Since all power is real power without reactive power jQ, we cannot use power factor PF to describe the energy-transferring process.

Since DC components exist without harmonics in input and output voltage, THD is not available to be used to describe the energy-transferring process and waveform distortion.

To simplify the research and analysis, we usually assume the condition without power losses during power-transferring process to investigate power DC/DC converters. Consequently, the efficiency $\eta = 1$ is 100% for most of the description of power DC/DC investigation. Otherwise, efficiency η must be considered for special investigations regarding the power losses.

In general conditions, all four factors are not available to apply in the analysis of power DC/DC converters. This situation makes the designers of power DC/DC converters confusing for very long time. People would like to find other new parameters to describe the characteristics of power DC/DC converters.

Energy storage in power DC/DC converters has been paid attention long time ago. Unfortunately, there is no clear concept to describe the phenomena and reveal the relationship between the stored energy and the characteristics of power DC/DC converters. We have theoretically defined a new concept, "energy factor (EF)", and researched the relationship between EF and the mathematical modeling of power DC/DC converters. EF is a new concept in power electronics and conversion technology, which thoroughly differs from the traditional concepts such as power factor (PF), power transfer efficiency (η), total harmonic distortion (THD) and ripple factor (RF). EF and the sub-sequential other parameters can illustrate the system stability, reference response and interference recovery. This investigation is very helpful for system design and DC/DC converters characteristics foreseeing.

Assuming the instantaneous input voltage and current of a DC/DC converter are, $v_1(t)$ and $i_1(t)$, and their average values are V_1 and I_1, respectively. The instantaneous output voltage and current of a DC/DC converter are, respectively, $v_2(t)$ and $i_2(t)$, and their average values are V_2 and I_2, respectively. The switching frequency is f, the switching period is $T = 1/f$, the conduction duty cycle is k and the voltage transfer gain is $M = V_2/V_1$.

2.2 PUMPING ENERGY (*PE*)

All power DC/DC converters have pumping circuit to transfer the energy from the source to some energy-storage passive elements, e.g. inductors and capacitors. The pumping energy (*PE*) is used to count the input energy in a switching period T. Its calculation formula is:

$$PE = \int_0^T P_{\text{in}}(t)\mathrm{d}t = \int_0^T V_1 i_1(t)\mathrm{d}t = V_1 I_1 T \qquad (2.1)$$

where $I_1 = \int_0^T i_1(t)\mathrm{d}t$ is the average value of the input current if the input voltage V_1 is constant. Usually, the input average current I_1 depends on the conduction duty cycle.

2.2.1 ENERGY QUANTIZATION

In switching power circuits the energy is not continuously flowing from source to actuator. The energy delivered in a switching period T from source to actuator is likely an energy quantum. Its value is the *PE*.

2.2.2 ENERGY QUANTIZATION FUNCTION

From Equation (2.1) it can be seen that the energy quantum (*PE*) is the function of switching frequency f or period T, conduction duty cycle k, input voltage v_1 and current i_1. Since the variables T, k, v_1 and i_1 can vary on time, *PE* is the time function. Usually, in a steady state the variables T, k, v_1 and i_1 cannot vary, consequently *PE* is a constant value in a steady state.

2.3 STORED ENERGY (*SE*)

Energy storage in power DC/DC converters has been paid attention long time ago. Unfortunately, there is no clear concept to describe the phenomena and reveal the relationship between the stored energy and the characteristics of power DC/DC converters.

2.3.1 STORED ENERGY IN CONTINUOUS CONDUCTION MODE

If a power DC/DC converter works in the continuous conduction mode (CCM), then all inductor's currents and capacitor's voltages are continuous (not to be equal to zero).

Stored Energy (*SE*)

The stored energy in an inductor is:

$$W_{\text{L}} = \frac{1}{2}LI_{\text{L}}^2 \qquad (2.2)$$

The stored energy across a capacitor is.

$$W_C = \frac{1}{2}CV_C^2 \tag{2.3}$$

Therefore, if there are n_L inductors and n_C capacitors, the total stored energy in a DC/DC converter is:

$$SE = \sum_{j=1}^{n_L} W_{L_j} + \sum_{j=1}^{n_C} W_{C_j} \tag{2.4}$$

Usually, the stored energy (SE) is independent from the switching frequency f (as well as the switching period T). Since the inductor currents and the capacitor voltages rely on the conduction duty cycle k, the stored energy does also rely on the conduction duty cycle k. We use the stored energy (SE) as a new parameter in further description.

Capacitor–Inductor Stored Energy Ratio (*CIR*)

Most power DC/DC converters consist of inductors and capacitors. Therefore, we can define the capacitor–inductor stored energy ratio (*CIR*) as follows:

$$CIR = \frac{\displaystyle\sum_{j=1}^{n_C} W_{C_j}}{\displaystyle\sum_{j=1}^{n_L} W_{L_j}} \tag{2.5}$$

Energy Losses (*EL*)

Usually, most analyses applied in DC/DC converters are assuming no power losses, i.e. the input power is equal to the output power, $P_{in} = P_o$ or $V_1 I_1 = V_2 I_2$, so that pumping energy is equal to output energy in a period $PE = V_1 I_1 T = V_2 I_2 T$. It corresponds to the efficiency $\eta = V_2 I_2 T/PE = 100\%$.

Particularly, power losses always exist during the conversion process. They are caused by the resistance of the connection cables, resistance of the inductor and capacitor wire, and power losses across the semiconductor devices (diode, integrated gate bipolar transistors (IGBT), power metal-oxide semiconductor field effected transistors (MOSFET) and so on). We can sort them as the resistance power losses P_r, passive element power losses P_e and device power losses P_d. The total power losses are:

$$P_{loss} = P_r + P_e + P_d$$

and

$$P_{in} = P_O + P_{loss} = P_O + P_r + P_e + P_d = V_2 I_2 + P_r + P_e + P_d$$

Therefore,

$$EL = P_{\text{loss}} \times T = (P_r + P_e + P_d)T$$

The energy losses (EL) is in a period T:

$$EL = \int_0^T P_{\text{loss}}\, \mathrm{d}t = P_{\text{loss}}T \tag{2.6}$$

Since the output energy in a period T is $(PE - EL)T$, we can define the efficiency η to be:

$$\eta = \frac{P_O}{P_{\text{in}}} = \frac{P_{\text{in}} - P_{\text{loss}}}{P_{\text{in}}} = \frac{PE - EL}{PE} \tag{2.7}$$

If there are some energy losses ($EL > 0$), then the efficiency η is smaller than unity. If there are no energy losses during conversion process ($EL = 0$), then the efficiency η is equal to unity.

Stored Energy Variation on Inductors and Capacitors (*VE*)

The current flowing through an inductor has variation (ripple) Δi_L, the variation of stored energy in an inductor is:

$$\Delta W_L = \frac{1}{2}L(I_{\text{max}}^2 - I_{\text{min}}^2) = LI_L \Delta i_L \tag{2.8}$$

where

$$I_{\text{max}} = (I_L + \Delta i_L)/2 \quad \text{and} \quad I_{\text{min}} = (I_L - \Delta i_L)/2.$$

The voltage across a capacitor has variation (ripple) Δv_C, the variation of stored energy across a capacitor is:

$$\Delta W_C = \frac{1}{2}C(V_{\text{max}}^2 - V_{\text{min}}^2) = CV_C \Delta v_C \tag{2.9}$$

where

$$V_{\text{max}} = (V_C + \Delta v_C)/2 \quad \text{and} \quad V_{\text{min}} = (V_C - \Delta v_C)/2$$

In the steady state of CCM, the total variation of the stored energy (*VE*) is:

$$VE = \sum_{j=1}^{n_L} \Delta W_{L_j} + \sum_{j=1}^{n_C} \Delta W_{C_j} \tag{2.10}$$

2.3.2 STORED ENERGY IN DISCONTINUOUS CONDUCTION MODE (DCM)

If a power DC/DC converter works in the CCM, some component's voltage and current are discontinuous. In the steady state of the discontinuous conduction situation (DCM), some minimum currents through inductors and/or some minimum voltages across capacitors become zero. We define the **filling coefficients** m_L and m_C to describe the performance in DCM.

Usually, if the switching frequency f is high enough, the inductor's current is a triangle waveform. It increases and reaches I_{max} during the switching-on period kT, and it decreases and reaches I_{min} during the switching-off period $(1-k)T$. If it becomes zero at $t = t_1$ before next switching-on, we call the converter works in DCM. The waveform of the inductor's current is shown in Figure 2.1. The time t_1 should be in the range $kT < t_1 < T$, and the filling coefficient m_L is:

$$m_L = \frac{t_1 - kT}{(1-k)T} \tag{2.11}$$

where $0 < m_L < 1$. It means the inductor's current only can fill the time period $m_L(1-k)T$ during switch-off period. In this case, I_{min} is equal to zero and the average current I_L is:

$$I_L = \frac{1}{2} I_{max}[m_L + (1 - m_L)/k] \tag{2.12}$$

and

$$\Delta i_L = I_{max} \tag{2.13}$$

Therefore,

$$\Delta W_L = L I_L \Delta i_L = \frac{1}{2} L I_{max}^2 [m_L + (1 - m_L)/k] \tag{2.14}$$

Analogously, we define the filling coefficient m_C to describe the capacitor voltage discontinuity. The waveform is shown in Figure 2.2. Time t_2 should be $kT < t_2 < T$, and the filling coefficient m_C is:

$$m_C = \frac{t_2 - kT}{(1-k)T} \tag{2.15}$$

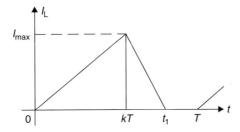

Figure 2.1 Discontinuous inductor current.

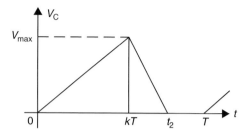

Figure 2.2   Discontinuous capacitor voltage.

where $0 < m_C < 1$. It means that the capacitor's voltage only can fill the time period $m_C(1-k)T$ during the switch-off period. In this case, V_{min} is equal to zero and the average voltage V_C is:

$$V_C = \frac{1}{2}V_{max}[m_C + (1-m_C)/k] \tag{2.16}$$

and

$$\Delta v_C = V_{max} \tag{2.17}$$

Therefore,

$$\Delta W_C = CV_C\Delta v_C = \frac{1}{2}CV_{max}^2[m_C + (1-m_C)/k] \tag{2.18}$$

We consider a converter working in DCM; it usually means only one or two energy-storage elements' voltage/current are discontinuous, and not all elements. We use the parameter VE_D to present the total variation of the stored energy:

$$VE_D = \sum_{j=1}^{n_{L-d}}\Delta W_{L_j} + \sum_{j=n_{L-d}+1}^{n_L}\Delta W_{L_j} + \sum_{j=1}^{n_{C-d}}\Delta W_{C_j} + \sum_{j=n_{C-d}+1}^{n_C}\Delta W_{C_j} \tag{2.19}$$

where n_{L-d} is the number of discontinuous inductor currents, and n_{C-d} is the number of discontinuous capacitor voltages. We have other chapters to discuss these cases. This formula form is very similar to Equation (2.10). For convenience, if there is no special necessity, we use Equation (2.10) to cover both CCM and CDM.

2.4 ENERGY FACTOR (*EF*)

As described in previous section the input energy in a period T is the pumping energy $PE = P_{in} \times T = V_{in}I_{in} \times T$. We now define that the energy factor (*EF*) is the ratio of the stored energy (*SE*) over the pumping energy (*PE*):

$$EF = \frac{SE}{PE} = \frac{SE}{V_1I_1T} = \frac{\sum\limits_{j=1}^{m}W_{L_j} + \sum\limits_{j=1}^{n}W_{C_j}}{V_1I_1T} \tag{2.20}$$

Energy factor (*EF*) is a very important factor of a power DC/DC converter. It is usually independent from the conduction duty cycle k, and proportional to the switching frequency f (inversely proportional to the) since the pumping energy (*PE*) is proportional to the switching period T.

2.5 VARIATION ENERGY FACTOR (*EF*$_V$)

We also define that the energy factor for the variation of stored energy (*EF*$_V$) is the ratio of the variation of stored energy over the pumping energy:

$$EF_V = \frac{VE}{PE} = \frac{VE}{V_1 I_1 T} = \frac{\sum\limits_{j=1}^{m} \Delta W_{L_j} + \sum\limits_{j=1}^{n} \Delta W_{C_j}}{V_1 I_1 T} \tag{2.21}$$

Energy factor (*EF*) and variation energy factor (*EF*$_V$) are available to be used to describe the characteristics of power DC/DC converters. The applications are listed in Section 2.7.

2.6 TIME CONSTANT, τ, AND DAMPING TIME CONSTANT, τ_d

We define the time constant, τ, and damping time constant, τ_d, of a power DC/DC converter in this section for the applications in Section 2.7.

2.6.1 TIME CONSTANT, τ

The **time constant**, τ, of a power DC/DC converter is a new concept to describe the transient process of a DC/DC converter. If there are no power losses in the converter, it is defined as:

$$\tau = \frac{2T \times EF}{1 + CIR} \tag{2.22}$$

This time constant (τ) is independent from switching frequency f (or period $T = 1/f$). It is available to estimate the converter responses for a unit-step function and impulse interference.

If there are power losses and $\eta < 1$, it is defined as:

$$\tau = \frac{2T \times EF}{1 + CIR}\left(1 + CIR\frac{1-\eta}{\eta}\right) \tag{2.23}$$

The time constant (τ) is still independent from switching frequency f (or period $T = 1/f$) and conduction duty cycle k. If there is no power loss and $\eta = 1$, then Equation (2.23) becomes Equation (2.22). Usually, the higher the power losses (the lower efficiency η), the larger the time constant τ since usually $CIR > 1$.

2.6.2 DAMPING TIME CONSTANT, τ_d

The **damping time constant**, τ_d, of a power DC/DC converter is a new concept to describe the transient process of a DC/DC converter. If there are no power losses, it is defined as:

$$\tau_d = \frac{2T \times EF}{1 + CIR} CIR \tag{2.24}$$

This damping time constant (τ_d) is independent from switching frequency f (or period $T = 1/f$). It is available to estimate the oscillation responses for a unit-step function and impulse interference.

If there are power losses and $\eta < 1$, then it is defined as:

$$\tau_d = \frac{2T \times EF}{1 + CIR} \frac{CIR}{\eta + CIR(1 - \eta)} \tag{2.25}$$

The damping time constant (τ_d) is also independent from switching frequency f (or period $T = 1/f$) and conduction duty cycle k. If there is no power loss and $\eta = 1$, then Equation (2.25) becomes Equation (2.24). Usually, the higher the power losses (the lower efficiency η), the smaller the damping time constant (τ_d) since usually $CIR > 1$.

2.6.3 TIME CONSTANT RATIO, ξ

The **time constant ratio**, ξ, of a power DC/DC converter is a new concept to describe the transient process of a DC/DC converter. If there are no power losses, it is defined as:

$$\xi = \frac{\tau_d}{\tau} = CIR \tag{2.26}$$

This time constant ratio is independent from switching frequency f (or period $T = 1/f$). It is available to estimate the oscillation responses for a unit-step function and impulse interference.

If there are power losses and $\eta < 1$, it is defined as:

$$\xi = \frac{\tau_d}{\tau} = \frac{CIR}{\eta \left(1 + CIR\frac{1-\eta}{\eta}\right)^2} \tag{2.27}$$

The time constant ratio is still independent from switching frequency f (or period $T = 1/f$) and conduction duty cycle k. If there is no power loss and $\eta = 1$, then Equation (2.27) becomes Equation (2.26). Usually, the higher the power losses (the lower efficiency η), the smaller the time constant ratio (ξ) since usually $CIR > 1$. From this analysis, most power DC/DC converters with lower power losses possess the output voltage oscillation when the converter operation state changed. Vice versa, power

DC/DC converters with high power losses will possess the output voltage smoothening when the converter operation state changed.

By cybernetic theory, we can estimate the unit-step function response using the ratio ξ. If the ratio ξ is equal to or smaller that 0.25 the corresponding unit-step function response has no oscillation and overshot. Vice versa, if the ratio ξ is greater that 0.25 the corresponding unit-step function response has oscillation and overshot. The higher the value of ratio ξ, the heavier the oscillation with higher overshot.

2.6.4 MATHEMATICAL MODELING FOR POWER DC/DC CONVERTERS

The mathematical modeling for all power DC/DC converters is:

$$G(s) = \frac{M}{1 + s\tau + s^2 \tau \tau_d} \tag{2.28}$$

where M is the voltage transfer gain ($M = V_2/V_1$); τ, the time constant in Equation (2.23); τ_d, the damping time constant in Equation (2.25) ($\tau_d = \xi\tau$) and s, the Laplace operator in the s-domain.

Using this mathematical model of power DC/DC converters, it is significantly easy to describe the characteristics of power DC/DC converters. In order to verify this theory, few converters are investigated to demonstrate the characteristics of power DC/DC converters and applications of the theory.

2.7 EXAMPLES OF APPLICATIONS

In order to demonstrate the parameters' calculation some examples are presented in this section. A buck converter, super-lift Luo-converter, boost converter, buck–boost converter and positive-output Luo-converter are used for this purpose.

2.7.1 A BUCK CONVERTER IN CCM

We will carefully discuss the mathematical model for buck converter in various conditions in this sub-section.

Buck Converter without Energy Losses ($r_L = 0 \, \Omega$)

A buck converter shown in Figure 2.3 has the components values: $V_1 = 40 \, \text{V}$, $L = 250 \, \mu\text{H}$ with resistance $r_L = 0 \, \Omega$, $C = 60 \, \mu\text{F}$, $R = 10 \, \Omega$, the switching frequency $f = 20 \, \text{kHz}$ ($T = 1/f = 50 \, \mu\text{s}$) and conduction duty cycle $k = 0.4$. This converter is stable and works in CCM.

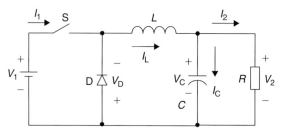

Figure 2.3   Buck converter.

Therefore, we have got the voltage transfer gain $M = 0.4$, i.e. $V_2 = V_C = MV_1 = 0.4 \times 40 = 16\,\mathrm{V}$, $I_L = I_2 = 1.6\,\mathrm{A}$, $P_{\mathrm{loss}} = 0\,\mathrm{W}$ and $I_1 = 0.64\,\mathrm{A}$. The parameter EF and others are listed below:

$$PE = V_1 I_1 T = 40 \times 0.64 \times 50\mu = 1.28\,\mathrm{mJ}$$

$$W_C = \frac{1}{2}CV_C^2 = 0.5 \times 60\mu \times 16^2 = 7.68\,\mathrm{mJ}$$

$$W_L = \frac{1}{2}LI_L^2 = 0.5 \times 250\mu \times 1.6^2 = 0.32\,\mathrm{mJ}$$

$$SE = W_L + W_C = 0.32 + 7.68 = 8\,\mathrm{mJ}$$

$$CIR = \frac{W_C}{W_L} = \frac{7.68}{0.32} = 24$$

$$EF = \frac{SE}{PE} = \frac{8}{1.28} = 6.25$$

$$EL = P_{\mathrm{loss}} \times T = 0\,\mathrm{mJ}$$

$$\eta = \frac{P_O}{P_O + P_{\mathrm{loss}}} = 1$$

$$\tau = \frac{2T \times EF}{1 + CIR}\left(1 + CIR\frac{1-\eta}{\eta}\right) = 25\,\mu\mathrm{s}$$

$$\tau_d = \frac{2T \times EF}{1 + CIR}\frac{CIR}{\eta + CIR(1-\eta)} = 625\,\mu\mathrm{s}$$

$$\xi = \frac{\tau_d}{\tau} = \frac{CIR}{\eta\left(1 + CIR\frac{1-\eta}{\eta}\right)^2} = 25 \gg 0.25$$

By cybernetic theory, since the damping time constant (τ_d) is much larger than the time constant (τ), the corresponding ratio (ξ) is $25 \gg 0.25$. The output voltage has heavy oscillation with high overshot. The corresponding transfer function is:

$$G(s) = \frac{M}{1 + s\tau + s^2\tau\tau_d} = \frac{M/\tau\tau_d}{(s + s_1)(s + s_2)} \tag{2.29}$$

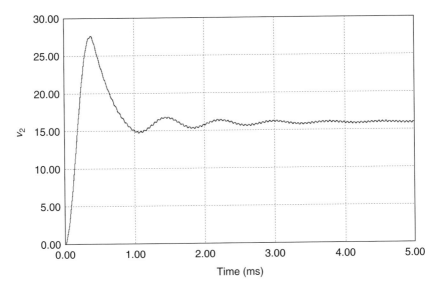

Figure 2.4   Buck converter unit-step function response.

where

$$s_1 = \sigma + j\omega \quad \text{and} \quad s_2 = \sigma - j\omega$$

with

$$\sigma = \frac{1}{2\tau_d} = \frac{1}{1200\mu s} = 833\,\text{Hz} \quad \text{and} \quad \omega = \frac{\sqrt{4\tau\tau_d - \tau^2}}{2\tau\tau_d} = \frac{\sqrt{60,000 - 625}}{30,000\mu}$$

$$= \frac{243.67}{30,000\mu} = 8122\,\text{rad/s}$$

The unit-step function response is:

$$v_2(t) = 16[1 - e^{(-t/0.0012)}(\cos 8122t - 0.1026\sin 8122t)]\,\text{V} \qquad (2.30)$$

The unit-step function response (transient process) has oscillation progress with damping factor (σ) and frequency (ω). The simulation is shown in Figure 2.4.

The impulse interference response is:

$$\Delta v_2(t) = 0.205 U e^{-t/0.0012} \sin 8122t \qquad (2.31)$$

where U is the interference signal. The impulse response (interference recovery process) has oscillation progress with damping factor (σ) and frequency (ω). The simulation is shown in Figure 2.5.

In order to verify the analysis, calculation and simulation results, we constructed a test rig with same conditions. The corresponding test results are shown in Figures 2.6 and 2.7.

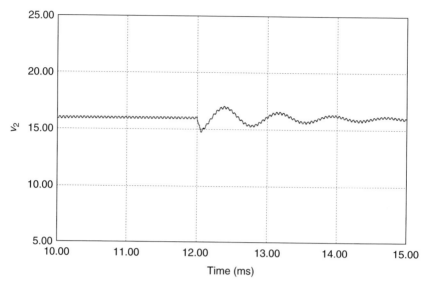

Figure 2.5   Buck converter impulse response.

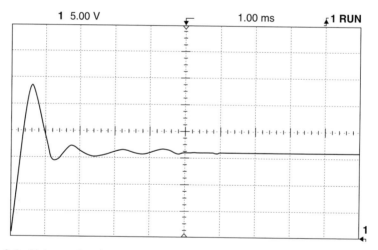

Figure 2.6   Unit-step function responses of buck converter (experiment).

Buck Converter with Small Energy Losses ($r_L = 1.5\Omega$)

A buck converter shown in Figure 2.3 has the components values: $V_1 = 40\,\text{V}$, $L = 250\,\mu\text{H}$ with resistance $r_L = 1.5\,\Omega$, $C = 60\,\mu\text{F}$, $R = 10\,\Omega$, the switching frequency $f = 20\,\text{kHz}$ ($T = 1/f = 50\,\mu\text{s}$) and conduction duty cycle $k = 0.4$. This converter is stable and works in CCM.

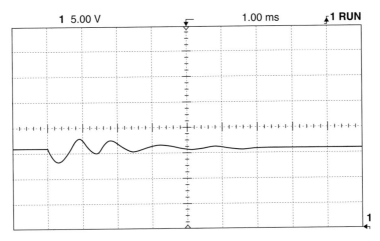

1 5.00 V				1.00 ms		⬩1 RUN

Figure 2.7   Impulse responses of buck converter (experiment).

Therefore, we have got the voltage transfer gain $M = 0.35$, i.e. $V_2 = V_C = MV_1 = 0.35 \times 40 = 14\,\text{V}$, $I_L = I_2 = 1.4\,\text{A}$, $P_{\text{loss}} = I_L^2 \times r_L = 1.4^2 \times 1.5 = 2.94\,\text{W}$ and $I_1 = 0.564\,\text{A}$. The parameter EF and others are listed below:

$$PE = V_1 I_1 T = 40 \times 0.564 \times 50\mu = 1.128\,\text{mJ}$$

$$W_C = \frac{1}{2}CV_C^2 = 0.5 \times 60\mu \times 14^2 = 5.88\,\text{mJ}$$

$$W_L = \frac{1}{2}LI_L^2 = 0.5 \times 250\mu \times 1.4^2 = 0.245\,\text{mJ}$$

$$SE = W_L + W_C = 0.245 + 5.88 = 6.125\,\text{mJ}$$

$$CIR = \frac{W_C}{W_L} = \frac{5.88}{0.245} = 24$$

$$EF = \frac{SE}{PE} = \frac{6.125}{1.128} = 5.43$$

$$EL = P_{\text{loss}} \times T = 2.94 \times 50 = 0.147\,\text{mJ}$$

$$\eta = \frac{P_O}{P_O + P_{\text{loss}}} = 0.87$$

$$\tau = \frac{2T \times EF}{1 + CIR}\left(1 + CIR\frac{1 - \eta}{\eta}\right) = 99.6\,\mu\text{s}$$

$$\tau_d = \frac{2T \times EF}{1 + CIR}\frac{CIR}{\eta + CIR(1 - \eta)} = 130.6\,\mu\text{s}$$

$$\xi = \frac{\tau_d}{\tau} = \frac{CIR}{\eta\left(1 + CIR\frac{1-\eta}{\eta}\right)^2} = 1.31 \gg 0.25$$

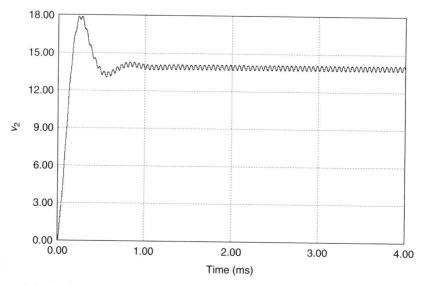

Figure 2.8 Buck converter unit-step function response ($r_L = 1.5\,\Omega$).

By cybernetic theory, since the damping time constant (τ_d) is much larger than the time constant (τ), the corresponding ratio (ξ) is $1.31 >> 0.25$. The output voltage has heavy oscillation with high overshot. The corresponding transfer function is:

$$G(s) = \frac{M}{1 + s\tau + s^2\tau\tau_d} = \frac{M/\tau\tau_d}{(s + s_1)(s + s_2)} \tag{2.32}$$

where

$$s_1 = \sigma + j\omega \quad \text{and} \quad s_2 = \sigma - j\omega$$

with

$$\sigma = \frac{1}{2\tau_d} = \frac{1}{261.2\,\mu s} = 3833\,\text{Hz} \quad \text{and} \quad \omega = \frac{\sqrt{4\tau\tau_d - \tau^2}}{2\tau\tau_d} = \frac{\sqrt{52{,}031 - 9920}}{26{,}015.5}$$

$$= \frac{205.2}{26{,}015.5\mu} = 7888\,\text{rad/s}$$

The unit-step function response is:

$$v_2(t) = 14[1 - e^{-t/0.000261}(\cos 7888t - 0.486\sin 7888t)]\,\text{V} \tag{2.33}$$

The unit-step function response (transient process) has oscillation progress with damping factor (σ) and frequency (ω). The simulation is shown in Figure 2.8.

The impulse interference response is:

$$\Delta v_2(t) = 0.975U e^{-t/0.000261}\sin 7888t \tag{2.34}$$

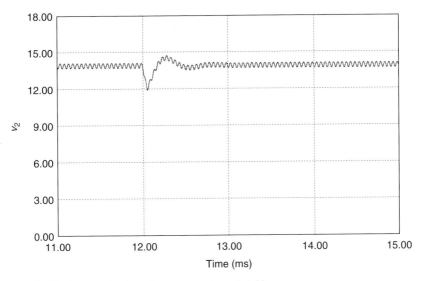

Figure 2.9 Buck converter impulse response ($r_L = 1.5 \, \Omega$).

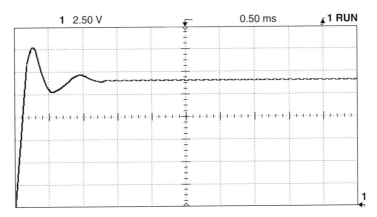

Figure 2.10 Unit-step function responses of buck converter ($r_L = 1.5 \, \Omega$ experiment).

where U is the interference signal. The impulse response (interference recovery process) has oscillation progress with damping factor (σ) and frequency (ω). The simulation is shown in Figure 2.9.

In order to verify the analysis, calculation and simulation results, we constructed a test rig with same conditions. The corresponding test results are shown in Figures 2.10 and 2.11.

Buck Converter with Energy Losses ($r_L = 4.5 \, \Omega$)

A buck converter shown in Figure 2.3 has the components values: $V_1 = 40 \, V$, $L = 250 \, \mu H$ with resistance $r_L = 4.5 \, \Omega$, $C = 60 \, \mu F$, $R = 10 \, \Omega$, the switching frequency

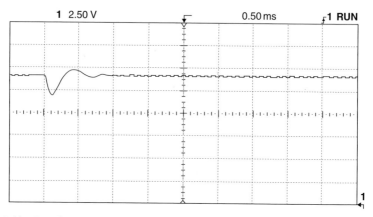

Figure 2.11   Impulse responses of buck converter ($r_L = 1.5\,\Omega$ experiment).

$f = 20\,\text{kHz}$ ($T = 1/f = 50\,\mu\text{s}$) and conduction duty cycle $k = 0.4$. This converter is stable and works in CCM.

Therefore, we have got the voltage transfer gain $M = 0.2756$, i.e. $V_2 = V_C = MV_1 = 0.2756 \times 40 = 11\,\text{V}$, $I_L = I_2 = 1.1\,\text{A}$, $P_{\text{loss}} = I_L^2 \times r_L = 1.1^2 \times 4.5 = 5.445\,\text{W}$ and $I_1 = 0.4386\,\text{A}$. The parameter EF and others are listed below:

$$PE = V_1 I_1 T = 40 \times 0.4386 \times 50\mu = 0.877\,\text{mJ}$$

$$W_C = \frac{1}{2}CV_C^2 = 0.5 \times 60\mu \times 11^2 = 3.63\,\text{mJ}$$

$$W_L = \frac{1}{2}LI_L^2 = 0.5 \times 250\mu \times 1.1^2 = 0.151\,\text{mJ}$$

$$SE = W_L + W_C = 0.151 + 3.63 = 3.781\,\text{mJ}$$

$$CIR = \frac{W_C}{W_L} = \frac{3.63}{0.151} = 24$$

$$EF = \frac{SE}{PE} = \frac{3.781}{0.877} = 4.31$$

$$EL = P_{\text{loss}} \times T = 5.445 \times 50 = 0.2722\,\text{mJ}$$

$$\eta = \frac{P_O}{P_O + P_{\text{loss}}} = 0.689$$

$$\tau = \frac{2T \times EF}{1 + CIR}\left(1 + CIR\frac{1-\eta}{\eta}\right) = 203.2\,\mu\text{s}$$

$$\tau_d = \frac{2T \times EF}{1 + CIR}\frac{CIR}{\eta + CIR(1-\eta)} = 50.8\,\mu\text{s}$$

$$\xi = \frac{\tau_d}{\tau} = \frac{CIR}{\eta\left(1 + CIR\frac{1-\eta}{\eta}\right)^2} = 0.25$$

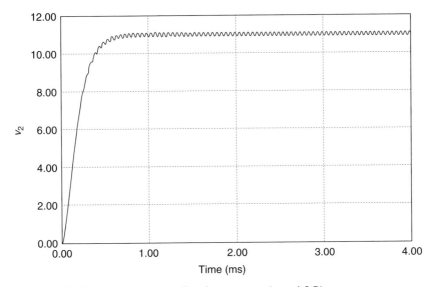

Figure 2.12 Buck converter unit-step function response ($r_L = 4.5\,\Omega$).

By cybernetic theory, since the damping time constant (τ_d) is the critical value, the corresponding ratio (ξ) is equal to 0.25. The output voltage has no oscillation. The corresponding transfer function is:

$$G(s) = \frac{M}{1 + s\tau + s^2\tau\tau_d} = \frac{M/\tau\tau_d}{(s+\sigma)^2} \tag{2.35}$$

where

$$\sigma = \frac{1}{2\tau_d} = \frac{1}{101.6\mu} = 9843\,\text{Hz}$$

The unit-step function response is:

$$v_2(t) = 11\left[1 - \left(1 + \frac{t}{0.0001016}\right)e^{-t/0.0001016}\right]\text{V} \tag{2.36}$$

The unit-step function response (transient process) has no oscillation progress with damping factor (σ). The simulation is shown in Figure 2.12.

The impulse interference response is:

$$\Delta v_2(t) = \frac{t}{0.0000508}Ue^{-t/0.0001016} \tag{2.37}$$

where U is the interference signal. The impulse response (interference recovery process) has no oscillation progress with damping factor (σ). The simulation is shown in Figure 2.13.

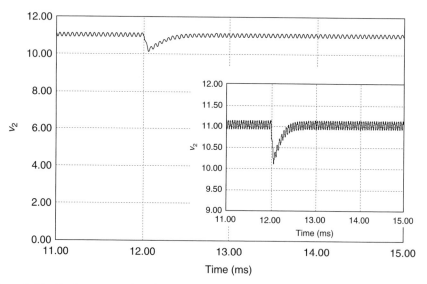

Figure 2.13   Buck converter impulse response ($r_\mathrm{L} = 4.5\,\Omega$).

Buck Converter with Large Energy Losses ($r_\mathrm{L} = 6\,\Omega$)

A buck converter shown in Figure 2.3 has the components values: $V_1 = 40\,\mathrm{V}$, $L = 250\,\mu\mathrm{H}$ with resistance $r_\mathrm{L} = 6\,\Omega$, $C = 60\,\mu\mathrm{F}$, $R = 10\,\Omega$, the switching frequency $f = 20\,\mathrm{kHz}$ ($T = 1/f = 50\,\mu\mathrm{s}$) and conduction duty cycle $k = 0.4$. This converter is stable and works in CCM.

Therefore, we have got the voltage transfer gain $M = 0.25$, i.e. $V_2 = V_\mathrm{C} = MV_1 = 0.25 \times 40 = 10\,\mathrm{V}$, $I_\mathrm{L} = I_2 = 1\,\mathrm{A}$, $P_\mathrm{loss} = I_\mathrm{L}^2 \times r_\mathrm{L} = 1^2 \times 6 = 6\,\mathrm{W}$ and $I_1 = 0.4\,\mathrm{A}$. The parameter EF and others are listed below:

$$PE = V_1 I_1 T = 40 \times 0.4 \times 50\mu = 0.8\,\mathrm{mJ}$$

$$W_\mathrm{C} = \frac{1}{2}CV_\mathrm{C}^2 = 0.5 \times 60\mu \times 10^2 = 3\,\mathrm{mJ}$$

$$W_\mathrm{L} = \frac{1}{2}LI_\mathrm{L}^2 = 0.5 \times 250\mu \times 1^2 = 0.125\,\mathrm{mJ}$$

$$SE = W_\mathrm{L} + W_\mathrm{C} = 0.125 + 3 = 3.125\,\mathrm{mJ}$$

$$CIR = \frac{W_\mathrm{C}}{W_\mathrm{L}} = \frac{3}{0.125} = 24$$

$$EF = \frac{SE}{PE} = \frac{3.125}{0.8} = 3.9$$

$$EL = P_\mathrm{loss} \times T = 6 \times 50 = 0.3\,\mathrm{mJ}$$

$$\eta = \frac{P_\mathrm{O}}{P_\mathrm{O} + P_\mathrm{loss}} = 0.625$$

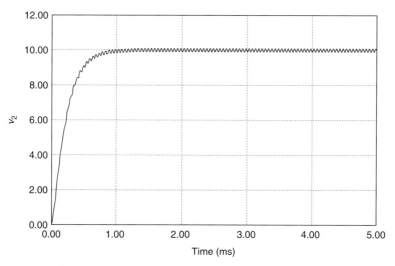

Figure 2.14 Buck converter unit-step function response ($r_L = 6\,\Omega$).

$$\tau = \frac{2T \times EF}{1 + CIR}\left(1 + CIR\frac{1 - \eta}{\eta}\right) = 240.3\,\mu s$$

$$\tau_d = \frac{2T \times EF}{1 + CIR}\frac{CIR}{\eta + CIR(1 - \eta)} = 38.9\,\mu s$$

$$\xi = \frac{\tau_d}{\tau} = \frac{CIR}{\eta\left(1 + CIR\frac{1-\eta}{\eta}\right)^2} = 0.162 < 0.25$$

By cybernetic theory, since the damping time constant (τ_d) is smaller than the time constant (τ), the corresponding ratio (ξ) is $0.162 < 0.25$. The output voltage has no oscillation. The corresponding transfer function is:

$$G(s) = \frac{M}{1 + s\tau + s^2\tau\tau_d} = \frac{M/\tau\tau_d}{(s + \sigma_1)(s + \sigma_2)} \tag{2.38}$$

with

$$\sigma_1 = \frac{\tau + \sqrt{4\tau\tau_d - \tau^2}}{2\tau\tau_d} = \frac{240.3 + 142.66}{18695.3\mu} = 20{,}500\,\text{Hz}$$

$$\sigma_2 = \frac{\tau - \sqrt{4\tau\tau_d - \tau^2}}{2\tau\tau_d} = \frac{240.3 - 142.66}{18695.3\mu} = 5200\,\text{Hz}$$

The unit-step function response is:

$$v_2(t) = 10(1 + 0.342e^{-20500t} - 1.342e^{-5200t})\,\text{V} \tag{2.39}$$

The unit-step function response (transient process) has no oscillation progress with damping factor (σ_2). The simulation is shown in Figure 2.14.

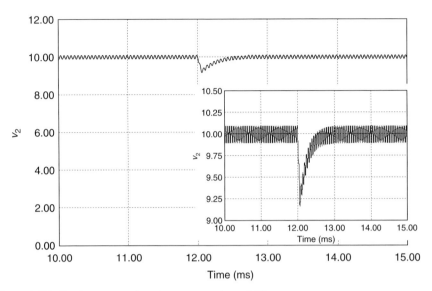

Figure 2.15 Buck converter impulse response ($r_L = 6\,\Omega$).

The impulse interference response is:

$$\Delta v_2(t) = 1.684 U(\mathrm{e}^{-20500t} - \mathrm{e}^{-5200t}) \tag{2.40}$$

where U is the interference signal. The impulse response (interference recovery process) has no oscillation progress with damping factor (σ_2). The simulation is shown in Figure 2.15.

2.7.2 A SUPER-LIFT LUO-CONVERTER IN CCM

Figure 2.16 shows a super-lift Luo-converter with the conduction duty $k = 0.5$. The components values are $V_1 = 20\,$V, $f = 50\,$kHz ($T = 20\,\mu$s), $L = 100\,\mu$H with resistance $r_L = 0.12\,\Omega$, $C_1 = 2500\,\mu$F, $C_2 = 800\,\mu$F and $R = 10\,\Omega$. This converter is stable and works in CCM.

Therefore, we have got the voltage transfer gain $M = 2.863$, i.e. the output voltage $V_2 = V_{C2} = 57.25\,$V, $V_{C1} = V_1 = 20\,$V, $I_1 = 17.175\,$A, $I_2 = 5.725\,$A, $I_L = 11.45\,$A and $P_{\mathrm{loss}} = I_L^2 \times r_L = 11.45^2 \times 0.12 = 15.73\,$W. The parameter EF and others are listed below:

$$PE = V_1 I_1 T = 20 \times 17.175 \times 20\mu = 6.87\,\mathrm{mJ}$$

$$W_L = \frac{1}{2} L I_L^2 = 0.5 \times 100\mu \times 11.45^2 = 6.555\,\mathrm{mJ}$$

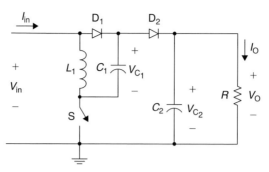

Figure 2.16 Super-lift Luo-converter.

$$W_{C_1} = \frac{1}{2}C_1V_{C_1}^2 = 0.5 \times 2500\mu \times 20^2 = 500\,\text{mJ}$$

$$W_{C_2} = \frac{1}{2}C_2V_{C_2}^2 = 0.5 \times 800\mu \times 57.25^2 = 1311\,\text{mJ}$$

$$SE = W_L + W_{C_1} + W_{C_2} = 6.555 + 500 + 1311 = 1817.6\,\text{mJ}$$

$$EF = \frac{SE}{PE} = \frac{1817.6}{6.87} = 264.6$$

$$EL = P_{\text{loss}}T = 15.73 \times 20 = 0.3146\,\text{mJ}$$

$$CIR = \frac{W_{C_1} + W_{C_2}}{W_L} = \frac{1811}{6.555} = 276.3$$

$$\eta = \frac{P_O}{P_O + P_{\text{loss}}} = \frac{327.76}{343.49} = 0.9542$$

$$\tau = \frac{2T \times EF}{1 + CIR}\left(1 + CIR\frac{1 - \eta}{\eta}\right) = \frac{40\mu \times 264.6 \times 13.26}{277.3} = 506\,\mu\text{s}$$

$$\tau_d = \frac{2T \times EF}{1 + CIR}\frac{CIR}{\eta + CIR(1 - \eta)} = \frac{40 \times 264.6 \times 20.3}{277.3} = 775\,\mu\text{s}$$

By cybernetic theory, since the damping time constant (τ_d) is much larger than the time constant (τ), the corresponding ratio (ξ) $= 775/506 = 1.53 >> 0.25$. The output voltage has heavy oscillation with high overshot. The transfer function of this converter has two poles ($-s_1$ and $-s_2$) that are located in the left-hand half plane (LHHP):

$$G(s) = \frac{M}{1 + s\tau + s^2\tau\tau_d} = \frac{M/\tau\tau_d}{(s + s_1)(s + s_2)} \tag{2.41}$$

where

$$s_1 = \sigma + j\omega \quad \text{and} \quad s_2 = \sigma - j\omega$$

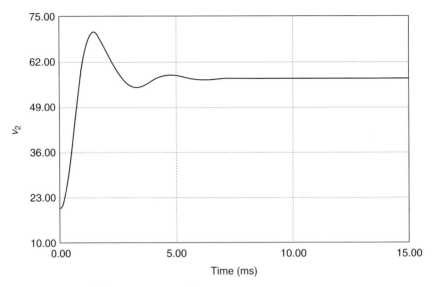

Figure 2.17   Super-lift Luo-converter unit-step responses.

with

$$\sigma = \frac{1}{2\tau_d} = \frac{1}{1.55\,\text{ms}} = 645\,\text{Hz} \quad \text{and} \quad \omega = \frac{\sqrt{4\tau\tau_d - \tau^2}}{2\tau\tau_d} = \frac{\sqrt{1{,}686{,}400 - 295{,}936}}{843{,}200}$$

$$= \frac{1197.2}{843{,}200\mu} = 1398\,\text{rad/s}$$

The unit-step function response is:

$$v_2(t) = 57.25[1 - e^{-t/0.00155}(\cos 1398t - 0.461 \sin 1398t)]\,\text{V} \qquad (2.42)$$

The unit-step function response (transient process) has oscillation progress with damping factor (σ) and frequency (ω) the simulation is shown in Figure 2.17.
    The impulse interference response is:

$$\Delta v_2(t) = 0.923 U e^{-t/0.00155} \sin 1398t \qquad (2.43)$$

where U is the interference signal. The impulse response (interference recovery process) has oscillation progress with damping factor (σ) and frequency (ω), and is shown in Figure 2.18.
    In order to verify the analysis, calculation and simulation results, we constructed a test rig with same conditions. The corresponding test results are shown in Figures 2.19 and 2.20.

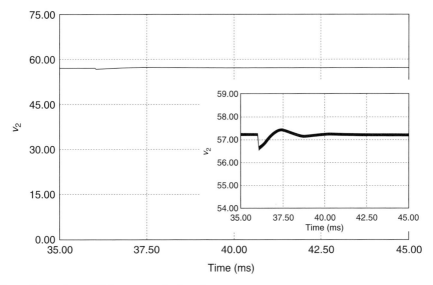

Figure 2.18　Super-lift Luo-converter impulse responses.

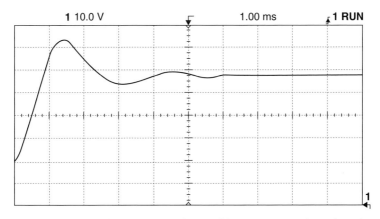

Figure 2.19　Unit-step function responses of super-lift Luo-converter (experiment).

2.7.3 A Boost Converter in CCM (no Power Losses)

A boost converter shown in Figure 2.21 has the components values: $V_1 = 40$ V, $L = 250\,\mu$H, $C = 60\,\mu$F, $R = 10\,\Omega$, the switching frequency $f = 20$ kHz ($T = 1/f = 50\,\mu$s) and conduction duty cycle $k = 0.6$. This converter is stable and works in CCM.

Therefore, we have got the voltage transfer gain $M = 1/(1-k) = 2.5$, i.e. $V_2 = V_C = V_1/(1-k) = 100$ V, $I_2 = 10$ A, and $I_1 = I_L = 25$ A. The parameter EF and

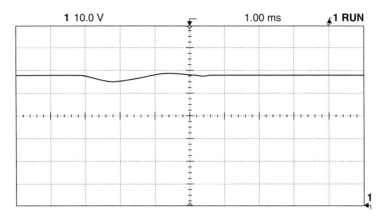

Figure 2.20   Impulse responses of super-lift Luo-converter (experiment).

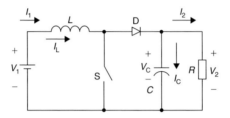

Figure 2.21   Boost converter.

others are listed below:

$$PE = V_1 I_1 T = 40 \times 25 \times 50\mu = 50\,\text{mJ}$$

$$W_{\text{L}} = \frac{1}{2}LI_{\text{L}}^2 = 0.5 \times 250\mu \times 25^2 = 78.125\,\text{mJ}$$

$$W_{\text{C}} = \frac{1}{2}CV_{\text{C}}^2 = 0.5 \times 60\mu \times 100^2 = 300\,\text{mJ}$$

$$SE = W_{\text{L}} + W_{\text{C}} = 78.125 + 300 = 378.125\,\text{mJ}$$

$$EF = \frac{SE}{PE} = \frac{378.125}{50} = 7.5625$$

$$CIR = \frac{W_{\text{C}}}{W_{\text{L}}} = \frac{300}{78.125} = 3.84$$

Since there are no power losses in the converter, $EL = 0$ and $\eta = 1$:

$$\tau = \frac{2T \times EF}{1 + CIR}\left(1 + CIR\frac{1-\eta}{\eta}\right) = \frac{100\mu \times 7.5625}{1 + 3.84} = 156.25\,\mu\text{s}$$

$$\tau_d = \frac{2T \times EF}{1 + CIR} \frac{CIR}{\eta + CIR(1 - \eta)} = \frac{100 \times 7.5625 \times 3.84}{4.84} = 600\,\mu s$$

$$\xi = \frac{\tau_d}{\tau} = \frac{CIR}{\eta \left(1 + CIR\frac{1-\eta}{\eta}\right)^2} = 3.84 > 0.25$$

By cybernetic theory, since the damping time constant (τ_d) is much larger than the time constant (τ), the corresponding ratio (ξ) is $3.84 > 0.25$. The output voltage has heavy oscillation with high overshot. The transfer function of this converter has two poles ($-s_1$ and $-s_2$) that are located in the LHHP:

$$G(s) = \frac{M}{1 + s\tau + s^2\tau\tau_d} = \frac{M/\tau\tau_d}{(s + s_1)(s + s_2)} \tag{2.44}$$

where

$$s_1 = \sigma + j\omega \quad \text{and} \quad s_2 = \sigma - j\omega$$

with

$$\sigma = \frac{1}{2\tau_d} = \frac{1}{1.2\,ms} = 833\,Hz \quad \text{and} \quad \omega = \frac{\sqrt{4\tau\tau_d - \tau^2}}{2\tau\tau_d} = \frac{\sqrt{375,000 - 24,414}}{187,500}$$

$$= \frac{592.1}{187,500\mu} = 3158\,rad/s$$

The unit-step function response is:

$$v_2(t) = 100[1 - e^{-t/0.0012}(\cos 3158t - 0.264 \sin 3158t)]\,V \tag{2.45}$$

The unit-step function response (transient process) has oscillation progress with damping factor (σ) and frequency (ω). The simulation is shown in Figure 2.22.

The impulse interference response is:

$$\Delta v_2(t) = 0.528Ue^{-t/0.0012} \sin 3158t \tag{2.46}$$

where U is the interference signal. The impulse response (interference recovery process) has oscillation progress with damping factor (σ) and frequency (ω), and is shown in Figure 2.23.

2.7.4 A BUCK–BOOST CONVERTER IN CCM (NO POWER LOSSES)

A boost converter shown in Figure 2.24 has the components values: $V_1 = 40\,V$, $L = 250\,\mu H$, $C = 60\,\mu F$, $R = 10\,\Omega$, the switching frequency $f = 20\,kHz$ ($T = 1/f = 50\,\mu s$) and conduction duty cycle $k = 0.6$. This converter is stable and works in CCM.

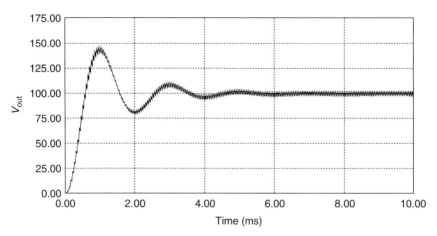

Figure 2.22   Boost converter unit-step responses.

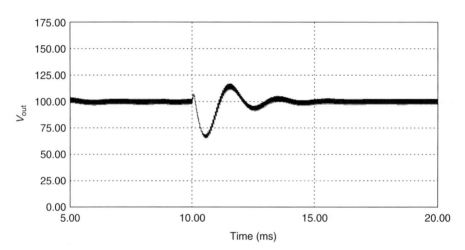

Figure 2.23   Boost converter impulse responses.

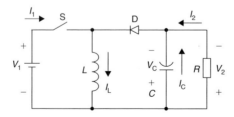

Figure 2.24   Buck–boost converter.

Therefore, we have got the voltage transfer gain $M = k/(1-k) = 1.5$, i.e. $V_2 = V_C = kV_1/(1-k) = 60\,\text{V}$, $I_2 = 6\,\text{A}$, $I_1 = 9\,\text{A}$ and $I_L = 15\,\text{A}$. The parameter EF and others are listed below:

$$PE = V_1 I_1 T = 40 \times 9 \times 50\mu = 18\,\text{mJ}$$

$$W_L = \frac{1}{2}LI_L^2 = 0.5 \times 250\mu \times 15^2 = 28.125\,\text{mJ}$$

$$W_C = \frac{1}{2}CV_C^2 = 0.5 \times 60\mu \times 60^2 = 108\,\text{mJ}$$

$$SE = W_L + W_C = 28.125 + 108 = 136.125\,\text{mJ}$$

$$CIR = \frac{W_C}{W_L} = \frac{108}{28.125} = 3.84$$

$$EF = \frac{SE}{PE} = \frac{136.125}{18} = 7.5625$$

Since there are no power losses in the converter, $EL = 0$ and $\eta = 1$:

$$\tau = \frac{2T \times EF}{1 + CIR}\left(1 + CIR\frac{1-\eta}{\eta}\right) = \frac{100\mu \times 7.5625}{1 + 3.84} = 156.25\,\mu\text{s}$$

$$\tau_d = \frac{2T \times EF}{1 + CIR}\frac{CIR}{\eta + CIR(1 - \eta)} = \frac{100 \times 7.5625 \times 3.84}{4.84} = 600\,\mu\text{s}$$

$$\xi = \frac{\tau_d}{\tau} = \frac{CIR}{\eta\left(1 + CIR\frac{1-\eta}{\eta}\right)^2} = 3.84 > 0.25$$

By cybernetic theory, since the damping time constant (τ_d) is much larger than the time constant (τ), the corresponding ratio (ξ) is $3.84 > 0.25$. The output voltage has heavy oscillation with high overshot. The transfer function of this converter has two poles ($-s_1$ and $-s_2$) that are located in the LHHP:

$$G(s) = \frac{M}{1 + s\tau + s^2\tau\tau_d} = \frac{M/\tau\tau_d}{(s + s_1)(s + s_2)} \tag{2.47}$$

where

$$s_1 = \sigma + j\omega \quad \text{and} \quad s_2 = \sigma - j\omega$$

with

$$\sigma = \frac{1}{2\tau_d} = \frac{1}{1.2\,\text{ms}} = 833\,\text{Hz} \quad \text{and} \quad \omega = \frac{\sqrt{4\tau\tau_d - \tau^2}}{2\tau\tau_d} = \frac{\sqrt{375,000 - 24,414}}{187,500}$$

$$= \frac{592.1}{187,500\mu} = 3158\,\text{rad/s}$$

The unit-step function response is:

$$v_2(t) = 60[1 - e^{-t/0.0012}(\cos 3158t - 0.264 \sin 3158t)] \text{ V} \qquad (2.48)$$

The unit-step function response (transient process) has oscillation progress with damping factor (σ) and frequency (ω). The simulation is shown in Figure 2.25.

The impulse interference response is:

$$\Delta v_2(t) = 0.528 U e^{-t/0.0012} \sin 3158t \qquad (2.49)$$

where U is the interference signal. The impulse response (interference recovery process) has oscillation progress with damping factor (σ) and frequency (ω), and is shown in Figure 2.26.

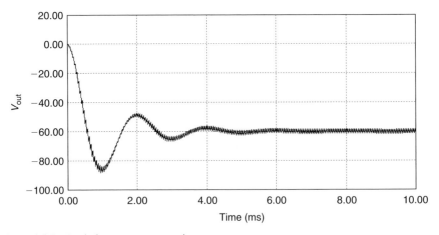

Figure 2.25   Buck–boost converter unit-step responses.

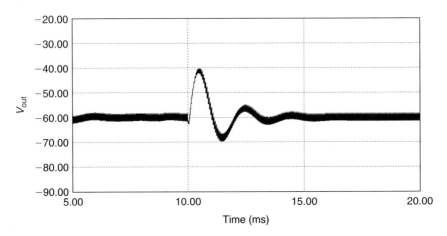

Figure 2.26   Buck–boost converter impulse responses.

2.7.5 POSITIVE-OUTPUT LUO-CONVERTER IN CCM (NO POWER LOSSES)

Figure 2.27 shows a positive-output Luo-converter with the conduction duty k. The components values are $V_1 = 20\,\text{V}$, $f = 50\,\text{kHz}$ $(T = 20\,\mu\text{s})$, $L_1 = L_2 = 1\,\text{mH}$, $k = 0.5$, $C_1 = C_2 = 20\,\mu\text{F}$ and $R = 10\,\Omega$. This converter is stable and works in CCM.

Therefore, we have got the voltage transfer gain $M = k/(1-k) = 1$, i.e. the output voltage $V_2 = V_{C2} = kV_1/(1-k) = 40\,\text{V}$, $V_{C1} = V_1 = 40\,\text{V}$, $I_1 = 4\,\text{A}$, $I_2 = 4\,\text{A}$ and $I_{L_1} = I_{L_2} = 4\,\text{A}$. The parameter EF and others are listed below:

$$PE = V_1 I_1 T = 40 \times 4 \times 20\mu = 3.2\,\text{mJ}$$

$$W_{C_1} = \frac{1}{2}C_1 V_{C_1}^2 = 0.5 \times 20\mu \times 40^2 = 16\,\text{mJ}$$

$$W_{C_2} = \frac{1}{2}C_2 V_{C_2}^2 = 0.5 \times 20\mu \times 40^2 = 16\,\text{mJ}$$

$$W_{L_1} = \frac{1}{2}L_1 I_{L_1}^2 = 0.5 \times 1m \times 4^2 = 8\,\text{mJ}$$

$$W_{L_2} = \frac{1}{2}L_2 I_{L_2}^2 = 0.5 \times 1m \times 4^2 = 8\,\text{mJ}$$

$$SE = W_{L_1} + W_{L_2} + W_{C_1} + W_{C_2} = 16 + 32 = 48\,\text{mJ}$$

$$EF = \frac{SE}{PE} = \frac{48}{3.2} = 15$$

$$CIR = \frac{W_{C_1} + W_{C_2}}{W_{L_1} + W_{L_2}} = \frac{32}{16} = 2$$

Since there are no power losses in the converter, $EL = 0$ and $\eta = 1$:

$$\tau = \frac{2T \times EF}{1 + CIR}\left(1 + CIR\frac{1-\eta}{\eta}\right) = \frac{40\mu \times 15}{3} = 200\,\mu\text{s}$$

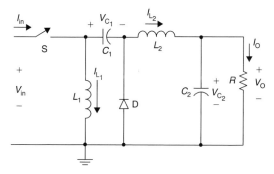

Figure 2.27 Positive-output Luo-converter.

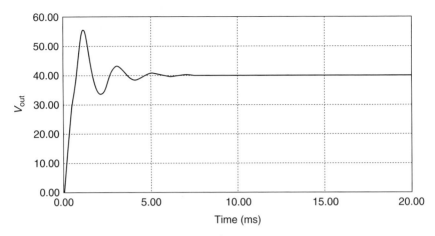

Figure 2.28 Positive-output Luo-converter unit-step responses.

$$\tau_d = \frac{2T \times EF}{1 + CIR} \frac{CIR}{\eta + CIR(1 - \eta)} = \frac{40\mu \times 15 \times 2}{3} = 400 \, \mu s$$

$$\xi = \frac{\tau_d}{\tau} = \frac{CIR}{\eta \left(1 + CIR\frac{1-\eta}{\eta}\right)^2} = 2 > 0.25$$

By cybernetic theory, since the damping time constant (τ_d) is much larger than the time constant (τ), the corresponding ratio (ξ) is $2 > 0.25$. The output voltage has no oscillation and overshot. The transfer function of this converter has two real poles ($-s_1$ and $-s_2$) that are located in the LHHP:

$$G(s) = \frac{M}{1 + s\tau + s^2 \tau \tau_d} = \frac{M/\tau \tau_d}{(s + s_1)(s + s_2)} \tag{2.50}$$

where

$$s_1 = \sigma + j\omega \quad \text{and} \quad s_2 = \sigma - j\omega$$

with

$$\sigma = \frac{1}{2\tau_d} = \frac{1}{0.8 \, ms} = 1250 \, Hz \quad \text{and} \quad \omega = \frac{\sqrt{4\tau\tau_d - \tau^2}}{2\tau\tau_d} = \frac{\sqrt{320,000 - 40,000}}{160,000}$$

$$= \frac{529.2}{160,000\mu} = 3307 \, rad/s$$

The unit-step function response is:

$$v_2(t) = 40[1 - e^{-t/0.0008}(\cos 3307t - 0.378 \sin 3307t)] \, V \tag{2.51}$$

The unit-step function response (transient process) has oscillation progress with damping factor (σ) and frequency (ω). The simulation is shown in Figure 2.28.

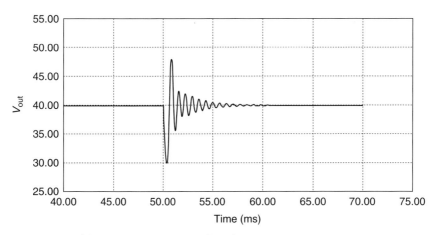

Figure 2.29 Positive-output Luo-converter impulse responses.

The impulse interference response is:

$$\Delta v_2(t) = 0.756 U e^{-t/0.0008} \sin 3307t \tag{2.52}$$

where U is the interference signal. The impulse response (interference recovery process) has oscillation progress with damping factor (σ) and frequency (ω), and is shown in Figure 2.29.

2.8 SMALL SIGNAL ANALYSIS

We analyzed the characteristics of power DC/DC converters in large signal operation in Section 2.6.4. We analyze the characteristics of power DC/DC converters in small signal operation in this section. It will verify that the transfer function (2.28) is generally correct for both large and small signal analyses, and it describes the native characteristics of a power DC/DC converter.

If the conduction duty cycle (k) changes from k_1 to k_2 ($\Delta k = k_2 - k_1$) in a small increment to the new value ($k_2 = k_1 + \Delta k$), the pumping energy PE has to change correspondingly in an increment to the new value ($PE + \Delta PE$). Analogously, the inductor currents and capacitor voltages have to change correspondingly, and the stored energy SE changes to ($SE + \Delta SE$):

$$\Delta PE = \int_0^T V_1 i_1(t)|_{k=k_2} \, dt - \int_0^T V_1 i_1(t)|_{k=k_1} \, dt = V_1(I_{1-k_2} - I_{1-k_1})T = V_1 \Delta I_1 T \tag{2.53}$$

The stored energy in an inductor is:

$$\Delta W_L = \frac{1}{2} L(I_{L-k_2}^2 - I_{L-k_1}^2) \tag{2.54}$$

The stored energy across a capacitor is

$$\Delta W_{\mathrm{C}} = \frac{1}{2} C (V_{\mathrm{C}-k_2}^2 - V_{\mathrm{C}-k_1}^2) \tag{2.55}$$

Therefore, if there are n_{L} inductors and n_{C} capacitors the total stored energy in a DC/DC converter is:

$$\Delta SE = \sum_{j=1}^{n_{\mathrm{L}}} \Delta W_{\mathrm{L}j} + \sum_{j=1}^{n_{\mathrm{C}}} \Delta W_{\mathrm{C}j} \tag{2.56}$$

We define the energy factor EF in small signal operation as:

$$EF = \frac{\Delta SE}{\Delta PE} = \frac{\sum_{j=1}^{m} \Delta W_{\mathrm{L}j} + \sum_{j=1}^{n} \Delta W_{\mathrm{C}j}}{V_1 \Delta I_1 T} \tag{2.57}$$

Correspondingly, the capacitor/inductor stored energy ratio (CIR) is:

$$CIR = \frac{\sum_{j=1}^{n_{\mathrm{C}}} \Delta W_{\mathrm{C}j}}{\sum_{j=1}^{n_{\mathrm{L}}} \Delta W_{\mathrm{L}j}} \tag{2.58}$$

The energy losses increment (ΔEL) in a period T is defined as:

$$\Delta EL = \Delta P_{\mathrm{loss}} \times T \tag{2.59}$$

so that the efficiency η is:

$$\eta = \frac{\Delta PE - \Delta EL}{\Delta PE} \tag{2.60}$$

Although the **time constant** (τ), **damping time constant** (τ_{d}) and **time constant ratio** (ξ) are not changed, they are still defined in the same forms as:

$$\tau = \frac{2T \times EF}{1 + CIR} \left(1 + CIR \frac{1 - \eta}{\eta} \right) \tag{2.61}$$

$$\tau_{\mathrm{d}} = \frac{2T \times EF}{1 + CIR} \frac{CIR}{\eta + CIR(1 - \eta)} \tag{2.62}$$

$$\xi = \frac{\tau_{\mathrm{d}}}{\tau} = \frac{CIR}{\eta \left(1 + CIR \frac{1 - \eta}{\eta} \right)^2} \tag{2.63}$$

The transfer function is not changed:

$$G(s) = \frac{M}{1 + s\tau + s^2\tau\tau_d} = \frac{M}{1 + s\tau + \xi s^2\tau^2} \qquad (2.64)$$

In order to verify this theory and offer examples to readers, we prepare two converters: a buck converter and super-lift Luo-converter to demonstrate the characteristics of power DC/DC converters and applications of the theory.

2.8.1 A BUCK CONVERTER IN CCM WITHOUT ENERGY LOSSES ($r_L = 0$)

A buck converter shown in Figure 2.3 has the components values: $V_1 = 40\,\text{V}, L = 250\,\mu\text{H}$ with resistance $r_L = 0\,\Omega$, $C = 60\,\mu\text{F}$, $R = 10\,\Omega$, the switching frequency $f = 20\,\text{kHz}$ ($T = 1/f = 50\,\mu\text{s}$) and conduction duty cycle (k) changing from 0.4 to 0.5. This converter is stable and works in CCM.

Therefore, we have got the voltage transfer gain $M = 0.5$, i.e. $V_2 = V_C = MV_1 = 0.5 \times 40 = 20\,\text{V}$, $I_L = I_2 = 2\,\text{A}$, $P_{\text{loss}} = 0\,\text{W}$ and $I_1 = 1\,\text{A}$. The increments are $\Delta V_2 = 4\,\text{V}$, $\Delta I_2 = \Delta I_L = 0.4\,\text{A}$, $\Delta I_1 = 0.34\,\text{A}$. The parameter EF and others are listed below:

$$\Delta PE = V_1 \Delta I_1 T = 40 \times 0.36 \times 50\mu = 0.72\,\text{mJ}$$

$$\Delta W_C = \frac{1}{2}C(V_{C-0.5}^2 - V_{C-0.4}^2) = 4.32\,\text{mJ}$$

$$\Delta W_L = \frac{1}{2}L(I_{L-0.5}^2 - I_{L-0.4}^2) = 0.5 \times 250\mu \times (2^2 - 1.6^2) = 0.18\,\text{mJ}$$

$$\Delta SE = \Delta W_L + \Delta W_C = 4.5\,\text{mJ}$$

$$\Delta EL = 0\,\text{mJ}$$

$$\eta = \frac{\Delta PE - \Delta EL}{\Delta PE} = 1$$

$$EF = \frac{\Delta SE}{\Delta PE} = \frac{4.5}{0.72} = 6.25$$

$$CIR = \frac{\Delta W_C}{\Delta W_L} = \frac{4.32}{0.18} = 24$$

$$\tau = \frac{2T \times EF}{1 + CIR}\left(1 + CIR\frac{1 - \eta}{\eta}\right) = 25\,\mu\text{s}$$

$$\tau_d = \frac{2T \times EF}{1 + CIR}\frac{CIR}{\eta + CIR(1 - \eta)} = 600\,\mu\text{s}$$

From the above calculation and analysis we found out that the time constants are not changed. Therefore, the transfer function for small signal operation should not be

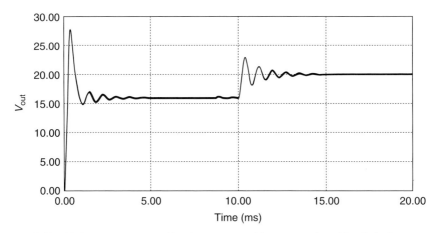

Figure 2.30   Unit-step responses of buck converter without power loss (simulation).

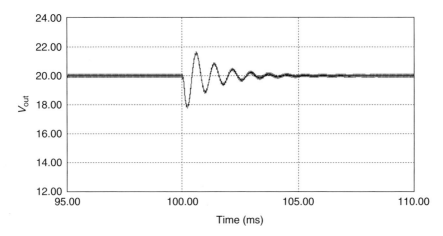

Figure 2.31   Impulse responses of buck converter without power loss (small signal).

changed, which is still Equation (2.29). Correspondingly, the unit-step response is:

$$v_2(t) = 16 + 4[1 - e^{-t/0.0012}(\cos 8122t - 0.1026 \sin 8122t)] \text{ V} \qquad (2.65)$$

The unit-step function response for large signal ($k = 0$–0.4) and small signal ($k = 0.4$–0.5) operation is shown in Figure 2.30 for comparison with each other.

The impulse response for small signal is described as:

$$\Delta v_2(t) = 0.205U e^{-t/0.0012} \sin 8122t \qquad (2.66)$$

where U is the interference signal. The small-signal impulse response is shown in Figure 2.31.

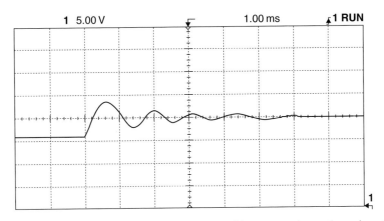

Figure 2.32   Unit-step responses of buck converter without power losses (experiment, small signal).

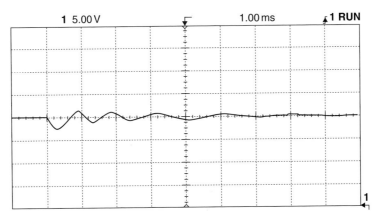

Figure 2.33   Small signal impulse responses of buck converter without power losses (experiment).

In order to verify this analysis and compare the simulation results to experimental results, a test rig was constructed. The conduction duty cycle (k) changes from 0.4 to 0.5. The experimental results for unit-step response ($k = 0.4$–0.5) and impulse interference response ($k = 0.5$–0.4) are shown in Figures 2.32 and 2.33. We can see that both the simulation and experimental results are identical.

2.8.2 BUCK-CONVERTER WITH SMALL ENERGY LOSSES ($r_L = 1.5\,\Omega$)

A buck converter shown in Figure 2.3 has the components values: $V_1 = 40\,\text{V}$, $L = 250\,\mu\text{H}$ with resistance $r_L = 1.5\,\Omega$, $C = 60\,\mu\text{F}$, $R = 10\,\Omega$, the switching frequency

$f = 20\,\text{kHz}\ (T = 1/f = 50\,\mu\text{s})$ and conduction duty cycle (k) changing from 0.4 to 0.5. This converter is stable and works in CCM.

We have got the voltage transfer gain $M = 0.435$, i.e. $V_2 = V_C = MV_1 = 0.435 \times 40 = 17.4\,\text{V}$, $I_L = I_2 = 17.4\,\text{A}$, $P_{\text{loss}} = I_L^2 r_L = 1.74^2 \times 1.5 = 4.54\,\text{W}$, $I_1 = 0.871\,\text{A}$, and $\Delta V_2 = 3.4\,\text{V}$, $\Delta I_2 = \Delta I_L = 0.34\,\text{A}$, $\Delta P_{\text{loss}} = (I_{L-0.5}^2 - I_{L-0.4}^2) r_L = (1.74^2 - 1.4^2)1.5 = 1.6\,\text{W}$ and $\Delta I_1 = 0.313\,\text{A}$. The parameters are listed below:

$$\Delta PE = V_1 \Delta I_1 T = 40 \times 0.313 \times 50\mu = 0.626\,\text{mJ}$$

$$\Delta W_C = \frac{1}{2}C(V_{C-0.5}^2 - V_{C-0.5}^2) = 3.284\,\text{mJ}$$

$$\Delta W_L = \frac{1}{2}L(I_{L-0.5}^2 - I_{L-0.4}^2) = 0.136\,\text{mJ}$$

$$\Delta SE = \Delta W_L + \Delta W_C = 3.42\,\text{mJ}$$

$$\Delta EL = \Delta P_{\text{loss}} \times T = 1.6 \times 50 = 0.08\,\text{mJ}$$

$$\eta = \frac{\Delta PE - \Delta EL}{\Delta PE} = \frac{0.546}{0.626} = 0.87$$

$$EF = \frac{\Delta SE}{\Delta PE} = \frac{3.42}{0.626} = 5.43$$

$$CIR = \frac{\Delta W_C}{\Delta W_L} = \frac{3.284}{0.136} = 24$$

$$\tau = \frac{2T \times EF}{1 + CIR}\left(1 + CIR\frac{1 - \eta}{\eta}\right) = 100\,\mu\text{s}$$

$$\tau_d = \frac{2T \times EF}{1 + CIR}\frac{CIR}{\eta + CIR(1 - \eta)} = 130.6\,\mu\text{s}$$

From the above calculation and analysis we found out that the time constants are not changed. Therefore, the transfer function for small signal operation should not be changed, which is still Equation (2.32). Correspondingly, the unit-step response is:

$$v_2(t) = 14 + 3.4[1 - e^{-t/0.000261}(\cos 7888t - 0.486 \sin 7888t)]\,\text{V} \qquad (2.67)$$

The unit-step function response for large ($k = 0$–0.4) and small signal ($k = 0.4$–0.5) operation is shown in Figure 2.34 for comparison with each other. The impulse response for small signal is described as:

$$\Delta v_2(t) = 0.973 U e^{-t/0.000261} \sin 7888t \qquad (2.68)$$

where U is the interference signal. The small-signal impulse response is shown in Figure 2.35.

In order to verify this analysis and compare the simulation results to experimental results, a test rig was constructed. The conduction duty cycle (k) changes from 0.4 to 0.5.

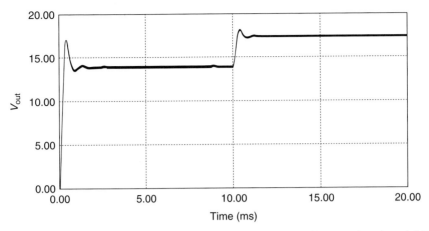

Figure 2.34 Unit-step responses (simulation) of buck converter with power loss ($r_L = 1.5\,\Omega$).

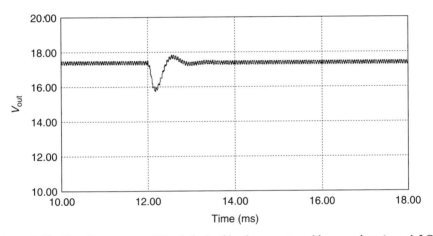

Figure 2.35 Impulse responses (simulation) of buck converter with power loss ($r_L = 1.5\,\Omega$, small signal).

The experimental results for unit-step response ($k = 0.4$–0.5) and impulse interference response ($k = 0.5$–0.4) are shown in Figures 2.36 and 2.37.

We can see that both the simulation and experimental results are identical.

2.8.3 SUPER-LIFT LUO-CONVERTER WITH ENERGY LOSSES ($r_L = 0.12\,\Omega$)

Figure 2.16 shows a super-lift Luo-converter with the conduction duty (k) changing from 0.5 to 0.6. The components values are $V_1 = 20\,\text{V}$, $f = 50\,\text{kHz}$ ($T = 20\,\mu\text{s}$),

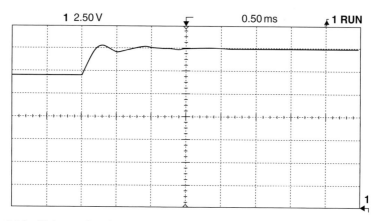

Figure 2.36 Unit-step function responses (experiment, small signal) of buck converter with power loss ($r_L = 1.5\,\Omega$).

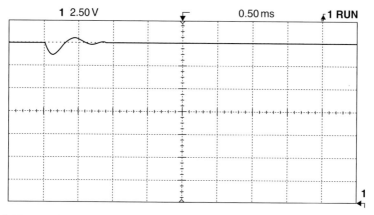

Figure 2.37 Impulse responses (experiment, small signal) of buck converter with power loss ($r_L = 1.5\,\Omega$).

$L = 100\,\mu\text{H}$ with resistance $r_L = 0.12\,\Omega$, $C_1 = 2500\,\mu\text{F}$, $C_2 = 800\,\mu\text{F}$ and $R = 10\,\Omega$. This converter is stable and works in CCM.

We then obtain $V_2 = 65.09\,\text{V}$, $I_2 = 6.509\,\text{A}$, $I_1 = 22\,\text{A}$, $I_L = 14.91\,\text{A}$, $P_{\text{loss}} = I_L^2 \times r_L = 14.91^2 \times 0.12 = 26.67\,\text{W}$, $V_{C_1} = V_1 = 20\,\text{V}$, $V_{C_2} = V_2 = 65.09\,\text{V}$, and $\Delta V_2 = 7.74\,\text{V}$, $\Delta I_2 = 0.784\,\text{A}$, $\Delta I_1 = 4.825\,\text{A}$, $\Delta I_L = 3.46\,\text{A}$, $\Delta P_{\text{loss}} = (I_{L-0.6}^2 - I_{L-0.5}^2)\,r_L = 7.4\,\text{W}$, $\Delta V_{C_1} = \Delta V_1 = 0\,\text{V}$ and $\Delta V_{C_2} = \Delta V_2 = 7.84\,\text{V}$. The parameters are:

$$\Delta PE = V_1 \Delta I_1 T = 1.93\,\text{mJ}$$

$$\Delta W_L = 1.385\,\text{mJ}$$

$$\Delta W_{C_1} = 0\,\text{mJ}$$

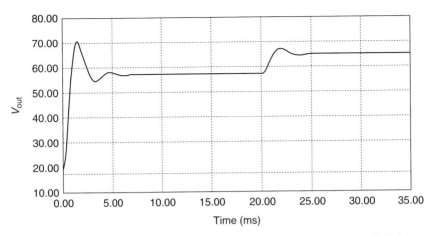

Figure 2.38 Unit-step responses (simulation) of super-lift Luo-converter ($r_L = 0.12\,\Omega$).

$$\Delta W_{C_2} = 383.68\,\text{mJ}$$

$$\Delta SE = \Delta W_L + \Delta W_{C_1} + \Delta W_{C_2} = 385.06\,\text{mJ}$$

$$EF = \frac{\Delta SE}{\Delta PE} = \frac{385.06}{1.9} = 203$$

$$\Delta EL = \Delta P_{\text{loss}} \times T = 7.4 \times 20 = 0.148\,\text{mJ}$$

$$\eta = \frac{\Delta PE - \Delta EL}{\Delta PE} = \frac{1.93 - 0.148}{1.93} = 0.923$$

$$CIR = \frac{\Delta W_C}{\Delta W_L} = \frac{383.68}{1.38} = 277$$

$$\tau = \frac{2T \times EF}{1 + CIR}\left(1 + CIR\frac{1-\eta}{\eta}\right) = 543\,\mu\text{s}$$

$$\tau_d = \frac{2T \times EF}{1 + CIR}\frac{CIR}{\eta + CIR(1-\eta)} = 768\,\mu\text{s}$$

From the above calculation and analysis we found out that the time constants are not changed. Therefore, the transfer function (2.33) for small signal operation should not be changed. Correspondingly, the unit-step response is:

$$v_2(t) = 57.25 + 7.8[1 - e^{-t/1.55}(\cos 1398t - 0.461 \sin 1398t)]\,\text{V} \qquad (2.69)$$

The unit-step function response for large signal ($k = 0$–0.5) and small signal ($k = 0.5$–0.6) operation are shown in Figure 2.38. The impulse response for small signal is described as:

$$\Delta v_2(t) = 0.923 U e^{-t/0.00155} \sin 1398t \qquad (2.70)$$

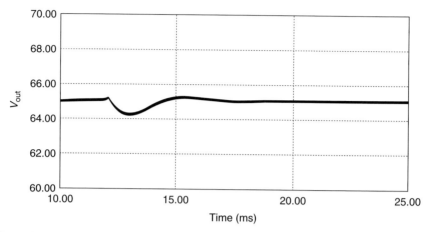

Figure 2.39 Impulse responses (simulation) of super-lift Luo-converter ($r_L = 0.12\,\Omega$, small signal).

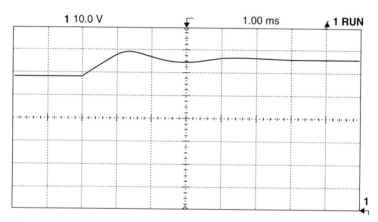

Figure 2.40 Unit-step function responses (experiment, small signal) of super-lift Luo-converter ($r_L = 0.12\,\Omega$).

where U is the interference signal. The small signal impulse response is shown in Figure 2.39.

In order to verify this analysis and compare the simulation results to experimental results, a test rig was constructed. The components values are $V_1 = 20\,\text{V}$, $f = 50\,\text{kHz}$ ($T = 20\,\mu\text{s}$), $k = 0.5$, $L = 100\,\mu\text{H}$ (with $r_L = 0.12\,\Omega$), $C_1 = 750\,\mu\text{F}$, $C_2 = 200\,\mu\text{F}$ and $R = 10\,\Omega$.

The experimental results for unit-step (small signal: $k = 0.5$–0.6) response and impulse interference (small signal: $k = 0.6$–0.5) responses are shown in Figures 2.40 and 2.41. We can see that both the simulation and experimental results are identical.

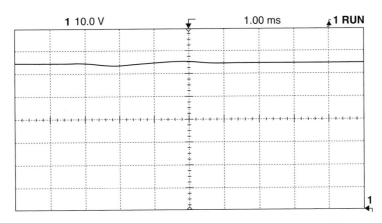

Figure 2.41 Impulse responses (experiment, small signal) of super-lift Luo-converter ($r_L = 0.12\,\Omega$).

FURTHER READING

1. Sira-Ramirez H., Sliding motions in bilinear switched networks, *IEEE Trans CAS*, Vol. 34, No. 8, August 1987, pp. 919–933.
2. Sira-Ramirez H. and Ilic M., Exact linearization in switched mode DC to DC power converters, *Int J Contr*, Vol. 50, No. 2, August 1989, pp. 511–524.
3. Sira-Ramirez H., A geometric approach to pulse-width modulated control in nonlinear dynamical systems, *IEEE Trans Automat Contr*, Vol. 34, No. 2, February 1989, pp. 184–187.
4. Czarkowski D. and Kazimierczuk M. K., Energy conservation approach to modelling PWM dc–dc converters, *IEEE Trans Aero Electron Sys*, Vol. 29, No. 3, July 1993, pp. 1059–1063.
5. Sira-Ramirez H. and Miguel Rios-Bolivar, Sliding mode control of dc-to-dc power converters via extended linearization, *IEEE Trans CAS*, Vol. 41, No. 10, October 1994, pp. 652–661.
6. Sira-Ramirez H., Ortega R., Perez-Moreno R. and Garcia-Esteban M., A sliding mode controller–observer for DC-to-DC power converters: a passivity approach, *Proc IEEE-DAC '95*, Vol. 4, 1995, pp. 3379–3384.
7. Kazimierczuk M. K. and Cravens II R., Open and closed-loop dc and small-signal characteristics of PWM buck–boost converter for CCM, *J Circuit, Sys Comp*, Vol. 5, No. 3, September 1995, pp. 261–3003.
8. Kazimierczuk M. K. and Cravens II R., Closed-loop characteristics of voltage-mode controlled PWM boost converter with an integral-lead controller, *J Circuit, Sys Comp*, Vol. 4, No. 4, December 1994, pp. 429–458.
9. Kazimierczuk M. K. and Cravens II R., Experimental results for the small-signal study of the PWM boost converter with an integral-lead controller, *J Circuit, Sys Comp*, Vol. 5, No. 4, December 1995, pp. 747–755.
10. Dariusz Czarkowski, Pujara L. R. and Marian K. Kazimierczuk, Robust stability of state-feedback control of PWM DC–DC push–pull converter, *IEEE Trans IE*, Vol. 42, No. 1, February 1995, pp. 108–111.

11. Wong R. C., Owen H. A. and Wilson T. G., An efficient algorithm for the time-domain simulation of regulated energy-storage dc-to-dc converters, *IEEE-Trans PEL*, Vol. 2, 1987, pp. 154–168.

12. Cheng K. W. E., Storage energy for classical switched mode power converters, *IEE-Proc EPA*, Vol. 150, No. 4, 2003, pp. 439–446.

13. Lee Y. S., A systemic and unified approach to modeling switches in switch-mode power supplies, *IEEE-Trans IE*, Vol. 32, 1985, pp. 445–448.

14. Middlebrook R. and Cúk S., A general unified approach to modeling switching-converter power stages, *J Electron*, Vol. 42, No. 6, 1977, pp. 521–550.

15. Erickson R. W. and Maksimovic D., *Fundamentals of Power Electronics*, 2nd edn., Kluwer Academic Publishers, Boulder, CO, USA, 2001.

16. Smedley K. M. and Cuk S., One-cycle control of switching converters, *IEEE Trans PEL*, Vol. 10, No. 6, November 1995, pp. 625–633.

17. Luo F. L. and Ye H., Energy factor and mathematical modeling for power DC/DC converters, *IEE-Proc EPA*, Vol. 152, No. 2, 2005, pp. 233–248.

18. Luo F. L. and Ye H., Mathematical modeling for power DC/DC converters, *Proc IEEE Int Conf POWERCON'2004*, Singapore, 21–24 November 2004, pp. 323–328.

19. Luo F. L. and Ye H., *Advanced DC/DC Converters*, CRC Press LLC, Boca Rotan, FL, USA, 2003.

20. Luo F. L. and Ye H., Positive output super-lift Luo-converters, *Proc IEEE Int Conf PESC'2002*, Cairns, Australia, 23–27 June 2002, pp. 425–430.

21. Luo F. L. and Ye H., Positive output super-lift converters, *IEEE-Trans PEL*, Vol. 18, No. 1, January 2003, pp. 105–113.

22. Luo F. L. and Ye H., Negative output super-lift Luo-converters, *Proc IEEE Int Conf PESC '2003*, Acapulco, Mexico, 15–19 June 2003, pp. 1361–1366.

23. Luo F. L. and Ye H., Negative output super-lift converters, *IEEE-Trans PEL*, Vol. 18, No. 5, September 2003, pp. 1113–1121.

24. Luo F. L., Positive output Luo-converters: voltage lift technique, *IEE-EPA Proc*, Vol. 146, No. 4, July 1999, pp. 415–432.

25. Luo F. L., Negative output Luo-converters: voltage lift technique, *IEE-EPA Proc*, Vol. 146, No. 2, March 1999, pp. 208–224.

26. Luo F. L., Double output Luo-converters: advanced voltage lift technique, *IEE-EPA Proc*, Vol. 147, No. 6, November 2000, pp. 469–485.

27. Luo F., Ye H. and Rashid M. H., Multiple-quadrant Luo-converters, *IEE-EPA Proc*, Vol. 148, No. 1, January 2002, pp. 9–18.

28. Luo F. L. and Ye H., Positive output cascade boost converters, *IEE-EPA Proc*, Vol. 151, No. 5, September 2004, pp. 590–606.

APPENDIX A – A SECOND-ORDER TRANSFER FUNCTION

A typical second-order transfer function in the *s*-domain is shown below:

$$G(s) = \frac{M}{1 + s\tau + s^2 \tau \tau_d} = \frac{M}{1 + s\tau + \xi s^2 \tau^2} \tag{2A.1}$$

where M is the voltage transfer gain; τ, The time constant; τ_d, The damping time constant ($\tau_d = \xi\tau$) and s, The Laplace operator in the *s*-domain.

We now discuss various situations for the transfer function in detail.

A.1 *Very Small Damping Time Constant*

If the damping time constant is very small (i.e. $\tau_d \ll \tau, \xi \ll 1$) and it can be ignored, the value of the damping time constant (τ_d) is omitted (i.e. $\tau_d = 0, \xi = 0$). The transfer function (2A.1) is downgraded to the first order as:

$$G(s) = \frac{M}{1 + s\tau} \tag{2A.2}$$

The unit-step function response in the time domain is:

$$g(t) = M(1 - e^{-t/\tau}) \tag{2A.3}$$

The transient process (settling time) is nearly 3 times of the time constant, i.e. 3τ, to produce $g(t) = g(3\tau) = 0.95$ M. The response in time domain is shown in Figure 2A.1 with $\tau_d = 0$.

The impulse interference response is:

$$\Delta g(t) = U e^{-t/\tau} \tag{2A.4}$$

where U is the interference signal. The interference recovering progress is nearly 3 times of the time constant, 3τ, and shown in Figure 2A.2 with $\tau_d = 0$.

A.2 *Small Damping Time Constant*

If the damping time constant is small (i.e. $\tau_d < \tau/4, \xi < 0.25$) and it cannot be ignored, the value of the damping time constant (τ_d) is not omitted. The transfer function (2A.1) retained the second-order function with two real poles $-\sigma_1$ and $-\sigma_2$ as:

$$G(s) = \frac{M}{1 + s\tau + s^2 \tau \tau_d} = \frac{M/\tau\tau_d}{(s + \sigma_1)(s + \sigma_2)} \tag{2A.5}$$

where

$$\sigma_1 = \frac{\tau + \sqrt{\tau^2 - 4\tau\tau_d}}{2\tau\tau_d} \quad \text{and} \quad \sigma_2 = \frac{\tau - \sqrt{\tau^2 - 4\tau\tau_d}}{2\tau\tau_d}$$

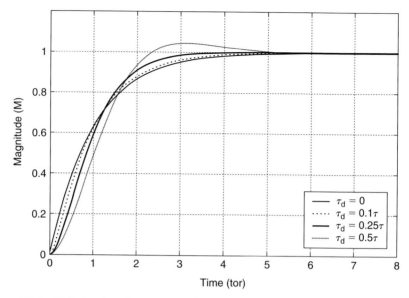

Figure 2A.1   Unit-step function responses ($\tau_d = 0$, 0.1τ, 0.25τ and 0.5τ).

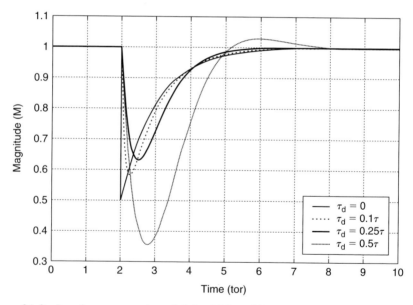

Figure 2A.2   Impulse responses ($\tau_d = 0$, 0.1τ, 0.25τ and 0.5τ).

There are two real poles in the transfer function, and assuming $\sigma_1 > \sigma_2$. The unit-step function response in the time domain is:

$$g(t) = M(1 + K_1 e^{-\sigma_1 t} + K_2 e^{-\sigma_2 t}) \tag{2A.6}$$

where

$$K_1 = -\frac{1}{2} + \frac{\tau}{2\sqrt{\tau^2 - 4\tau\tau_d}} \quad \text{and} \quad K_2 = -\frac{1}{2} - \frac{\tau}{2\sqrt{\tau^2 - 4\tau\tau_d}}$$

The transient process is nearly 3 times of the time value $1/\sigma_1$, $3/\sigma_1 < 3\tau$. The response process is quick without oscillation. The corresponding waveform in time domain is shown in Figure 2A.1 with $\tau_d = 0.1\tau$.

The impulse interference response is:

$$\Delta g(t) = \frac{U}{\sqrt{1 - 4\tau_d/\tau}} (e^{-\sigma_2 t} - e^{-\sigma_1 t}) \tag{2A.7}$$

where U is the interference signal. The transient process is nearly 3 times of the time value $1/\sigma_1$, $3/\sigma_1 < 3\tau$. The response waveform in time domain is shown in Figure 2A.2 with $\tau_d = 0.1\tau$.

A.3 Critical Damping Time Constant

If the damping time constant is equal to the critical value (i.e. $\tau_d = \tau/4$), the transfer function (2A.1) retained the second-order function with two equaled real poles $\sigma_1 = \sigma_2 = \sigma$ as:

$$G(s) = \frac{M}{1 + s\tau + s^2\tau\tau_d} = \frac{M/\tau\tau_d}{(s + \sigma)^2} \tag{2A.8}$$

where

$$\sigma = \frac{1}{2\tau_d} = \frac{2}{\tau}$$

There are two-folded real poles in the transfer function. This expression describes the characteristics of the DC/DC converter. The unit-step function response in the time domain is:

$$g(t) = M\left[1 - \left(1 + \frac{2t}{\tau}\right)e^{-2t/\tau}\right] \tag{2A.9}$$

The transient process is nearly 2.4 times of the time constant $\tau(2.4\tau)$. The response process is quick without oscillation. The response waveform in time domain is shown in Figure 2A.1 with $\tau_d = 0.25\tau$.

The impulse interference response is:

$$\Delta g(t) = \frac{4U}{\tau} t\, e^{-2t/\tau} \tag{2A.10}$$

where U is the interference signal. The transient process is still nearly 2.4 times of the time constant, 2.4τ. The response waveform in time domain is shown in Figure 2A.2 with $\tau_d = 0.25\tau$.

A.4 *Large Damping Time Constant*

If the damping time constant is large (i.e. $\tau_d > \tau/4, \xi > 0.25$), the transfer function (2A.1) is a second-order function with a couple of conjugated complex poles $-s_1$ and $-s_2$ in the LHHP in s-domain:

$$G(s) = \frac{M}{1 + s\tau + s^2\tau\tau_d} = \frac{M/\tau\tau_d}{(s + s_1)(s + s_2)} \tag{2A.11}$$

where

$$s_1 = \sigma + j\omega \quad \text{and} \quad s_2 = \sigma - j\omega$$

with

$$\sigma = \frac{1}{2\tau_d} \quad \text{and} \quad \omega = \frac{\sqrt{4\tau\tau_d - \tau^2}}{2\tau\tau_d}$$

There is a couple of conjugated complex poles $-s_1$ and $-s_2$ in the transfer function. This expression describes the characteristics of the DC/DC converter. The unit-step function response in the time domain is:

$$g(t) = M\left[1 - e^{-t/2\tau_d}\left(\cos\omega t - \frac{1}{\sqrt{4\tau_d/\tau - 1}}\sin\omega t\right)\right] \tag{2A.12}$$

The transient response has oscillation progress with damping factor (σ) and frequency (ω). The corresponding waveform in time domain is shown in Figure 2A.1 with $\tau_d = 0.5\tau$ and in Figure 2A.3 with $\tau, 2\tau, 5\tau$ and 10τ.

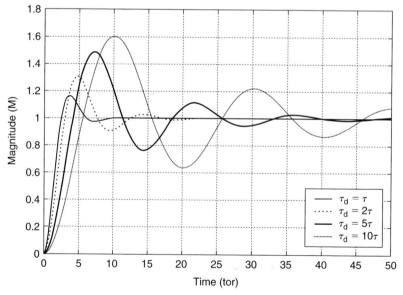

Figure 2A.3 Unit-step function responses ($\tau_d = \tau, 2\tau, 5\tau$ and 10τ).

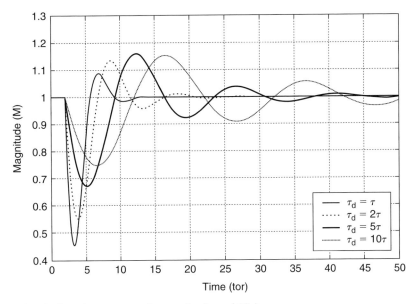

Figure 2A.4 Impulse responses ($\tau_d = \tau$, 2τ, 5τ and 10τ).

The impulse interference response is:

$$\Delta g(t) = \frac{U}{\sqrt{\frac{\tau_d}{\tau} - \frac{1}{4}}} e^{-t/2\tau_d} \sin(\omega t) \tag{2A.13}$$

where U is the interference signal. The recovery process is a curve with damping factor (σ) and frequency (ω). The response waveform in time domain is shown in Figure 2A.2 with $\tau_d = 0.5\tau$ and in Figure 2A.4 with τ, 2τ, 5τ and 10τ.

APPENDIX B – SOME CALCULATION FORMULAE DERIVATIONS

B.1 *Transfer Function of Buck Converter*

With reference to Figure 2.3, we obtain the output voltage $v_2(t)$ from input source voltage $v_1(t) = V_1$ using the voltage division formula:

$$v_1(t) = \begin{cases} V_1 & 0 \le t < kT \\ 0 & kT \le t < T \end{cases}$$

Correspondingly, the transfer function in the s-domain is:

$$G(s) = \frac{v_2(s)}{v_1(s)} = M \frac{\frac{R(1/sC)}{R+(1/sC)}}{sL + \frac{R(1/sC)}{R+(1/sC)}} = \frac{M}{1 + s\frac{L}{R} + s^2 LC} \tag{2B.1}$$

where M is the voltage transfer gain in the steady state.

This transfer function is in the second-order form. It is available for other fundamental converters which consist of two passive energy-storage elements (one inductor L and one capacitor C) and load R such as boost converter and buck–boost converter. The voltage transfer gain is $M = k$ for buck converter, $M = 1/(1 - k)$ for boost converter, $M = k/(1 - k)$ for buck–boost converter.

B.2 *Transfer Function of Super-Lift Luo-converter*

With reference to Figure 2.16, we obtain the output voltage $v_2(t)$ from input source voltage $v_1(t) = V_1$ $(0 \le t < kT)$ using the voltage division formula. Correspondingly, the transfer function in the s-domain is:

$$G(s) = \frac{v_2(s)}{v_1(s)} = M \frac{\frac{R(1/sC_2)}{R+(1/sC_2)}}{sL + \frac{1}{sC_1} + \frac{R(1/sC_2)}{R+(1/sC_2)}} = \frac{MsRC_1}{1 + sR(C_1 + C_2) + s^2 LC_1 + s^3 RLC_1 C_2} \tag{2B.2}$$

where M is the voltage transfer gain in the steady state. $M = (2 - k)/(1 - k)$ for the elementary circuit of positive-output super-lift Luo-converter. This is a third-order transfer function in the s-domain.

B.3 *Simplified Transfer function of Super-Lift Luo-converter*

With reference to Figure 2.16, we obtain the output voltage $v_2(t)$ from input source voltage $v_1(t) = V_1$ using the voltage division formula. Correspondingly, the transfer function in the s-domain is shown in Equation (2B.2). If the capacitance C_2 is very small it can be ignored, the item involving the C_2 can be deleted in Equation (2B.2). Therefore, we obtain the simplified transfer function (2B.3) as given below:

$$G(s) = \frac{MsRC_1}{1 + sRC_1 + s^2LC_1} \tag{2B.3}$$

This is a second-order transfer function in the s-domain with two "poles" and one "zero", which means that there is offset from beginning.

In the other case, if C_1 is very large and $1/SC_1 = 0$, we may obtain in other form as:

$$G(s) = \frac{M}{1 + s\frac{L}{R} + s^2LC_2} \tag{2B.4}$$

This is a second-order transfer function in the s-domain with two "poles". There is no offset from beginning.

B.4 *Time Constants τ and τ_d, and Ratio ξ*

The deviation of time constants τ and τ_d, and ratio ξ can be referred to the transfer function of buck converter with power losses ($r_L \neq 0$).

$$G(s) = \frac{M}{1 + \frac{r}{R} + s\left(Cr + \frac{L}{R}\right) + s^2LC} = \frac{M\eta}{1 + s\left(Cr + \frac{L}{R}\right)\eta + s^2LC\eta} \tag{2B.5}$$

$$\begin{aligned}
\tau &= \eta\left(Cr + \frac{L}{R}\right) = \frac{CRr}{R+r} + \eta\frac{L}{R} = (1-\eta)\frac{2W_C}{P_O} + \eta\frac{2W_L}{P_O} \\
&= \frac{2T(1-\eta)EF \times CIR}{\eta(1 + CIR)} + \frac{2T \times EF}{1 + CIR} = \frac{2T \times EF}{1 + CIR}\left(1 + CIR\frac{1-\eta}{\eta}\right)
\end{aligned} \tag{2B.6}$$

$$\begin{aligned}
\tau_d &= \frac{CL\eta}{\eta\left(Cr + \frac{L}{R}\right)} = \frac{CL\eta}{\frac{2T \times EF}{1+CIR}\left(1 + CIR\frac{1-\eta}{\eta}\right)} = \frac{\left(\frac{2T \times EF}{1+CIR}\right)^2\frac{CIR}{\eta}}{\frac{2T \times EF}{1+CIR}\left(1 + CIR\frac{1-\eta}{\eta}\right)} \\
&= \frac{2T \times EF}{1 + CIR}\frac{CIR}{\eta + CIR(1 - \eta)}
\end{aligned} \tag{2B.7}$$

$$\xi = \frac{\tau_d}{\tau} = \frac{CIR}{\eta\left(1 + CIR\frac{1-\eta}{\eta}\right)^2} \tag{2B.8}$$

From Equation (2.23)

$$
\tau = \frac{2T \times EF}{1 + CIR}\left(1 + CIR\frac{1-\eta}{\eta}\right) = \frac{2T\,(SE/PE)}{1 + CIR}\left(1 + CIR\frac{1-\eta}{\eta}\right)
$$

$$
= \frac{2 \times SE/P_{in}}{1 + CIR}\left(1 + CIR\frac{1-\eta}{\eta}\right)
$$

Since the stored energy (SE), CIR, input power (P_{in}) and the efficiency (η) are dependent on the working state, but independent from the switching frequency (f) and conduction duty cycle (k), the time constant (τ) is independent from the switching frequency (f) and conduction duty cycle (k).

From Equation (2.25)

$$
\tau_d = \frac{2T \times EF}{1 + CIR}\frac{CIR}{\eta + CIR(1-\eta)} = \frac{2T(SE/PE)}{1 + CIR}\frac{CIR}{\eta + CIR(1-\eta)}
$$

$$
= \frac{2 \times SE/P_{in}}{1 + CIR}\frac{CIR}{\eta + CIR(1-\eta)}
$$

Analogously, the time constant ratio (ξ) is independent from the switching frequency (f) and conduction duty cycle (k).

Usually the stored energy is proportional to the input power. Therefore, when the working state changes from one steady state to a new one, the time constant (τ), the damping time constant (τ_d) and the time constant ratio (ξ) are not changed. They are the parameters to rely on the circuit structure and power losses. Readers can try changing the k and/or f to repeat the exercises in Section 2.5. You can find the time constant (τ), the damping time constant (τ_d) and the time constant ratio (ξ) are not changed.

Chapter 3

Basic Mathematics of Digital Control Systems

Digital control systems are described by digital control theory. Some necessary fundamental knowledge on digital control theory is introduced in this Chapter as mathematical tools, which are used in further Chapters.

3.1 INTRODUCTION

Today, computers are more advanced and are almost inevitable in many industries. As computers operate on digital signals, the need for handling digital signals also increased proportionally. High-speed processing capabilities of modern computers attracted applications that make use of digital signals, which further accelerate the development of the use of digital signals. Hence, digital control systems have gradually become more prominent in today's industries. Accompanying the growth of digital signals, the use of switching circuits also increases tremendously for industrial applications. These switching circuits transfer energy from the source to the load in switching status. For example, the energy from the source is transferred to the load in discrete format. In the 1980's Dr. F. L. Luo paid attention to this phenomenon, which differs from the traditional analog control methods. Note that any system that involves the switching circuits must be in the discrete-data control system and the sampling interval is the switching period T, where $T = 1/f_C$ in which f_C is the switching frequency.

Signals are variations that "transport" information from one place to another. There are two main types of signals: the *analog signals* and the *digital signals*. Analog signals, also known as *continuous-time signals* can take any value and are defined at every instant of time. Whereas digital signals, also known as *discrete-data signals*, are only defined at finite number of levels and points in time. Real-time signals are analog in nature and have to be converted to digital signals for further processing and storage purposes.

Digital control systems have many advantages over analog control systems. Analog control systems are based on circuitries, whose hardware components' properties are affected by manufacturers' tolerance and external factors such as temperature. As digital control systems are mainly software based, they are almost completely unaffected by these problems. Digital control systems are smaller in size and consume less power than their analog counterparts. Digital control systems are also highly reproducible and have virtually unlimited programmability. The greatest advantage of digital technology is the flexibility that allows modification to be done.

Digital computers are used for simulation and computation of control systems dynamics for analysis and design of complex control systems. This eased the hassle of laboratory work that is tedious and expensive. Computer simulations also allow the designers to check or present the results obtained by analytical means. In addition, digital computers can also be used as controllers or processors.

3.1.1 BASIC MODULATION METHODS

A signal in discrete-time state is not a continuous function. It is a pulse-train corresponding to certain parameters. The pulse-train usually has certain repeating period T called sampling period, and amplitude. There are two typical modulation methods applied for the discrete signals. They are pulse-amplitude modulation (PAM) and pulse-width modulation (PWM) methods.

Pulse-Amplitude Modulation

Figure 3.1 shows the continuous-data (analog) signal $f(t)$ to be sampled, and the corresponding sampled output signal $f_p^*(t)$ is a train of finite-width pulses whose amplitude are modulated by the input $f(t)$. The carrier signal $p(t)$ is the sampling control signal which is a train of periodic pulses, each with unity amplitude and the sampling (switching) frequency f_C, and the sampling period (or sampling interval) $T = 1/f_C$. The p is the sample time (or sample width) that is assumed $p < T$. This sampler is called the uniform-rate sampler.

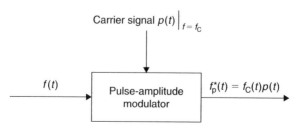

Figure 3.1   Pulse-amplitude modulator.

Typical input and output waveforms of a uniform-rate sampler are shown in Figure 3.2. This sampling method is called the pulse-amplitude modulation (PAM). The corresponding output pulse-train has the pulses with same width p and different amplitudes:

$$f_p^*(t) = f(t)p(t) \tag{3.1}$$

This modulation method will be discussed in the next section in detail.

Pulse-Width Modulation

Another typical sampling is the pulse-width modulation (PWM). The corresponding modulator is called the pulse-width modulator. Typical input and output waveforms of a pulse-width modulator are shown in Figure 3.3. The output pulse-train has the pulses with same amplitude and different widths, which correspond to the input signal at the sampling instants. This modulation method is very popular in most industrial applications.

The amplitude modulation ratio m_a is arranged in certain area, which is usually yielded by a uniformed-amplitude triangle (carrier) signal with the amplitude $V_{\text{tri-m}}$. The maximum amplitude of input signal is assumed $V_{\text{in-m}}$. We define the amplitude

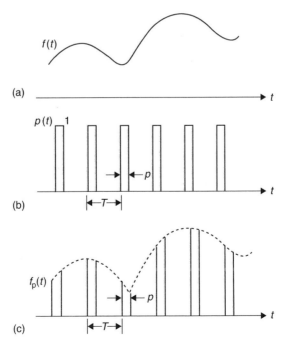

Figure 3.2 Typical input and output waveforms of a uniform-rate sampler: (a) input, (b) carrier and (c) output signals.

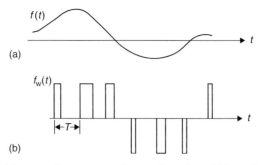

Figure 3.3 Typical input and output waveforms of a pulse-width modulator: (a) input and (b) output signals.

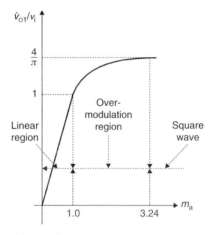

Figure 3.4 Voltage control by varying m_a.

modulation ratio m_a as follows:

$$m_a = \frac{V_{\text{in-m}}}{V_{\text{tri-m}}} \tag{3.2}$$

The input signal at most time points is smaller than its amplitude. The modulation ratio is defined as

$$m = \frac{V_{\text{in}}}{V_{\text{tri-m}}}$$

We also define the frequency modulation ratio m_f as follows:

$$m_f = \frac{f_{\text{tri-m}}}{f_{\text{in-m}}} \tag{3.3}$$

Since the value of the input signal is always smaller than or equal to the maximum amplitude $V_{\text{in-m}}$, the modulation ratio m is always smaller than or equal to the amplitude modulation ratio m_a. The voltage control by varying m_a is shown in Figure 3.4.

Linear Range (m_a ≤ 1.0)

The condition $(\hat{V}_{Ao})_1 = m_a(V_d/2)$ determines the linear region. It is a sinusoidal PWM where the amplitude of the fundamental frequency voltage varies linearly with the amplitude modulation ratio m_a. The PWM pushes the harmonics into a high-frequency range around the switching frequency and its multiples. However, the maximum available amplitude of the fundamental frequency component may not be as high as the desired.

Overmodulation (3.24 > m_a > 1.0)

The condition $\frac{V_d}{2} < (\hat{V}_{Ao})_1 < \frac{4}{\pi}\frac{V_d}{2}$ determines the overmodulation region. When the amplitude of the fundamental frequency component in the output voltage is increased beyond 1.0, it reaches overmodulation. In overmodulation range, the amplitude of the fundamental frequency voltage no longer varies linearly with m_a.

Overmodulation causes the output voltage to contain many more harmonics in the sidebands as compared with the linear range. The harmonics with dominant amplitudes in the linear range may not be dominant during overmodulation.

Square Wave (Sufficiently Large m_a > 3.24)

The inverter voltage waveform degenerates from a pulse-width-modulated waveform into a square wave.

3.1.2 BASIC ELEMENTS OF A DISCRETE-DATA CONTROL SYSTEM

For convenience, only the mathematical modeling of the uniform-rate sampling operation is discussed in this section. Once we have established the input–output relation of the uniform-rate sampler, the analysis can be easily extended to some of the other types of non-uniform-rate sampling.

The carrier signal $p(t)$ is expressed as follows:

$$p(t) = \sum_{k=-\infty}^{\infty} [u(t - kT) - u(t - kT - p)] \quad p < T \tag{3.4}$$

where $u(t)$ is the unit-step function. We assumed that the sampling operation begins from $t = -\infty$ and the leading edge of the pulse at $t = 0$. The output of the sampler is written as:

$$f_p^*(t) = f(t)p(t) \tag{3.5}$$

where

$$p(t) = \begin{cases} 1 & kT \leq t < kT + p \\ 0 & kT + p \leq t < (k+1)T \end{cases} \tag{3.6}$$

or

$$p(t) = \sum_{k=-\infty}^{\infty} [u(t - kT) - u(t - kT - p)] \quad \text{where } p < T \qquad (3.7)$$

Substituting Equation (3.7) in Equation (3.5), we get,

$$f_p^*(t) = f(t) \sum_{k=-\infty}^{\infty} [u(t - kT) - u(t - kT - p)] \quad p < T \qquad (3.8)$$

Equation (3.8) gives a time domain description of the input–output relation of the uniform-rate finite-pulse-width sampler. Since the unit-pulse-train $p(t)$ is a periodic function with period T, it can be represented by a Fourier series,

$$p(t) = \sum_{k=-\infty}^{\infty} C_n e^{jn\omega_C t} \qquad (3.9)$$

where ω_C is the sampling angular frequency in rad/s, and is equal to $2\pi f_C$ or $2\pi/T$; C_n denotes the complex series coefficients and is given by:

$$C_n = \frac{1}{T} \int_0^T p(t) e^{-jn\omega_C t} \, dt \qquad (3.10)$$

Considering $p(t) = 1$ for $0 \le t \le p$, Equation (3.10) becomes:

$$C_n = \frac{1}{T} \int_0^p e^{-jn\omega_C t} dt = \frac{1 - e^{-jn\omega_C p}}{jn\omega_C T} \qquad (3.11)$$

Using well-known trigonometric identities, C_n is written as:

$$C_n = \frac{p}{T} \frac{\sin(n\omega_C p/2)}{n\omega_C p/2} e^{-jn\omega_C p/2} \qquad (3.12)$$

Substituting Equation (3.12) in Equation (3.9), we get,

$$p(t) = \sum_{n=-\infty}^{\infty} \frac{p}{T} \frac{\sin(n\omega_C p/2)}{n\omega_C p/2} e^{-jn\omega_C p/2} e^{jn\omega_C t} \qquad (3.13)$$

Considering Equation (3.9), we rewrite Equation (3.5) as follows:

$$f_p^*(t) = \sum_{n=-\infty}^{\infty} C_n f(t) e^{jn\omega_C t} \qquad (3.14)$$

The Fourier transform of $f_p^*(t)$ is obtained as:

$$f_p^*(j\omega) = \Im\left[f_p^*(t)\right] = \int_{-\infty}^{\infty} f_p^*(t)\,e^{-j\omega t}\,dt \tag{3.15}$$

Using the complex shifting theorem of the Fourier transform which states that:

$$\Im\left[e^{jn\omega_C t}f(t)\right] = F(j\omega - jn\omega_C) \tag{3.16}$$

the Fourier transform of $f_p^*(t)$ is written as:

$$F_p^*(j\omega) = \sum_{n=-\infty}^{\infty} C_n F(j\omega - jn\omega_C) \tag{3.17}$$

Since n changes from $-\infty$ to ∞, the sign minus $(-)$ can be written in plus $(+)$. So that:

$$|F_p^*(j\omega)| = \left| \sum_{n=-\infty}^{\infty} C_n F(j\omega + jn\omega_C) \right| \le \sum_{n=-\infty}^{\infty} |C_n||F(j\omega + jn\omega_C)| \tag{3.18}$$

or

$$\left|F_p^*(j\omega)\right| \le \sum_{n=-\infty}^{\infty} \frac{p}{T} \left| \frac{\sin(n\omega_C p/2)}{n\omega_C p/2} \right| |F(j\omega + jn\omega_C)| \tag{3.19}$$

and

$$C_0 = \lim_{n \to 0} C_n = \frac{p}{T} \tag{3.20}$$

Then we get the amplitude spectra of input and output signals of a finite-pulse-width sampler shown in Figure 3.5.

3.2 DIGITAL SIGNALS AND CODING

Examining the waveform conjectured as $|F_p^*(j\omega)|$ in Figure 3.5(c), we can see that the sampling operation retains the fundamental component of $F(j\omega)$, but in addition, the sampler output also contains the harmonics components, $F(j\omega + jn\omega_C)$, for $n = \pm 1, \pm 2, \ldots$. The frequency $\omega_C/2$ is called *folding frequency*, and the frequency ω_N in Figure 3.5(b) and (c) is called *Nyquist frequency*. The conditions below must be satisfied:

$$\frac{\omega_C}{2} < \omega_N$$

and/or

$$T > \frac{T_N}{2}$$

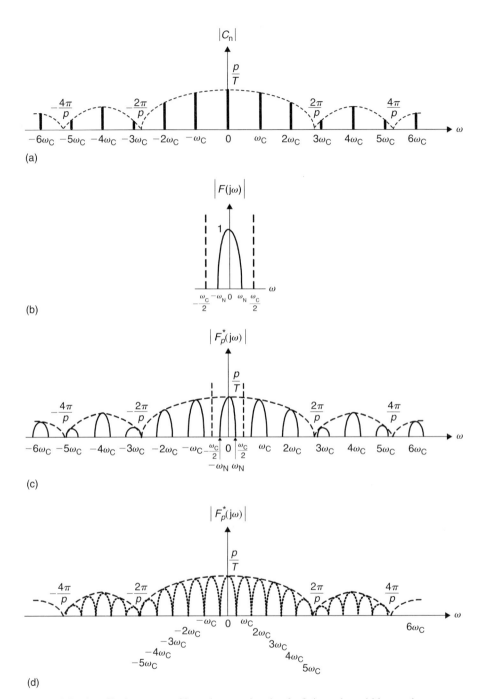

Figure 3.5   Amplitude spectra of input/output signals of a finite-pulse-width sampler.

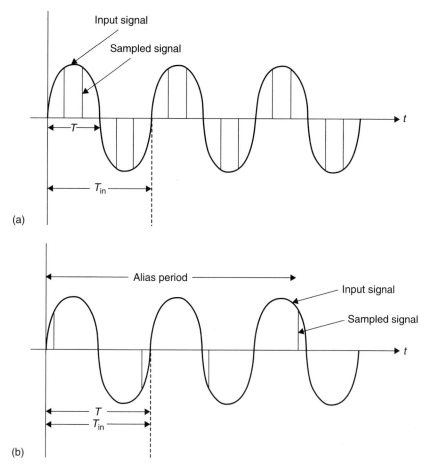

Figure 3.6 Normal sampling and aliasing.

It means that the folded frequency must be smaller than the Nyquist frequency. Figure 3.6 shows the input signal and the sampling signal. In Figure 3.6(a) the sampling interval $T = 1/f_C$ is smaller than the half-cycle of the input signal. The output signal of the sampler is described by the theoretical analysis in Section 3.1. Vice versa, if the sampling interval $T = 1/f_C$ is greater than the half-cycle of the input signal as shown in Figure 3.6(b), the output signal will have a different frequency from the input frequency (the phenomenon is called *aliasing*), and the output signal frequency is called the *alias frequency*. The period of the output is called the *alias period*.

The requirement that the sampling frequency ω_C be at least twice as large as the highest frequency (Nyquist frequency) contained in the signal $f(t)$ is formally known as the Shannon's sampling theorem.

3.3 SHANNON'S SAMPLING THEOREM

Sampling is of great practical importance as it has many applications in engineering and physics. Many people have discovered or rediscovered the sampling theorem in the recent centuries.

3.3.1 BRIEF INTRODUCTION TO NYQUIST SAMPLING THEORY

Nyquist sampling theorem states that an analog signal can be perfectly re-created from its sample values if the sampling interval is chosen correctly. For example, a signal with a maximum frequency of ω Hz must be sampled at least 2ω times per second to reconstruct the original signal from the samples.

3.3.2 SHANNON SAMPLING THEOREM

The Shannon sampling theorem was discovered by Claude E. Shannon, and is also known as the Nyquist criterion. Shannon formalized the Nyquist sampling theory by stating that any band-limited signal can be reconstructed from its samples provided the sampling frequency is at least twice the highest frequency in the signal. Whenever this condition is not fulfilled, *aliasing* occurs. This theory was described in a manuscript in 1940, but it was not published until 1949, that too after the end of World War II.

In depth, Shannon's sampling theorem states that a function of time $f(t)$ which contains no frequency components greater than ω_C in **rad/s** (band limited) can be reconstructed by the values of $f(t)$ at any set of sampling points that are spaced apart by $T < \pi/\omega_C$ seconds. The continuous-time frequency band-limited signal $f(t)$ can be reconstructed from the sampled signal by using:

$$f(t) = \sum_{k=-\infty}^{+\infty} e(kT) \frac{\sin v(t, kT)}{v(t, kT)}$$

where $v(t, kT) \equiv (\omega_S - \omega_S kT)/2$.
The Shannon sampling theorem is stated below:

> *If a signal contains no frequency higher than ω_C in **rad/s**, it is completely characterized by the values of the signal measured at instants of time separated by $T = \pi/\omega_N$ in **second**.*

It means that the sampling frequency ω_C must be greater than twice of the input signal frequency, the output signal can be successfully sampled. The sampling frequency condition is $\omega_S > 2\omega_C$, where $\omega_C = \omega_S/2$ is called the Nyquist frequency. Below the Nyquist sampling frequency, signal frequency information is lost. At Nyquist sampling frequency, amplitude data are lost.

The principal impact of the Shannon sampling theorem on information theory is that it allows the replacement of a continuous band-limited signal by a discrete sequence of its samples without loss of any information. Also it specifies the lowest rate of sampling to reproduce the original signal.

3.4 SAMPLE-AND-HOLD DEVICES

Sample-and-hold (S/H) device is the important equipment (device) for the sampling and holding process of the digitized operation. We introduce some devices for this operation.

3.4.1 DIGITAL WORDS AND CODES

Digital words and codes are normally used to represent digital signals. The information carried by the digital codes is generally in the form of discrete bits. The numerical value of the digital word or code then represents the magnitude of the information in the variable the word represents.

The digital signal stored in a digital computer is made up of binary number of zeros and ones. Each binary digit is referred to as a *bit*. One bit itself is too small to carry the full information needed. Hence, the bits are strung together to form larger and more useful information units.

The accuracy of a digital computer depends on the ability to store and manipulate digital signals as indicated by its word length.

3.4.2 SAMPLING PROCESS

The S/H device makes a fast acquisition (sample) of an analog signal and then holds the signal at a constant value until the next acquisition is made. This device converts an analog or continuous-time signal into a digital or pulse-modulated signal.

S/H device is an important component in the digital and sampled-data control systems. Its fundamental block diagram is shown in Figure 3.7. Hence, it is important to ensure that the system is realistic and is mathematically simple for analytical purposes. Normally, S/H is performed by a single unit normally known as sample-and-hold. However, it is better to treat the sampling and holding operations separately when doing analysis.

S/H devices are commonly used in digital systems. They are normally used to maintain fast-moving signals during conversion operations. It is also used to store multiplexer outputs while the signal is being converted and to detect signal's peak.

Figure 3.7 Functional block diagram of an S/H unit.

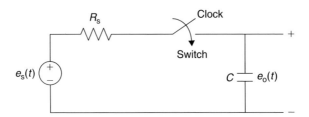

Figure 3.8   Circuit illustration of S/H.

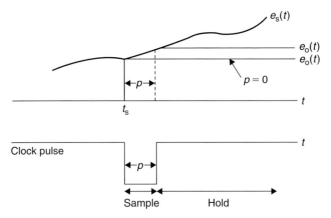

Figure 3.9   Ideal output waveform of an S/H device.

Figure 3.8 illustrates the simplest form of an S/H application. The opening and closing of the switch is controlled by a sample command. When the switch is closed, the S/H output samples and tracks the input signal $e_s(t)$. When the switch is opened, the output is held at the voltage that the capacitor is charged to. The time interval where the switch is closed is known as the sampling duration. The resistance is non-zero in practice. The capacitor will charge toward the sampled input signal with a time constant $R_S C$. The operation of the sampler is not instantaneous as it needs time to respond to the S/H command. Figure 3.9 shows the ideal output waveform of an S/H device.

A hold device maintains the value of the pulse for a prescribed time duration. During the hold mode, a typical output signal of an S/H is characterized by several sources of time delays and imperfect holding. Figure 3.10 shows the input signal and corresponding output of a practical S/H with finite time delay. The following terms used in Figure 3.10 can be briefed as follows:

T_a: Acquisition time
The duration from the instant when SAMPLE command is given to the time when the output of the S/H enters and remains within a specified error band around the input signal.

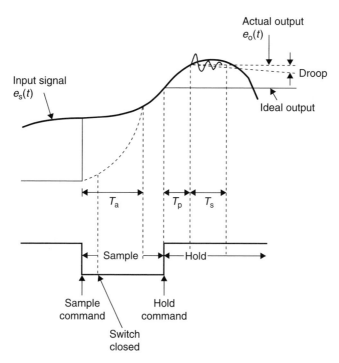

Figure 3.10 Input signal and corresponding output of a practical S/H with finite time delay.

T_p: *Aperture time*
The time between the start of the HOLD command and the time the switch or sampler is opened.

T_s: *Settling time*
The time needed for the transient oscillation to settle to within a certain percent of error band.

In digital systems, the S/H operation is often controlled by a periodic clock. The input and output signals of S/H with a uniform-periodic sampling rate are shown in Figure 3.11. The time duration between the sample commands is called the sampling period T. Usually, the S/H is available in one unit. However, it is better to treat the sampling and holding operations separately when doing analysis.

Figure 3.12 shows the block diagram approximation of the S/H with output filter. This equivalent block diagram isolates the S/H functions and the effects of all the delay times and transient operations, where T_d is the pure time delay, the time which approximates the acquisition time and the aperture time delays. The filter represents the finite time constant and dynamics of the buffer application.

The sampler can be regarded as a pulse-amplitude modulator which has a pulse or sampling duration of p. The hold device simply holds the sampled signal during the holding periods. Usually, the time-delay element and filter are equipped in the S/Hs, and the ideal S/H is shown in Figure 3.13.

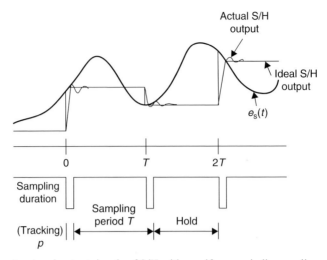

Figure 3.11   Input and output signals of S/H with a uniform-periodic sampling rate.

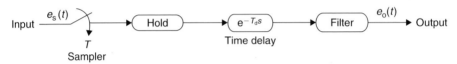

Figure 3.12   Block diagram of the S/H with output filter.

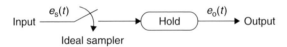

Figure 3.13   Ideal S/H.

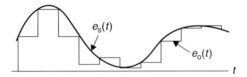

Figure 3.14   Input and output signals of an ideal S/H.

In practice, majority of the S/H operations have very small sampling duration p as compared with the sampling period T and the significant time constant of the input analog signal. The time delay of the S/H is also comparatively small so that it is negligible. If $p \ll T$, the time delay due to S/H is small. It has zero sampling duration when $p = 0$. In this case, the sampler is called an ideal sampler. The input and output signals of an ideal S/H is shown in Figure 3.14.

The most common type of modulation in the S/H operation (shown in Figure 3.15) is the PAM. In the figure, p is the sampling pulse duration and T is the sampling period.

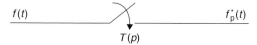

Figure 3.15 Block diagram representation of a periodic/uniform-rate sampler with finite sampling duration.

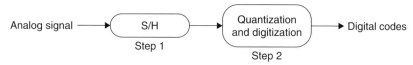

Figure 3.16 Block diagram for A/D conversion.

3.5 ANALOG-TO-DIGITAL CONVERSION

Analog-to-digital (A/D) conversion is an important operation to convert the particular physical parameter into a digital control system. Real-world signals are *analog* in nature and have to be converted to *digital* signals for signal processing and storage purposes.

3.5.1 A/D CONVERSION PROCESS

A/D conversion occurs in two steps as shown in Figure 3.16.

Step 1: Sample and hold

Sampling normally occurs at regular time intervals known as sampling periods. At each sampling point, the analog signal is sampled and the sampled value is held steady until the next sampling point. This process is called sample-and-hold (S/H). Sampling must be fast enough to capture the most rapid changes in the signal being sampled. If the sampling is too slow, important signal information may be lost. This problem is known as aliasing.

Step 2: Quantization and digitization

This second step can begin at the completion of sample acquisition. The hold interval normally gives enough time for this step to be completed. As soon as possible after each sampling instant, the converter selects a quantization level that approximates the S/H value as closely as possible and then assign a binary code that identifies the quantization level.

A/D conversion may cause quantization errors. The larger the number of bits used, the smaller the errors will be. The accuracy of a digital computer depends on the ability to

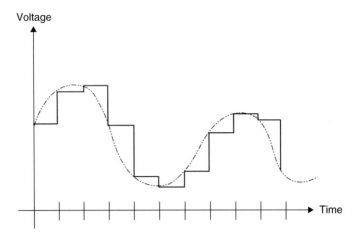

Figure 3.17   S/H signal for an A/D.

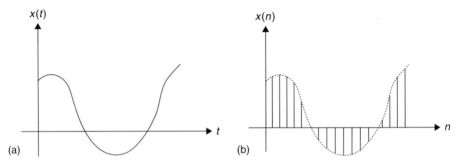

Figure 3.18   Typical A/D signal. (a) Analog signal: continuous in magnitude and time, and (b) digital signal: discrete in magnitude and time n (means nT).

store and manipulate digital signals as indicated by its word length. The sampling and holding signal from an analog signal to digital signal is shown in Figure 3.17.

If an analog signal $x(t)$ (shown in Figure 3.18(a)) is converted to the corresponding digital signal $x(n)$ (shown in Figure 3.18(b)), then the instantaneous value of $x(n)$ is not exactly equal to the instantaneous value of $x(t)$. The digital signal is quantized in discrete values after digitization. The quantization and digitization are shown in Figure 3.19.

3.5.2 A/D CONVERTERS

An A/D converter is generally more expensive and has slower response for the same conversion accuracy. There are a large number of A/D circuits available in the market.

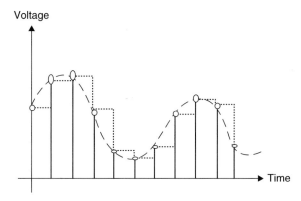

Figure 3.19 Quantization and digitization.

However, they may be classified into three categories based on their architectures and performance:

1. *Serial A/D converters*: They convert analog input to the equivalent digital output 1 LSB (least-significant bit) at a time (i.e. from MSB (most-significant bit) to LSB) or 1 bit at a time (per clock). *Examples*: single slope, dual slope (integrating), successive approximation, delta-sigma.
2. *Flash A/D converter*: All the bits are determined simultaneously in a single step or in one clock cycle.
3. *Subranging A/D converter*: They combine both serial and parallel techniques as a compromise between the serial and parallel A/D converters. *Example*: two-step, multistep, pipeline.

$$\text{Resolutions: serial} > \text{flash} > \text{Subranging}$$
$$\text{Speed: subranging} > \text{flash} > \text{serial}$$

3.6 DIGITAL-TO-ANALOG CONVERSION

Digital-to-analog (D/A) conversion is an important operation to convert the calculated output signal in a digital control system to the particular physical actuator.

3.6.1 D/A CONVERSION PROCESS

A D/A conversion occurs in two steps as shown in Figure 3.20:

Step 1: Convert to analog level

The D/A conversion process is to convert each digital code into analog voltage level that is proportional to the size of the digital number.

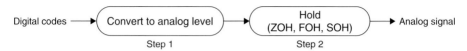

Figure 3.20 Block diagram of D/A conversion.

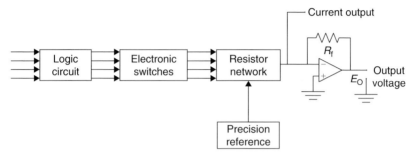

Figure 3.21 Basic elements of a D/A converter.

Step 2: Hold the signal

The height at each sample corresponds to the analog voltage obtained from the digital codes. A holder is used to maintain the analog voltage level for the duration of a sampling period.

3.6.2 D/A CONVERTERS

The basic elements of a D/A converter are shown in Figure 3.21. The function of the logic circuit is to control the switching of the precision reference voltage or current source to the proper input terminals of the resistor network as a function of the digital value of each digital input bit.

D/A converters may be classified into serial and parallel types. The output of a serial D/A converter is obtained serially and the full output is either obtained in N clocks or in $2^N - 1$ clocks for an N-bit D/A converter. The output of a parallel D/A converter is obtained in one clock cycle.

The structures of D/A converter are relatively simpler compared to A/D converter. Most state-of-the-art D/A converter are based on complementary metal-oxide semiconductor (CMOS). There are a large number of architectures for D/A converter.

Since D/A converter converts a digital signal into an analog signal of corresponding magnitude, it is regarded as a device consisting of a decoding and an S/H unit.

A decoder decodes the digital word into a number of an amplitude-modulated pulse. The transfer relation of the decoder is simply a constant gain, which ideally equals to unity. The block diagram representation of a D/A converter is shown in Figure 3.22.

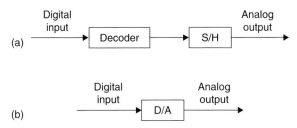

Figure 3.22 Block diagram representation of a D/A converter: (a) block diagram and (b) typical mark of a D/A converter.

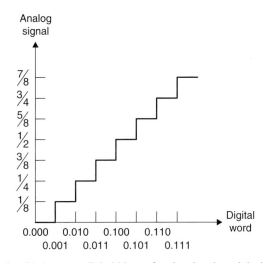

Figure 3.23 Relationship between digital binary fractional code and decimal numbers.

3.6.3 A/D AND D/A CONVERSION ERRORS

Digital computers are used for simulation and computation of control systems dynamics for analysis and design of complex control systems. This eased the hassle of laboratory work that is tedious and expensive. Computer simulations also allow the designer to check or present the results obtained by analytical means. In addition, digital computers can also be used as controllers or processors.

The accuracy of a digital computer depends on the ability to store and manipulate digital signals indicated by its word length. Figure 3.23 shows the relationship between digital binary fractional code and decimal numbers. Real-time signals are analog in nature and have to be converted to digital signals for further processing and storage purposes. Hence, it is important to study their conversion processes.

The conversion error depends on the sampling operation. Figure 3.24 shows the maximum error by full step sampling. The digital-coded signal is shown in Figure 3.24(a).

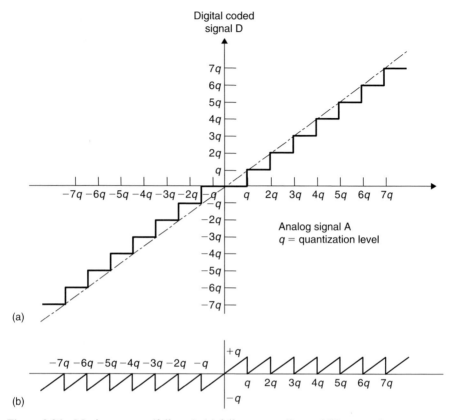

Figure 3.24   Maximum error (full step): (a) full-step sampling and (b) conversion error.

This sampling operation introduces the error to be equal to $+q$ or $-q$ in Figure 3.24(b). Another sampling operation, half-step sampling method, is shown in Figure 3.25. It shows the minimum error. The digital-coded signal is shown in Figure 3.25(a). This sampling operation introduces the error to be equal to $+q/2$ or $-q/2$ in Figure 3.25(b).

3.7 ENERGY QUANTIZATION

Computers use groups of bits to represent numbers. The number of bits used limits the number of values that can be represented by the computer. An analog sample is coded by choosing the closest quantization level available.

When N bits are used, 2^N possible values can be represented by the computer. The larger the number of bits used, the more closely the digital signal corresponds to the analog signal, but the more time consuming the calculations will be.

When an analog signal with a certain range of values is coded using N bits, each sample must be coded to one of the 2^N levels. The gap between levels is called quantization

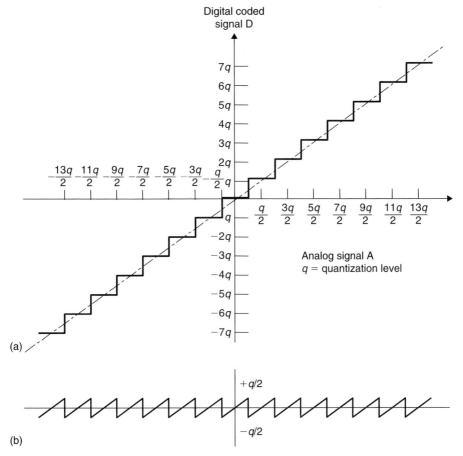

Figure 3.25 Minimum error (half step): (a) half-step sampling and (b) conversion error.

step:

$$\frac{\text{Quantization step}}{\text{Resolution of a quantizer } Q} = \frac{R}{2^N} \tag{3.21}$$

where R is the full-scale analog range and N is the number of bits. The quantization step grows smaller as the number of bits increases. Quantization error that occurs is the difference between the quantized value and the actual value of sample:

$$\text{Quantization error} = \text{quantized value} - \text{actual value} \tag{3.22}$$

The possible maximum error is one full quantization step. The errors can be reduced if the quantization levels are shifted to lie symmetrically around the diagonal. When this

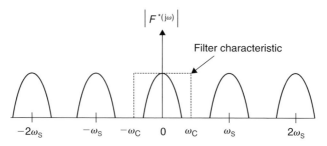

Figure 3.26   Reconstruction of continuous data from digital data using an ideal low pass filter.

occurs, the quantization errors will be reduced by half. Quantization of unipolar data varies from negative minimum to positive maximum.

Quantization error affects bipolar, the same way it affects unipolar. For bipolar, symmetry around zero is maintained to keep errors small. Since the quantization errors are proportional to the quantization step, errors can be reduced by increasing the number of bits used to represent each sample.

In D/A and A/D conversions, it is important to understand the MSB and the LSB, and the weight of each in a digitally coded word in the conversion process. The practical D/A and D/A converters based on the natural binary codes make use of the fractional codes.

3.8 INTRODUCTION TO RECONSTRUCTION OF SAMPLED SIGNALS

A data-reconstruction device also known as a *filter* is often used to interface between digital and analog components. The hold circuit in the S/H device is the most common filtering device in the discrete-data systems. Filtering devices or data-reconstruction devices are used to sieve out high-frequency harmonic components in a signal resulted from the sampling operation. Although the S/H device comes in a single unit, for mathematical simplification, only the hold device is modeled.

Firstly, assume that an ideal sampler is sampling at a sampling frequency ω_C, which is at least twice as large as the maximum frequency component of the continuous input signal being sampled. Figure 3.26 shows the reconstruction of continuous data from digital data using an ideal low-pass filter.

The amplitude characteristics of an ideal filter are shown in Figure 3.27. The amplitude gain is unity.

Perfect reconstruction of the continuous signal is based on the assumption that $f(t)$ is band limited. Hence, it is impossible to recover a totally perfect continuous signal once it is sampled. The best way to reconstruct a signal is to approximate the original time function as closely as possible.

The hold device is the simplest form of a general data-reconstruction problem. The problem of data reconstruction can be regarded as a given sequence of numbers $f(0)$,

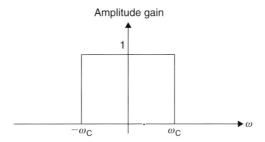

Figure 3.27 Amplitude characteristic of an ideal filter.

$f(T), f(2T), \ldots, f(kT), \ldots$. An analog signal $f(t)$, where $t > 0$ is to be reconstructed from the information contained in the sequence. This data-reconstruction process may be regarded as an extrapolation process, since the analog signal is to be constructed based on information available only at past sampling instants.

Power series expansion of $f(t)$ in the interval between the sampling instants kT and $(k+1)T$ is used to generate a desired approximation.

The approximation is:

$$f_k(t) = f(kT) + f^{(1)}(kT)(t - kT) + \frac{f^{(2)}(kT)}{2!}(t - kT)^2 + \cdots \quad (3.23)$$

or

$$f_k(t) = \sum_{n=0}^{\infty} \frac{f^{(n)}(kT)}{n!}(t - kT)^n \quad (3.24)$$

The higher the order of the derivation, the larger will be the number of delayed pulses required. In general, the number of delayed pulse data required to approximate the value of $f^{(n)}(kT)$ is $n + 1$. Thus, the extrapolating device consists of a series of time delays, and the number of delays depends on the accuracy of the estimate of the time function $f(t)$ during the time interval from kT to $(k+1)T$.

Although utilizing a higher-order derivative produces a more accurate extrapolation, it causes a reduction on the stability of the closed-loop control systems and it also makes the circuitry more complicated and expensive.

3.9 DATA CONVERSION: THE ZERO-ORDER HOLD

The most widely used holding device is the zero-order hold (ZOH). This is because it is less complicated and less expensive. A simple ZOH is shown in Figure 3.28.

It is called zero-order extrapolator as its polynomial used is of the zeroth order. It holds the value of the sampled value $f(kT)$ for $kT \le t < (k+1)T$ until the next sample $f[(k+1)T]$ arrives:

$$f_k(t) = f(kT) \quad (3.25)$$

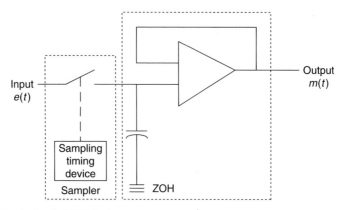

Figure 3.28   A simple ZOH device.

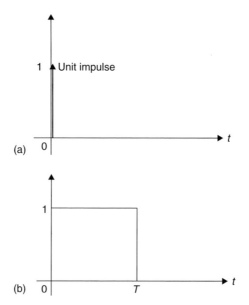

Figure 3.29   Responses of a ZOH: (a) unit-impulse input to and (b) impulse response of ZOH.

This equation is used for approximation of $f(t)$ during the time interval $kT \leq t < (k+1)T$. If a unit impulse input signal is applied to a ZOH, the impulse response is shown in Figure 3.29.

The ZOH is a linear device as it satisfies the principle of superposition. The impulse response of a ZOH is expressed as:

$$g_{h0}(t) = u_s(t) - u_s(t-T) \tag{3.26}$$

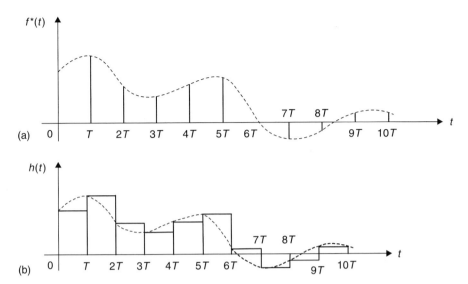

Figure 3.30 Input and output waveforms of a ZOH in the time domain: (a) input signal $f(t)$ and sampled signal $f^*(t)$, and (b) output waveform of ZOH.

where $u_s(t)$ is the unit-step function. The response of the ZOH to the unit-impulse input is equivalent to the difference between two unit-step function. By taking a Laplace transform, the transfer function is obtained as:

$$G_{h0}(s) = \frac{1 - e^{-Ts}}{s} \tag{3.27}$$

Figure 3.30 shows the output waveform of a ZOH with a pulse sequence input. The output waveforms clearly indicate that the accuracy of the ZOH greatly depends on the magnitude of the sampling frequency ω_S. As the sampling frequency increases to infinity or sampling period T approaches zero, the output of the ZOH $h(t)$ approaches the analog signal $f(t)$.

The ZOH is a data-reconstruction or data-filtering device. Hence it is useful to examine its frequency domain characteristics. By replacing s by $j\omega$ in the transfer function, we get,

$$G_{h0}(j\omega) = \frac{1 - e^{-Tj\omega}}{j\omega} \tag{3.28}$$

It can be conditioned as:

$$G_{h0}(j\omega) = \frac{2e^{-j\omega T/2}(e^{j\omega T/2} - e^{-j\omega T/2})}{j2\omega} = \frac{2\sin(\omega T/2)}{\omega}e^{-j\omega T/2} \tag{3.29}$$

or

$$G_{h0}(j\omega) = T\frac{\sin(\omega T/2)}{\omega T/2}e^{-j\omega T/2} = \frac{2\pi}{\omega_S}\frac{\sin(\pi\omega/\omega_S)}{\pi\omega/\omega_S}e^{-j(\pi\omega/\omega_S)} \tag{3.30}$$

Since T is the sampling period in seconds, and $T = 2\pi/\omega_S$ where ω_S is the sampling frequency in rad/s.

The magnitude of $G_{h0}(j\omega)$ is:

$$|G_{h0}(j\omega)| = \frac{2\pi}{\omega_S}\left|\frac{\sin(\pi\omega/\omega_S)}{\pi\omega/\omega_S}\right| \tag{3.31}$$

The phase of $G_{h0}(j\omega)$ is:

$$\angle G(j\omega_S) = \angle\sin(\pi\omega/\omega_S) - \pi\omega/\omega_S \tag{3.32}$$

The change of sign from $+$ to $-$ can be regarded as a phase change of $-180°$.

3.10 THE FIRST-ORDER HOLD

The first-order hold (FOH) uses the first two terms of the power series to extrapolate the time function $f(t)$ over the time interval $kT \le t < (k+1)T$. The equation for the FOH is:

$$f_k(t) = f(kT) + f^{(1)}(kT)(t - kT) \tag{3.33}$$

where the first-order derivative of $f(t)$ at $t = kT$ is approximated as:

$$f^{(1)}(kT) = \frac{f(kT) - f[(k-1)T]}{T} \tag{3.34}$$

Substituting Equation (3.34) in Equation (3.33) gives:

$$f_k(t) = f(kT) + \frac{f(kT) - f[(k-1)T]}{T}(t - kT) \tag{3.35}$$

The output of the FOH between two consecutive sampling instants is a ramp function. The slope of the ramp is equal to the difference of $f(kT)$ and $f[(k+1)T]$.

By applying a unit impulse at $t=0$ as input, an impulse response of the FOH is obtained. The corresponding output is obtained by setting $k = 0, 1, 2, \ldots$ for the various time intervals. For $k = 0$,

$$f_0(t) = f(0) + \frac{f(0) - f(-T)}{T}t \quad 0 \le t \le T \tag{3.36}$$

For a unit-impulse input, $f(0) = 1$ and $f(-T) = 0$, the impulse response of the FOH for $0 \le t \le T$ is:

$$g_{h1}(t) = 1 + \frac{t}{T}$$

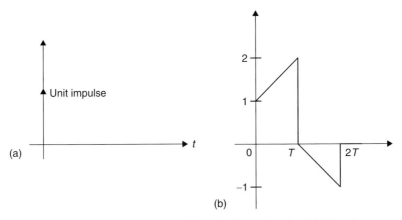

Figure 3.31 Responses of an FOH: (a) unit-impulse input signal and (b) impulse response of the FOH.

For $k = 1$,

$$f_1(t) = f(T) + \frac{f(T) - f(0)}{T}(t - T)$$ (3.37)

The impulse response of the FOH over time interval $T \leq t \leq 2T$ is:

$$g_{h1}(t) = -\frac{t - T}{T}$$

If a unit-impulse input signal applied to a FOH, the impulse response is shown in Figure 3.31.

The impulse response of the FOH for $t > 2T$ is zero, since $f(t) = 0$ for $t > 2T$. Functionally, the impulse response in Figure 3.31(b) can be written as:

$$g_{h1}(t) = u_S(t) + \frac{t}{T}u_S(t - T) - \frac{2(t - T)}{T}u_S(t - T)$$

$$+ \frac{(t - 2T)}{T}u_S(t - 2T)u_S(t - 2T) + u_S(t - 2T)$$ (3.38)

The transfer function of the FOH is obtained by taking the Laplace transform of the last equation:

$$G_{h1}(s) = \frac{1 + Ts}{T}\left[\frac{1 - e^{-Ts}}{s}\right]^2$$ (3.39)

or simply,

$$G_{h1}(s) = \frac{1 + Ts}{T}[G_{h0}(s)]^2$$ (3.40)

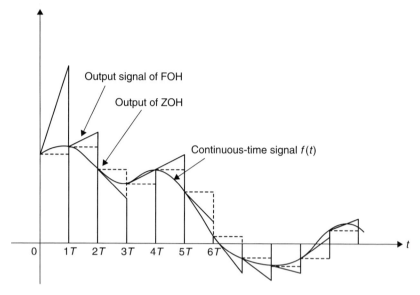

Figure 3.32   Reconstruction of a continuous-time signal by means of an FOH.

The frequency response of the FOH:

$$G_{h1}(j\omega) = \frac{1 + Tj\omega}{T} \left[\frac{1 - e^{Tj\omega}}{j\omega} \right]^2 \tag{3.41}$$

The magnitude and phase response of $G_{h1}(j\omega)$ are obtained as:

$$|G_{h0}(j\omega)| = \frac{2\pi}{\omega_s} \sqrt{1 + \frac{4\pi^2\omega^2}{\omega_s^2} \left[\frac{\sin \pi\omega/\omega_s}{\pi\omega/\omega_s} \right]^2} \tag{3.42}$$

$$\angle G_{h1}(j\omega) = \tan^{-1}\left(\frac{2\pi\omega}{\omega_s} \right) - \frac{2\pi\omega}{\omega_s} \tag{3.43}$$

The reconstruction of a continuous-time signal by means of an FOH is shown in Figure 3.32.

3.11 THE SECOND-ORDER HOLD

The Second-order hold(SOH) uses the first three terms of the power series to extrapolate the time function $f(t)$ over the time interval $kT \le t < (k + 1)T$. The equation for the

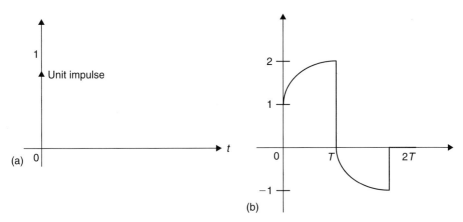

Figure 3.33 Responses of an SOH: (a) unit-impulse input to and (b) impulse response of SOH.

SOH is:

$$f_k(t) = f(kT) + f^{(1)}(kT)(t - kT) + \frac{f^{(2)}(kT)}{2!}(T - kT)^2 \qquad (3.44)$$

The output of the second order hold between two consecutive sampling instants may be a parabola function. The simplest of the curve is a square-law to the difference of $f(kT)$ and $f[(k + 1)T]$.

By applying a unit impulse at $t = 0$ as input, an impulse response of the second-order hold is obtained. The corresponding output is obtained by setting $k = 0, 1, 2....$ for the various time intervals. For $k = 0$, a unit impulse input, $f(0) = 1$ and $f(-T) = 0$, the impulse response of the second order hold for $0 \le t \le T$ is

$$g_{h2}(t) = 2 - \left(1 + \frac{t}{T}\right)^2.$$

For $k = 1$, the impulse response of the second order hold over time interval $T \le t \le 2T$ is

$$g_{h2}(t) = -1 + \left(\frac{t}{T} - 2\right)^2.$$

If a unit impulse input signal is applied to an SOH, the impulse response is shown in Figure 3.33.

Although using a higher-order derivative produces a more accurate extrapolation, it causes a reduction on the stability of the closed-loop control systems and it also makes the circuitry more complicated and expensive. Generally, say that the impulse response of an SOH is a parabola function as shown in Figure 3.33.

The reconstruction of a continuous-time signal by means of an SOH is shown in Figure 3.34.

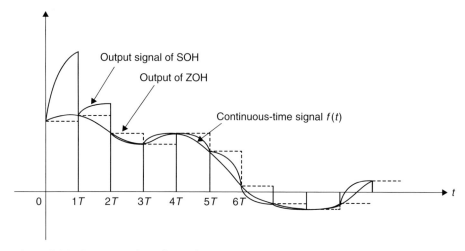

Figure 3.34 Reconstruction of a continuous-time signal by means of an SOH.

The typical second-order transfer function in the s-domain is shown as follows:

$$G(s) = \frac{M}{1 + s\tau + s^2\tau\tau_d} = \frac{M}{1 + s\tau + \xi s^2\tau^2} \tag{3.45}$$

where M is the voltage-transfer gain; τ, the time constant; τ_d, the damping time constant (in which $\tau_d = \xi\tau$) and s is the Laplace operator in s-domain.

In order for the typical second-order transfer function to be applicable to the digital modeling for power DC/DC converters, a z-transformation is necessary. The second-order transfer function undergoes a transformation based on Table 3.1. There are a total of four transfer functions that will be derived to describe the SOH as an analysis in Chapter 2.

3.11.1 VERY SMALL DAMPING TIME CONSTANT τ_d

For a very small damping time constant (i.e. $\tau_d \ll \tau, \xi \ll 1$) that can be ignored, the value of the damping time constant is omitted (i.e. $\tau_d = 0$, $\xi = 0$). The typical second-order transfer function (6.1) is then reduced to first order,

$$G(s) = \frac{M}{1 + s\tau} \tag{3.46}$$

To transform Equation (3.46) in order to describe the digital modeling for power DC/DC converters, the transformation from the z-transform table (see Table 3.1) is used:

$$F(s) = \frac{1}{s + a} \iff F(z) = \frac{z}{z + e^{-aT}} \tag{3.47}$$

Equation (3.46) is rearranged so as to apply the transformation,

$$G(s) = M \frac{1/\tau}{s + 1/\tau} = \frac{M}{\tau} \frac{1}{s + 1/\tau} \tag{3.48}$$

Applying the transformation, the mathematical modeling for the SOH for very small damping time constant is:

$$G(z) = \frac{M}{\tau} \frac{z}{z + e^{-aT}}$$

where

$$a = \frac{1}{\tau} \tag{3.49}$$

Simplifying,

$$G(z) = \frac{M}{\tau} \frac{z}{z + e^{-T/\tau}} = \frac{Mz}{\tau(z + e^{-T/\tau})} \tag{3.50}$$

3.11.2 SMALL DAMPING TIME CONSTANT $\tau_d < \tau/4$

If the damping time constant is small (i.e. $\tau_d < \tau/4$, $\xi < 0.25$) and cannot be ignored, then the value of the damping time constant is not omitted. The typical transfer function (3.45) is retained as second order with two real poles $-\sigma_1$ and $-\sigma_2$:

$$G(s) = \frac{M}{1 + s\tau + s^2\tau\tau_d} = \frac{M/\tau\tau_d}{(s + \sigma_1)(s + \sigma_2)} \tag{3.51}$$

where $\sigma_1 = (\tau - \sqrt{\tau^2 - 4\tau\tau_d})/2\tau\tau_d$ and $\sigma_2 = (\tau + \sqrt{\tau^2 - 4\tau\tau_d})/2\tau\tau_d$.

To transform Equation (3.51) in order to describe the digital modeling for power DC/DC converters, the transformation from the z-transform table (see Table 3.1) is used:

$$F(s) = \frac{1}{(s + a)(s + b)} \Leftrightarrow F(z) = \frac{1}{(b - a)} \left[\frac{z}{z - e^{-aT}} - \frac{z}{z - e^{-bT}} \right] \tag{3.52}$$

Equation (3.52) is rearranged so as to apply the transformation,

$$G(s) = \frac{M}{\tau\tau_d} \frac{1}{(s + \sigma_1)(s + \sigma_2)} \tag{3.53}$$

Applying the transformation, the mathematical modeling for the SOH for small damping time constant is:

$$G(z) = \frac{M}{\tau\tau_d(\sigma_2 - \sigma_1)} \left[\frac{z}{z - e^{-\sigma_1 T}} - \frac{z}{z - e^{-\sigma_2 T}} \right] \tag{3.54}$$

Expanding and simplifying Equation (3.54), we get,

$$
\begin{aligned}
G(z) &= \frac{M}{\sqrt{\tau^2 - 4\tau\tau_d}} \left[\frac{z}{z - e^{-\sigma_1 T}} - \frac{z}{z - e^{-\sigma_2 T}} \right] \\
&= \frac{Mz}{\sqrt{\tau^2 - 4\tau\tau_d}} \left[\frac{e^{-\sigma_1 T} - e^{-\sigma_2 T}}{(z - e^{-\sigma_1 T})(z - e^{-\sigma_2 T})} \right] \\
&= \frac{Mz \left(e^{-\left(\frac{\tau - \sqrt{\tau^2 - 4\tau\tau_d}}{2\tau\tau_d}\right)T} - e^{-\left(\frac{\tau + \sqrt{\tau^2 - 4\tau\tau_d}}{2\tau\tau_d}\right)T} \right)}{\sqrt{\tau^2 - 4\tau\tau_d} \left(z - e^{-\left(\frac{\tau - \sqrt{\tau^2 - 4\tau\tau_d}}{2\tau\tau_d}\right)T} \right) \left(z - e^{-\left(\frac{\tau + \sqrt{\tau^2 - 4\tau\tau_d}}{2\tau\tau_d}\right)T} \right)}
\end{aligned}
\tag{3.55}
$$

as $\sigma_2 - \sigma_1 = \left(\sqrt{\tau^2 - 4\tau\tau_d}\right)/\tau\tau_d$.

3.11.3 CRITICAL DAMPING TIME CONSTANT $\tau_d = \tau/4$

For a damping time constant that is equal to the critical value (i.e. $\tau_d = \tau/4$), the typical second-order transfer function (3.45) is then retained as second order with two equal real poles $-\sigma_1 = -\sigma_2 = -\sigma$,

$$
G(s) = \frac{M}{1 + s\tau + s^2 \tau\tau_d} = \frac{M/\tau\tau_d}{(s + \sigma)^2}
\tag{3.56}
$$

where $\sigma = 1/2\tau_d = 2/\tau$.

To transform Equation (3.56) in order to describe the digital modeling for power DC/DC converters, the transformation from the z-transform table (see Table 3.1) is used:

$$
F(s) = \frac{1}{(s + a)^2} \quad \Leftrightarrow \quad F(z) = \frac{Tz\,e^{-aT}}{(z - e^{-aT})^2}
\tag{3.57}
$$

Equation (3.56) is rearranged so as to apply the transformation,

$$
G(s) = \frac{M}{\tau\tau_d} \frac{1}{(s + \sigma)^2}
\tag{3.58}
$$

Applying the transformation, the mathematical modeling for the SOH for critical damping time constant is:

$$
G(z) = \frac{M}{\tau\tau_d} \frac{Tz\,e^{-\sigma T}}{(z - e^{-\sigma T})^2}
\tag{3.59}
$$

Simplifying Equation (3.59),

$$G(z) = \frac{M}{\tau\tau_d} \frac{Tz\,e^{-(2/\tau)^T}}{(z - e^{-(2/\tau)T})^2} \tag{3.60}$$

3.11.4 LARGE DAMPING TIME CONSTANT $\tau_d > \tau/4$

If the damping time constant is large (i.e. $\tau_d > \tau/4$, $\xi > 0.25$), the typical transfer function (3.45) is retained as second order with a couple of conjugated complex poles s_1 and s_2 in the left-hand half-plane in s-domain as:

$$G(s) = \frac{M}{1 + s\tau + s^2\tau\tau_d} = \frac{M/\tau\tau_d}{(s + s_1)(s + s_2)} \tag{3.61}$$

where $s_1 = \sigma + j\omega$ and $s_2 = \sigma - j\omega$, in which $\sigma = 1/2\tau_d$ and $\omega = \left(\sqrt{4\tau\tau_d - \tau^2}\right)/2\tau\tau_d$.

To transform Equation (3.61) in order to describe the digital modeling for power DC/DC converters, the transformation from the z-transform table (see Table 3.1) is used:

$$F(s) = \frac{\omega}{(s + a)^2 + \omega^2} \Leftrightarrow F(z) = \frac{z\,e^{-aT}\sin\omega T}{z^2 - 2z\,e^{-aT}\cos\omega T + e^{-2aT}} \tag{3.62}$$

Equation (3.61) is rearranged so as to apply the transformation,

$$
\begin{aligned}
G(s) &= \frac{M/\tau\tau_d}{(s + \sigma + j\omega)(s + \sigma - j\omega)} = \frac{M/\tau\tau_d}{(s + \sigma)^2 + \omega^2} = \frac{M}{\tau\tau_d\omega}\frac{\omega}{(s + \sigma)^2 + \omega^2} \\
&= \frac{2M}{\sqrt{4\tau\tau_d - \tau^2}}\frac{\omega}{(s + \sigma)^2 + \omega^2}
\end{aligned} \tag{3.63}
$$

Applying the transformation, the mathematical modeling for the SOH for large damping time constant is:

$$G(z) = \frac{2M}{\sqrt{4\tau\tau_d - \tau^2}}\frac{z\,e^{-aT}\sin\omega T}{z^2 - 2z\,e^{-aT}\cos\omega T + e^{-2aT}} \tag{3.64}$$

where $\sigma = a = 1/2\tau_d$. Expanding and simplifying Equation (3.64), we get,

$$
\begin{aligned}
G(z) &= \frac{2Mz\,e^{-aT}\sin\omega T}{\sqrt{4\tau\tau_d - \tau^2}(z^2 - 2z\,e^{-aT}\cos\omega T + e^{-2aT})} \\
&= \frac{2Mz\,e^{-T/2\tau_d}\sin\left(\frac{\sqrt{4\tau\tau_d - \tau^2}}{2\tau\tau_d}T\right)}{\sqrt{4\tau\tau_d - \tau^2}\left(z^2 - 2z\,e^{-T/2\tau_d}\cos\left(\frac{\sqrt{4\tau\tau_d - \tau^2}}{2\tau\tau_d}T\right) + e^{-T/\tau_d}\right)}
\end{aligned} \tag{3.65}
$$

Strictly speaking there are only three SOH mathematical modeling as the modeling for the very small damping SOH is reduced to first order.

3.12 THE LAPLACE TRANSFORM (THE *s*-DOMAIN)

The Laplace transform is useful to engineers in modeling a linear time-invariant analog system as a transfer function. The Laplace transform may also be used to find the time response of the system through simulations. The Laplace transform of a function $f(t)$ is defined as:

$$F(s) = L[f(t)] = \int_0^\infty f(t)\,e^{-st}\,dt \tag{3.66}$$

Therefore, the Laplace transform of the unit-step function is:

$$F(s) = L[f(t)] = \int_0^\infty 1(t)\,e^{-st}\,dt = \frac{1}{s} \tag{3.67}$$

The inverse Laplace transform of $F(s)$ is given as:

$$f(t) = \frac{1}{2\pi j} \int_{\sigma-j\infty}^{\sigma+j\infty} F(s)\,e^{st}\,ds \quad j = \sqrt{-1} \tag{3.68}$$

σ is determined by the singularities of $F(s)$. Normally, the inverse transformation is done using a table.

3.13 THE *z*-TRANSFORM (THE *z*-DOMAIN)

The *z*-transform is a useful mathematical tool to analyze and design discrete-data systems. A transformation from complex variable s to variable z:

$$z = e^{Ts} \Rightarrow s = \frac{1}{T}\ln z \tag{3.69}$$

where T is the sampling period in seconds and z is the complex variable whose real and imaginary parts are related to those of s through

$$\mathrm{Re}\,z = e^{T\sigma}\cos\omega T \tag{3.70}$$

$$\mathrm{Im}\,z = e^{T\sigma}\sin\omega T \tag{3.71}$$

with

$$s = \sigma + j\omega \tag{3.72}$$

The relation between s and z may be defined as the z-transformation.

$$F^* \left[s = \frac{1}{T} \ln z \right] = F(z) = \sum_{k=0}^{\infty} f(kT) z^{-k} \tag{3.73}$$

Hence $F(z)$ is the z-transform of $f(t)$:

$$F(z) = z\text{-transform of } f(t) = \mathbf{Z}[f(t)]$$

Since z-transform of $f(t)$ is obtained from the Laplace transform of $f^*(t)$ by performing the transformation $z = e^{Ts}$. Hence, any function $f(t)$ that is Laplace transformable also has a z-transform.

In short, the operation of taking the z-transform of a continuous-data function $f(t)$ involves three steps:

1. $f(t)$ is sampled by an ideal sampler to produce $f^*(t)$.
2. The Laplace transform of $f^*(t)$ gives,

$$F^*(s) = L[f^*(t)] = \sum_{k=0}^{\infty} f(kT) e^{-kTs} \tag{3.74}$$

3. Replace e^{Ts} by z in $F^*(s)$ to get,

$$F(z) = \sum_{k=0}^{\infty} f(kT) z^{-k} \quad \text{(infinite series)} \tag{3.75}$$

By replacing $e^{-Ts} = z^{-1}$, we get,

$$F(z) = \sum_{n=1}^{k} \frac{N(\xi_n)}{D'(\xi_n)} \frac{1}{1 - e^{\xi_n T} z^{-1}} \tag{3.76}$$

where $F(s) = N(s)/D(s)$ (finite number of simple poles).

If $F(s)$ has multiple poles $s_1, s_2, \ldots, s_k$ with multiplicity $m_1, m_2, \ldots, m_k$, respectively, the z-transform becomes:

$$F(z) = \sum_{n=1}^{k} \sum_{i=1}^{m_n} \frac{(-1)^{m_n-i} k_{ni}}{m_n - i} \left[\frac{d^{m_n-i}}{d_s^{m_n-i}} \frac{1}{1 - e^{-Ts}} \right] \Big|_{s=s-s_n} \Big|_{z=e^{Ts}} \tag{3.77}$$

where

$$K_{ni} = \frac{1}{(i-1)!} \left[\frac{d^{i-1}}{d_s^{i-1}} (s - s_n)^{m_n} F(s) \right] \Big|_{s=s_n}$$

The Laplace transform and z-transform is shown in Table 3.1.

Table 3.1

Laplace Transforms and z-Transforms

S/N	Laplace transform $F(s)$	z-transform $F(z)$
1	1	1
2	e^{-kTs}	z^{-k}
3	$\dfrac{1}{s}$	$\dfrac{z}{z-1}$
4	$\dfrac{1}{s^2}$	$\dfrac{Tz}{(z-1)^2}$
5	$\dfrac{2}{s^3}$	$\dfrac{T^2 z(z+1)}{(z-1)^3}$
6	$\dfrac{(k-1)!}{s^k}$	$\lim_{a\to 0}(-1)^{k-1}\dfrac{\partial^{k-1}}{\partial a^{k-1}}\left[\dfrac{z}{z-e^{-aT}}\right]$
7	$\dfrac{1}{s+a}$	$\dfrac{z}{z-e^{-aT}}$
8	$\dfrac{1}{(s+a)^2}$	$\dfrac{Tz\,e^{-aT}}{(z-e^{-aT})^2}$
9	$\dfrac{(k-1)!}{(s+a)^k}$	$(-1)^k\dfrac{\partial^k}{\partial_a^k}\dfrac{z}{z-e^{-aT}}$
10	$\dfrac{a}{s(s+a)}$	$\dfrac{z(1-e^{-aT})}{(z-1)(z-e^{-aT})}$
11	$\dfrac{1}{(s+a)(s+b)}$	$\dfrac{1}{(b-a)}\left[\dfrac{z}{z-e^{-aT}}-\dfrac{z}{z-e^{-bT}}\right]$
12	$\dfrac{a}{s^2(s+a)}$	$\dfrac{Tz}{(z-1)^2}-\dfrac{(1-e^{-aT})z}{a(z-1)(z-e^{-aT})}$
13	$\dfrac{a}{s^3(s+a)}$	$\dfrac{T^2 z}{(z-1)^3}+\dfrac{(aT-2)Tz}{2a(z-1)^2}+\dfrac{z}{a^2(z-1)}-\dfrac{z}{a^2(z-e^{-aT})}$
14	$\dfrac{a^2}{s(s+a)^2}$	$\dfrac{z}{z-1}-\dfrac{z}{z-e^{-aT}}-\dfrac{aT\,e^{-aT}z}{(z-e^{-aT})^2}$
15	$\dfrac{a^2}{s^2(s+a)^2}$	$\dfrac{1}{a}\left[\dfrac{(aT+2)z2z^2}{(z-1)^2}+\dfrac{2z}{z-e^{-aT}}+\dfrac{aT\,e^{-aT_z}}{(z-e^{-aT})^2}\right]$
16	$\dfrac{\omega}{s^2+\omega^2}$	$z\sin\omega T$
17	$\dfrac{s}{s^2+\omega^2}$	$\dfrac{z(z-\cos\omega T)}{z^2-2z\cos\omega T+1}$
18	$\dfrac{\omega}{s^2-\omega^2}$	$\dfrac{z\sinh\omega T}{z^2-2z\cosh\omega T+1}$
19	$\dfrac{s}{s^2-\omega^2}$	$\dfrac{z(z-\cosh\omega T)}{z^2-2z\cosh\omega T+1}$

(Continued)

Table 3.1

(*Continued*)

S/N	Laplace transform $F(s)$	z-transform $F(z)$
20	$\dfrac{\omega}{(s+a)^2+\omega^2}$	$\dfrac{z\,e^{-aT}\sin\omega T}{z^2-2z\,e^{-aT}\cos\omega T+e^{-2aT}}$
21	$\dfrac{a^2+\omega^2}{s[(s+a)^2+\omega^2]}$	$\dfrac{z}{z-1}-\dfrac{z^2-z\,e^{-aT}\sec\phi\,\cos(\omega T-\phi)}{z^2-2z\,e^{-aT}\cos\omega T+e^{-2aT}}$
22	$\dfrac{s+a}{(s+a)^2+\omega^2}$	$\dfrac{z^2-z\,e^{-aT}\cos\omega T}{z^2-2z\,e^{-aT}\cos\omega T+e^{-2aT}}$

Brief theory on mapping between s- and z-domains:

$$z = e^{Ts} \tag{3.78}$$

The mapping of the s-plane into the z-plane can be done by using $z = e^{Ts}$.
The one-sided Laplace transform of a sampled function $e^*(t)$ is:

$$E^*(s) = L[e^*(t)] = \sum_{k=0}^{\infty} e(kT)\varepsilon^{-kTs} \tag{3.79}$$

where $e^*(t) = 0$ for $t < 0$ (one-sided) and $k = 0, 1, 2, \dots$. Substitute $s + jl\omega_S$ for s where l is an integer.

$$E^*(s + jl\omega_S) = \sum_{k=0}^{\infty} e(kT)\varepsilon^{-kT(s+jl\omega)} = \sum_{k=0}^{\infty} e(kT)\varepsilon^{-kTs}\varepsilon^{-jkl\omega_S T} \tag{3.80}$$

since $\omega_S = 2\pi/T$ then $\varepsilon^{-jkl\omega_S T} = \varepsilon^{-j2\pi kl} = 1$

$$E^*(s + jl\omega_S) = \sum_{k=0}^{\infty} e(kT)\varepsilon^{-kTs} = E^*(s) \tag{3.81}$$

If $E(s)$ has a pole (or zero) at $s = s_1$, then $E^*(s)$ has poles (zeros) at $s = s_1 \pm jl\omega_S$, where $l = 0, \pm 1, \pm 2, \dots \pm \infty$.

This means that given any pole (or zero) $s = s_1$ in the s-plane, the sampled function $E^*(s)$ has the same value at all periodic frequency points $s_1 \pm jl\omega_S$.

Note that the primary poles (or zero) s_1 and its associated complementary poles (or zeros) are mapped onto the same point in the z-plane (aliasing effect).

FURTHER READING

1. Kuo B., *Digital Control Systems*, 2nd edn., Saunders College Publishing, New York, USA, 1992.
2. Houpis C. H. and Lamont G. B., *Digital Control Systems: Theory, Hardware, Software*, 2nd edn., McGraw-Hill, Inc., New York, USA, 1992.
3. Oldenbourg R. C. and Sartorius H., *The Dynamics of Automatic Control,* American Society of Mechanical Engineers, New York, USA, 1948.
4. Bucella T. and Ahmed I., Taking control with DSPs, *Mach Des*, October 1989, pp. 73–80.
5. Rabiner I. R. and Rader C. M. (Eds), *Digital Signal Processing*, IEEE Press, New York, USA, 1972.
6. Helms H. D. and Rabiner L. R. (Eds), *Literature in Digital Signal Processing: Terminology and Permuted Tule Index*, IEEE Press, New York, USA, 1972.
7. Openheim A. V. and Schafer R. W., *Digital Signal Processing*, Prentice-Hall, Englewood Cliffs, New Jersey, USA, 1975.
8. Sheingold D. M., *Analog–Digital Conversion Handbook,* Analog Devices, Inc., Norwood, Massachusetts, USA, 1972.
9. Schmid M., *Electronic Analog/Digital Conversion*, Van Nostrand Rheinhild, New York, USA, 1970.
10. Shannon C. E., Communication in the presence of noise, *Proc IRE*, Vol. 37, January 1949, pp. 10–21.
11. Shannon C. E., Oliver B. M. and Pierce J. R., The philosophy of pulse code modulation, *Proc IRE*, Vol. 36, November 1948, pp. 1324–1331.
12. Fogel L. J., A note of the sampling theorem, *IRE Trans Inform Theor*, Vol. 1, March 1955, pp. 47–48.
13. Linden D. A., A discussion of sampling theorem, *Proc IRE*, Vol. 47, July 1959, pp. 1219–1226.
14. Goff K. W., Dynamics in direct digital control, Part I, *ISA J*, Vol. 13, No. 11, November 1966, pp. 45–49.
15. Goff K. W., Dynamics in direct digital control, Part II, *ISA J*, Vol. 13, No. 12, December 1966, pp. 44–54.
16. Whitbeck R. F. and Hofmann L. G., *Analysis of Digital Flight Control Systems with Flying Qualities Applications*, Vol. II, Systems Technology, Inc., Hawthorne, CA, 1978.
17. Jury E. I., *Theory and Application of the z-Transform Method*, Wiley, New York, USA, 1964.
18. Jury E. I., *Inners and Stability of Dynamic Systems,* Wiley, New York, USA, 1974.
19. Muth E. J., *Transform Method with Applications to Engineering and Operation Research*, Prentice-Hall, Englewood Cliffs, NJ, 1977.
20. Jury E. I. and Bharucha B. H., Notes on the stability criterion for linear discrete systems, *IRE Trans Autom Control*, Vol. AC-6, February 1961, pp. 88–90.
21. Jury E. I. and Blanchard J., A stability test for linear discrete system in table form, *IEE Proc*, Vol. 49, No. 12, December 1961, pp. 1947–1948.
22. Raible R. E., A simplification of Jury's tabular form, *IEEE Trans AC*, Vol. 19, June 1974, pp. 248–250.

Chapter 4

Mathematical Modeling of Digital Power Electronics

All switching circuits including all AC/DC rectifiers, DC/AC inverters, DC/DC converters and AC/AC (AC/DC/AC) converters are working in the discrete-time state. Therefore, they have to be described by digital control theory rather than analog control. This chapter mainly describes the mathematical modeling for these four types of converters in digital control.

4.1 INTRODUCTION

Analog control theory and cybernetics describe the systems working in the continuous time process. All parameters in these systems vary from time to time continuously. The dynamic process can be described by differential equations and Laplace transform operations. If there is any element in a control system that is discrete-time circuit, then the whole system has to be treated as a discrete-time system since the signal transferring is no longer continuously varying.

All switching circuits are working in the discrete-time mode. They have not responded to the signal change in the continuous state. Therefore, any system involved in a switching circuit has to be treated as a discrete system.

Most digital control systems have the analog output functions. For example, digitally controlled variable-speed motor drive systems have analog output speed in continuous-time mode although their controllers are digital signal processor (DSP). The differences between analog and digital control systems are in the following aspects:

- transfer function's form,
- stability characteristics,
- unit-step responses and impulse responses.

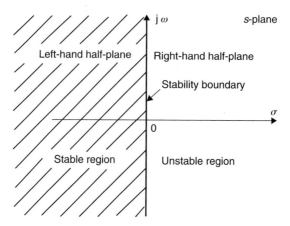

Figure 4.1   Stability boundary in the *s*-plane.

The transfer function of all analog control systems is described by the Laplace transform in the *s*-domain. The stability characteristics and criteria sates that if all poles of the transfer function are located in the left-hand half-plane (LHHP) of the *s*-plane, the system is stable. If any pole is located in the right-hand half-plane (RHHP) of the *s*-plane, the system is unstable. The stability boundary is shown in Figure 4.1. If any pole is located on the stability boundary, it means that the system performs at the critical stability state. Usually the system is considered unstable.

A stable analog control system has stable unit-step response with or without oscillation. Correspondingly, the impulse responses are also stable with or without oscillation. The system stability can be modified and improved by a closed-loop control and an optimization operation. A good analog control system has the following step response with the typical characteristics:

- fast response, i.e. the settling time is less than 4.7 times of the time constants;
- oscillation cycle number is not more than 2;
- the overshoot is not more than 5%.

These technical features are available for the impulse responses.

The transfer function of all digital control systems is described by either the Laplace transform in the *s*-domain and/or the *z*-transform in the *z*-domain. The stability characteristics and criteria sates that if all poles of the transfer function are located inside the unity-cycle of the *z*-plane, the system is stable. If any pole of the transfer function is located outside the unity-cycle of the *z*-plane, the system is unstable. The stability boundary (the unity-cycle in the *z*-plane) is shown in Figure 4.2. If any pole is located on the stability boundary, it means that the system performs at the critical stability state. Usually the system is considered unstable.

A stable digital control system has stable unit-step response with or without oscillation. Correspondingly, the impulse responses are also stable with or without oscillation.

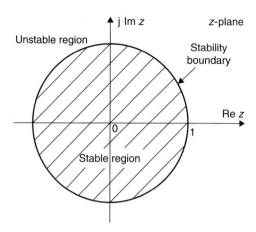

Figure 4.2 Stability boundary in the z-plane.

The system stability can be modified and improved by a closed-loop control and an optimization operation. A good analog control system has the following step response with the typical characteristics:

- fast response, i.e. the settling time can be only one-step delay;
- oscillation cycle number is not more than 2;
- the overshoot is not more than 5%.

These technical features are available for the impulse responses.

4.2 A ZERO-ORDER HOLD (ZOH) FOR AC/DC CONTROLLED RECTIFIERS

AC/DC rectifiers have many forms which are listed below:

1. single-phase half-wave controlled rectifier;
2. single-phase full-wave controlled rectifier;
3. three-phase half-wave controlled rectifier;
4. three-phase full-wave controlled rectifier;
5. double anti-star half-wave controlled rectifier with balanced inductor;
6. delta-star three-phase full-wave controlled rectifier;
7. four-quadrant operation controlled rectifiers:
    (a) four-quadrant operation controlled rectifiers with cycling current,
    (b) four-quadrant operation controlled rectifiers without cycling current.

The devices of all types of the AC/DC controlled rectifiers can be thyristors (silicon controlled rectifiers, SCRs), transistors, bipolar transistors (BTs), gate turn-off thyristors (GTOs) and Triacs. They are controlled by the corresponding firing pulse with

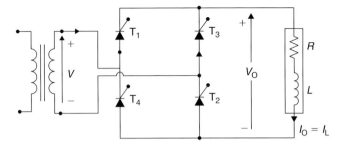

Figure 4.3 A single-phase fully controlled AC/DC rectifier.

certain firing angle α. In a sampling interval, usually in the commutation period the firing angle can be changed only once, i.e. the output voltage of an AC/DC controlled rectifier is changed period-by-period. Therefore, all types of the AC/DC controlled rectifiers are working in the discrete state.

In general situation, the load of the rectifiers is an L–R circuit with the time constant $\tau = L/R$. If the output current is continuous, then the average value of the output DC voltage is:

$$V_{\mathrm{d}} = V_{\mathrm{d\text{-}max}} \cos \alpha \qquad (4.1)$$

where α is the firing angle of the applied firing pulse. V_{d} is the output DC average voltage of the rectifier. $V_{\mathrm{d\text{-}max}}$ is the maximum output DC average voltage of the rectifier corresponding to the firing angle $\alpha = 0$.

Refer to the single-phase fully controlled AC/DC rectifier with an R–L load shown in Figure 4.3. The corresponding input and output voltage and current waveforms are shown in Figure 4.4. If the AC input power supply with the frequency $f = 50\,\mathrm{Hz}$ ($T = 1/f = 20\,\mathrm{ms}$), then each device is conducted in half a cycle, i.e. the conduction angle is 180° (or π rad) or in the interval of 10 ms. The current commutation happens twice a cycle.

The output voltage is out of control in a half-cycle once the firing pulse is applied. Therefore, it is the element to keep the output voltage in a period of $T/2 = 1/2f$. By per-unit system, the voltage transfer gain is unity (1) in a sampling interval $T/2 = 10\,\mathrm{ms}$. That is:

$$G(t) = \frac{V_{\mathrm{O}}}{V_{\mathrm{in}}}\Big|_{\text{per-unit}} = 1 \qquad (4.2)$$

Analogously, a single-phase half-wave AC/DC rectifier has the sampling interval to be 20 ms. A three-phase half-wave AC/DC rectifier has the sampling interval to be 6.67 ms (i.e. $T/3$) and a three-phase full-wave AC/DC rectifier has the sampling interval to be 3.33 ms (i.e. $T/6$). No matter how the multi-phase rectifier is, the output voltage is expressed by Equation (4.1), its voltage transfer gain in per-unit system is unity (1) as described in Equation (4.2).

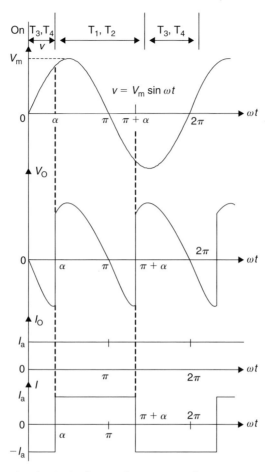

Figure 4.4 The input and output voltage and current waveforms.

4.2.1 TRADITIONAL MODELING FOR AC/DC CONTROLLED RECTIFIERS

Traditional modeling for AC/DC controlled rectifiers is a time-delay element in the s-domain. The delayed time is statistically as the sampling interval T or commutation period σ. For example, if the power supply frequency $f = 50$ Hz, a single-phase half-wave controlled rectifier has the time delay $T = \sigma = 20$ ms. The corresponding transfer function of this rectifier in per-unit system is:

$$G(s) = \mathrm{e}^{-Ts} = \mathrm{e}^{-\sigma s} \tag{4.3}$$

If the rectifier is used in a current control system or speed control system, and the current/speed responses are in the stage 0.1–1 s, we can consider the variable in a sampling interval of 20 ms is comparably small, i.e. $Ts = \sigma s \to 0$. Hence, Equation (4.3) can be rewritten as:

$$G(s) = e^{-\sigma s} = 1 - \sigma s = \frac{1}{1 + \sigma s} \tag{4.4}$$

This mathematical modeling is widely used in industrial applications.

4.2.2 A ZERO-ORDER HOLD FOR AC/DC CONTROLLED RECTIFIERS IN DIGITAL CONTROL

The mathematical modeling given in Equation (4.4) was very popular since 1960s to 1980s. Many large machinery systems have slow-time responses and use analog proportional-plus-integral (PI) controllers. This modeling is good enough to describe the system characteristics. Since 1980s the digital processors were applied in research and industrial application, where the systems performed faster. The rectifiers cannot be considered only a time-delayed element. Since its output voltage is out of control once the firing pulse was applied, it should be looked as a sample-and-hold element. Therefore, its mathematical modeling in per-unit digital control system should be a zero-order hold (ZOH) in both the s- and z-domains:

$$G(s) = \frac{1 - e^{-Ts}}{s} \tag{4.5}$$

$$G(z) = \mathbb{Z}[G(s)] = \mathbb{Z}\left[\frac{1 - e^{-Ts}}{s}\right] = \frac{z}{z - 1} - \frac{z}{(z - 1)}\frac{1}{z} = 1 \tag{4.6}$$

It means that the AC/DC controlled rectifier performs a sampling time delay in the s-domain and one-step delay (T) in a digital control system.

4.3 A FIRST-ORDER TRANSFER FUNCTION FOR DC/AC PULSE-WIDTH-MODULATION INVERTERS

Pulse-width-modulation (PWM) DC/AC inverters have many forms which are listed below:

1. DC/AC PWM inverter,
2. DC/AC single-phase PWM inverter,
3. multi-level PWM inverter,
4. multi-level single-phase PWM inverter,
5. vector multi-level PWM inverter,
6. space vector modulation (SVM) multi-level SPWM inverter.

The devices of all types of the DC/AC inverters can be a transistor, BT, GTO and power metal-oxide semiconductor field effected transistors (MOSFET). They are controlled by the PWM scheme with the certain carrier (chopping) frequency f_C. In a sampling interval $T = 1/f_C$, the pulse-width angle can be changed only once, i.e. the output voltage of an DC/AC PWM inverter is changed period by period. Therefore, all types of the DC/AC PWM inverters are working in discrete state.

In general situation, the input reference signal $v_{in}(t)$ is a sinusoidal waveform:

$$v_{in}(t) = V_m \sin \omega t \qquad (4.7)$$

The load of the inverters is an R–L circuit with time constant $\tau = L/R$, or the delayed angle is:

$$\phi = \tan^{-1} \frac{\omega L}{R} \qquad (4.8)$$

If the output current is continuous, the output AC voltage instantaneous value after the filter should be:

$$v_O(t) = V_m \sin(\omega t - \phi) \qquad (4.9)$$

where ϕ is the delay angle of the output voltage with reference to the input signal. $v_O(t)$ is the inverter output AC voltage instantaneous value after the filter. V_m is the amplitude of the output AC voltage of the inverter corresponding to the angle $(\omega t - \phi) = \pi/2$.

Refer to the single-phase PWM DC/AC inverter with an R–L load shown in Figure 4.5. The corresponding output voltage and current waveforms are shown in Figure 4.6 with the carrier frequency $f_C = 400$ Hz for indication although particular f_C may be very higher. If the input reference AC voltage signal with the frequency $f = 50$ Hz ($T = 1/f = 20$ ms) and the triangle waveform with carrier frequency $f_C = 400$ Hz ($T_C = 2.5$ ms), each device is conducted in a conduction period $= mT_C$, where m is the modulation ratio. The conduction period is less than a cycle T_C (sampling interval) since the modulation ratio m is usually smaller than unity (1) for linear modulation. The conduction angle is smaller than 360° (or 2π rad), or <2.5 ms. The current commutation happens once a chopping cycle.

The output voltage is out of control in a half-chopping cycle once the PWM pulse is applied. Therefore, it is the element to keep the output voltage in a sampling period

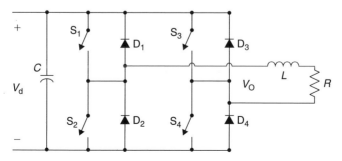

Figure 4.5   The single-phase PWM DC/AC inverter with an L–R circuit.

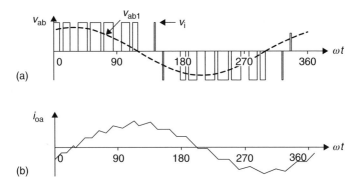

(a)

(b)

Figure 4.6   The output (a) voltage and (b) current waveforms.

of $T_C = 1/f_C$. By per-unit system, the voltage transfer gain is a linear element in a sampling interval $T_C = 2.5$ ms. That is,

$$G(t) = \frac{V_O}{V_{in}}|_{\text{per-unit}} = 1 - e^{-t/T_C} \tag{4.10}$$

Analogously, different chopping frequency only changes the sampling interval. The mathematical modeling is not changed. No matter how high or low the carrier frequency is, the output voltage is expressed by Equation (4.9), its voltage transfer gain per-unit system is a linear element as described in Equation (4.10).

4.3.1 TRADITIONAL MODELING FOR DC/AC PWM INVERTERS

Traditional modeling for DC/AC PWM inverters is a first-order element in the s-domain. The output voltage of a DC/AC PWM inverter is naturally a periodic pulse train with the repeating frequency f_O as requested. This output periodic pulse train has plenty of harmonics corresponding to both the requested frequency f_O and the carrier frequency f_C. Usually, the carrier frequency f_C must be much higher than the requested output frequency f_O to avoid yielding high total harmonic distortion (THD). Actually, no matter how higher the applied carrier frequency f_C is, THD cannot be zero.

   A low-pass filter must be set to filter the higher-order harmonics in order to obtain pure output sinusoidal waveform. The low-pass filter is usually an $R–C$ circuit with the time constant $\tau = R–C$ or an $R–L$ circuit with the time constant $\tau = L/R$. Normally, the time constant τ is much larger than the pulse width ($\tau \gg T = 1/f_C$) to avoid the parasitic power losses and additional distortion. The relations between them should be:

$$\frac{1}{f_O} \gg \tau \gg \frac{1}{f_C} \tag{4.11}$$

For example, if the requested output frequency $f_O = 50$ Hz, a modulation (chopping) frequency $f_C = 10$ kHz, usually the filter time constant τ will set in the range of

500 μs to 4 ms (typically 2 ms). The purpose to set the filter is to retain the fundamental harmonics in 50 Hz and filter the higher-order harmonics out, and then obtain very low THD. Sometimes, the time constant τ of the low-pass filter is not small to cause the fundamental harmonic in 50 Hz attenuated. The corresponding transfer function of this rectifier in per-unit system is:

$$G(s) = \frac{1}{1 + \tau s} \qquad (4.12)$$

From this point of view, the first-order transfer function is the feature of the first-order filter, but the characteristics of the DC/AC PWM inverters. The transfer function of the DC/AC PWM inverters is ignored. It is treated as a proportional element. Therefore, the traditional mathematical modeling is not really used to describe the DC/AC PWM inverters although it is generally agreed by most experts.

4.3.2 A FIRST-ORDER HOLD FOR DC/AC PWM INVERTERS IN DIGITAL CONTROL

The mathematical modeling seen in Equation (4.11) was very popular since 1960s to 1990s. Many large machinery systems have slow time responses and use analog PI controllers. This modeling is good enough to describe the system characteristics. On the other hand, if the inverters used to drive an induction motor, the stator circuit is a natural first-order filter with time constant $\tau = L/R$, which is usually lower than the power supply cycle $T = 1/f_O$. Carefully selecting the inverter carrier frequency f_C, the relation (4.11) is easily satisfied.

Since 1980s the digital processors have been applied in research and industrial application, and the systems perform faster. The DC/AC PWM inverters cannot be considered as only a proportional element. Since its output voltage is out of control once the pulse width is applied, it should be looked as a sample and linear-varying element. Therefore, its mathematical modeling in per-unit digital control system should be a first-order hold (FOH) in both s- and z-domains:

$$G(s) = \frac{1}{1 + Ts} \qquad (4.13)$$

$$G(z) = \mathbb{Z}[G(s)] = \mathbb{Z}\left[\frac{1}{1 + Ts}\right] = \frac{z}{z - 1/e} \qquad (4.14)$$

where T is the sampling interval $T = 1/f_C$. It means that the DC/AC PWM inverter performs a first-order inertial element with time constant T in the s-domain and a linear element in one step (T) in a digital control system in the z-domain. From Equation (4.14) the transfer function has one zero and one pole $z = 1/e$ inside the unit-cycle. Therefore, a DC/AC inverter is always a stable element.

4.4 A SECOND-ORDER TRANSFER FUNCTION FOR DC/DC CONVERTERS

DC/DC converters have many forms which are listed below:

1. fundamental converters such as buck, boost and buck–boost converters;
2. voltage-lift converters;
3. super-lift converters;
4. transformer-type converters;
5. other converters.

The devices of all types of the DC/DC converters can be transistor, BT, GTO and MOSFET. They are controlled by the PWM pulse with certain conduction duty cycle k. In a sampling interval/period T, the conduction duty cycle k can be changed only once, i.e. the output voltage of a DC/DC converter is changed period by period. Therefore, all types of the DC/DC converters are working in a discrete state.

In general situation, the load of the power DC/DC converters is a resistive load R. If the output current is continuous, the output DC voltage average value in a steady state is:

$$V_O = MV_I \tag{4.15}$$

or

$$M = \frac{V_O}{V_I}$$

where V_I and V_O are the input and output DC voltages, and M is the voltage transfer gain. If the switching frequency is f (the switching period $T = 1/f$) and the conduction duty cycle is k, the switching-on period is kT and switching-off period is $(1 - k)T$. The voltage transfer gain M is usually dependent to the conduction duty cycle k, and independent from the switching frequency f. For example, buck converter has the voltage transfer gain $M = k$, boost converter has the voltage transfer gain $M = 1/(1 - k)$ and buck–boost converter has the voltage transfer gain $M = k/(1 - k)$.

Refer to the single-ended primary inductance converter (SEPIC) shown in Figure 4.7. The inductor current i_{L_1} increases with slope $+V_C/L_1$ during switching

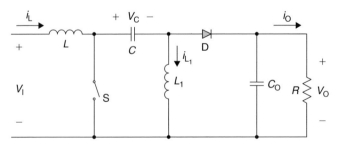

Figure 4.7   SEPIC.

on and decreases with slope $-V_O/L_1$ during switching off. Thus:

$$\frac{V_C}{L_1}kT = \frac{V_O}{L_1}(1-k)T$$

$$V_C = \frac{1-k}{k}V_O \tag{4.16}$$

The inductor current i_L increases with slope $+V_I/L$ during switching on and decreases with slope $-(V_C + V_O - V_I)/L$ during switching off. Thus:

$$\frac{V_I}{L}kT = \frac{V_C + V_O - V_I}{L}(1-k)T$$

$$V_O = \frac{k}{1-k}V_I \tag{4.17}$$

i.e.

$$M = \frac{V_O}{V_I} = \frac{k}{1-k}$$

Since the inductor L is in series connected to the source voltage, the inductor average current I_L is:

$$I_L = I_1$$

Since the inductor L_1 is in parallel connected to the capacitor C during switching off, the inductor average current I_{L_1} is ($I_{CO\text{-on}} = I_O$ and $I_{CO\text{-off}} = I_1$):

$$I_{L_1} = I_O$$

The variation of the current i_L is:

$$\Delta i_L = \frac{V_I}{L}kT$$

Therefore, the variation ratio of the current i_L is:

$$\xi = \frac{\Delta i_L/2}{I_L} = \frac{V_I}{2I_1 L}kT = \frac{k}{2M^2}\frac{R}{fL} \tag{4.18}$$

The variation of the current i_{L_1} is:

$$\Delta i_{L_1} = \frac{V_C}{L_1}kT$$

Therefore, the variation ratio of the current i_{L_1} is:

$$\xi_1 = \frac{\Delta i_{L_1}/2}{I_{L_1}} = \frac{V_C}{2I_O L_1} kT = \frac{1-k}{2} \frac{R}{fL_1} \tag{4.19}$$

The variation of the diode current i_D is:

$$\Delta i_D = \Delta i_L + \Delta i_{L_1} = \left(\frac{V_O}{L} + \frac{V_O}{L_1} \right)(1-k)T$$

We can define $L_{//} = L // L_1$:

$$\Delta i_D = \Delta i_L + \Delta i_{L_1} = \frac{V_O}{L_{//}}(1-k)T$$

and

$$I_D = I_L + I_{L_O} = I_1 + I_O = (M+1)I_O = \frac{1}{1-k} I_O$$

Therefore, the variation ratio of the diode current i_D is:

$$\zeta = \frac{\Delta i_D/2}{I_D} = \frac{V_O}{2I_O L_{//}}(1-k)^2 T = \frac{(1-k)^2}{2} \frac{R}{fL_{//}} \tag{4.20}$$

The variation of the voltage v_C is:

$$\Delta v_C = \frac{\Delta Q}{C} = \frac{I_1}{C}(1-k)T$$

Therefore, the variation ratio of the voltage v_C is:

$$\rho = \frac{\Delta v_C/2}{V_C} = \frac{I_1}{2CV_C}(1-k)T = \frac{kM}{2} \frac{1}{fRC} \tag{4.21}$$

The variation of the voltage v_{C_O} is:

$$\Delta v_{C_O} = \frac{\Delta Q_O}{C_O} = \frac{kTI_O}{C_O} = \frac{kI_O}{fC_O}$$

Therefore, the variation ratio of the voltage v_{C_O} is:

$$\varepsilon = \frac{\Delta v_{C_O}/2}{V_O} = \frac{kI_O}{2fC_O V_O} = \frac{k}{2fRC_O} \tag{4.22}$$

The output voltage is out of control in a period T once the duty cycle k is applied. There-fore, it is the element to keep the output voltage in a period $T = 1/f$. By per-unit system, the voltage transfer gain is unity (1) in a sampling interval. As discussed in Chapter 2, a power DC/DC converter is a second-order element, and its transfer function is:

$$G(s) = \frac{V_O}{V_I}\bigg|_{\text{per-unit}} = \frac{1}{1 + s\tau + s^2\tau\tau_d} \tag{4.23}$$

where τ is the time constant and τ_d is the damping time constant.

4.4.1 TRADITIONAL MODELING FOR DC/DC CONVERTERS

Traditional modeling for DC/DC converters is a complex element in the *s*-domain. The main method is the voltage division formula. Therefore, the transfer function with the orders is equal to the number of the passive energy-storage parts: inductors and capacitors. The simplest fundamental converters, such as buck, boost and buck–boost converters which have one inductor and one capacitor, possess a second-order transfer function. Other converters with multiple inductors and capacitors must have high-order transfer function. This problem has been discussed in Chapter 2. For example, a buck converter shown in Figure 2.1 has the transfer function as:

$$G(s) = \frac{M}{1 + s\frac{L}{R} + s^2 LC} = \frac{M}{1 + s\tau + s^2\tau\tau_d} \tag{4.24}$$

where M is the voltage transfer gain, τ is the time constant L/R and τ_d is the damping time constant RC.

For the DC/DC converters with two inductors plus two capacitors, their transfer function is in the fourth order. For example, the negative output Luo-converter elementary circuit shown in Figure 4.8 has the transfer function:

$$G(s) = \frac{M}{1 + s\frac{L_1+L_2}{R} + s^2(L_1C_1 + L_1C_2 + L_2C_2) + s^3\frac{L_1L_2C_1}{R} + s^4 L_1 L_2 C_1 C_2} \tag{4.25}$$

This mathematical modeling is very complex for industrial applications.

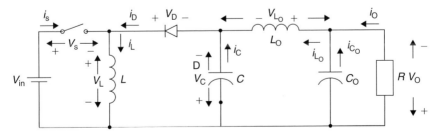

Figure 4.8   Elementary circuit of N/O Luo-converter.

4.4.2 A SECOND-ORDER HOLD FOR DC/DC CONVERTERS IN DIGITAL CONTROL

The traditional mathematical modeling was popular since 1940s till now. It is very difficult for the higher-order converters such as Luo-converters, super-lift converters, Cúk converter and SEPIC, and so on. In the period 2001–2004 Dr F. L. Luo and Dr H. Ye proposed a new method to model the power DC/DC converters. This new methodology was carefully described in Chapter 2. A second-order transfer function simulated all power DC/DC converters, the mathematical modeling in per-unit digital control system should be a second-order transfer function in the s-domain:

$$G(s) = \frac{1}{1 + s\tau + s^2\tau\tau_d} \tag{4.26}$$

where τ is the time constant (Equation (2.27)) and τ_d is the damping time constant (Equation (2.29)). In general situation if τ_d is greater than the critical value $\tau/4$, then there is a couple of conjugated complex poles s_1 and s_2. Equation (4.11) is rewritten as:

$$G(s) = \frac{1}{1 + s\tau + s^2\tau\tau_d} = \frac{1}{(s + s_1)(s + s_2)} \tag{4.27}$$

where

$$s_1 = \sigma + j\omega \quad \text{and} \quad s_2 = \sigma - j\omega$$

$$\sigma = \frac{1}{2\tau_d} \quad \text{and} \quad \omega = \frac{\sqrt{4\tau\tau_d - \tau^2}}{2\tau\tau_d}$$

Correspondingly, a power DC/DC converter is a second-order hold (SOH) in the z-domain:

$$G(z) = \mathbb{Z}[G(s)] = \mathbb{Z}\left[\frac{1}{(s + s_1)(s + s_2)}\right] = \frac{1}{s_1 - s_2}\left(\frac{z}{z - e^{-Ts_2}} - \frac{z}{z - e^{-Ts_1}}\right) \tag{4.28}$$

It means that the DC/DC converter performs a second-order response with oscillation in the s-domain and one-step delay (T) in a digital control system.

4.5 A FIRST-ORDER TRANSFER FUNCTION FOR AC/AC (AC/DC/AC) CONVERTERS

AC/AC (AC/DC/AC) converters have many forms which are listed below:

1. single phase-input single phase-output (SISO) amplitude modulated AC/AC converter,

2. multiphase-input multiphase-output (MIMO) amplitude modulated AC/AC converter,
3. SISO cycloconverter,
4. MIMO cycloconverter,
5. matrix AC/AC converter,
6. AC/DC/AC converters,
7. PWM converters.

The devices of all types of the AC/DC converters can be thyristor (SRC), transistor, BT, GTO and Triac. They are controlled by the corresponding firing pulse with certain firing angle α or the PWM scheme with the certain carrier (chopping) frequency f_c. In a sampling interval $T = 1/f_c$, the pulse-width angle can be changed only once, i.e. the output voltage of a DC/AC PWM inverter is changed period by period. Therefore, all types of the AC/AC PWM inverters are working in discrete state.

In general situation, the input reference signal $v_{in}(t)$ is a sinusoidal waveform:

$$v_{in}(t) = V_m \sin \omega t \tag{4.29}$$

The load of the inverters is an L–R circuit with time constant $\tau = L/R$, or the delayed angle is:

$$\phi = \tan^{-1} \frac{\omega L}{R} \tag{4.30}$$

If the output current is continuous, the output AC voltage instantaneous value after the filter should be:

$$v_O(t) = V_m \sin(\omega t - \phi) \tag{4.31}$$

where ϕ is the delay angle of the output voltage with reference to the input signal, which is calculated by Equation (4.30). $v_O(t)$ is the converter output AC voltage instantaneous value after the filter. V_m is the amplitude of the output AC voltage of the inverter corresponding to the angle $(\omega t - \phi) = \pi/2$.

Refer to the SISO AC/AC cycloconverter shown in Figure 4.9. The source AC voltage with frequency f_S is a sinusoidal waveform, and the output voltage should follow the

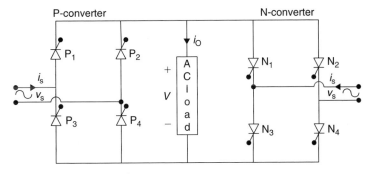

Figure 4.9 An SISO AC/AC cycloconverter.

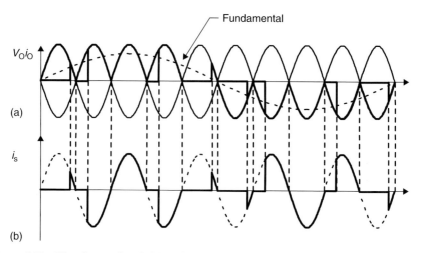

Figure 4.10   Waveforms of an SISO AC/AC cycloconverter (50–10 Hz) with R load. (a) Load voltage and load current; and (b) input supply current.

input reference voltage with frequency f. The corresponding input and output voltage and current waveforms are shown in Figure 4.10. If the input reference AC voltage signal with the frequency $f = 10$ Hz ($T = 1/f = 100$ ms), each bridge is conducted in a conduction period $= T/2 = 50$ ms.

The output voltage is out of control in a half-chopping cycle once the PWM pulse is applied. Therefore, it is the element to keep the output voltage in a period of $T_S/2 = 1/2f_S$. By per-unit system, the voltage transfer gain is a linear element in a sampling interval $T_S/2 = 10$ ms. That is,

$$G(t) = \frac{V_O}{V_{in}}|_{\text{per-unit}} = 1 - e^{-t/\tau} \tag{4.32}$$

Analogously, different chopping frequency only changes the sampling interval. The mathematical modeling is not changed. No matter how high or low the carrier frequency is, the output voltage is expressed by Equation (4.31), its voltage transfer gain in per-unit system is a linear element as described in Equation (4.32).

4.5.1 TRADITIONAL MODELING FOR AC/DC CONTROLLED RECTIFIERS

Traditional modeling for AC/AC converters is a first-order element in the s-domain. The output voltage of a AC/AC converter is naturally a periodic pulse train with the repeating frequency f_O as requested. This output periodic pulse train has plenty of harmonics corresponding to both the requested frequency f_O and the carrier frequency f_C. Usually, the carrier frequency f_C must be much higher than the requested output frequency f_O to avoid yielding high THD. Actually, no matter how higher the carrier frequency f_C applied is, THD cannot be zero.

A low-pass filter must be set to filter the higher-order harmonics in order to obtain pure output sinusoidal waveform. The low-pass filter is usually an R–C circuit with the time constant $\tau = RC$ or an R–L circuit with the time constant $\tau = L/R$. Normally, the time constant τ is much larger than the pulse width ($\tau \gg T = 1/f_C$) to avoid the parasitic power losses and additional distortion. The relations between them should be:

$$\frac{1}{f_O} \gg \tau \gg \frac{1}{f_C} \tag{4.33}$$

For example, if the output requested frequency $f_O = 50$ Hz, a modulation (chopping) frequency $f_C = 10$ kHz, usually then the filter time constant τ sets in the range of 500 μs to 4 ms (typically 2 ms). The purpose to set the filter is to retain the fundamental harmonic in 50 Hz and filter the higher-order harmonics out, and then obtain very low THD. Sometimes, the time constant τ of the low-pass filter is not small to cause the fundamental harmonic in 50 Hz attenuated. The corresponding transfer function of this rectifier in per-unit system is:

$$G(s) = \frac{1}{1 + \tau s} \tag{4.34}$$

From this point of view, the first-order transfer function is the feature of the first-order filter, but the characteristics of the AC/AC PWM inverters. The transfer function of the AC/AC PWM inverters is ignored. It is treated as a proportional element. Therefore, the traditional mathematical modeling is not really used to describe the AC/AC PWM inverters, although it is generally agreed by most experts and it is widely used in industrial applications.

4.5.2 A FOH for AC/AC Converters in Digital Control

The mathematical modeling given in Equation (4.14) was very popular since 1960s to 1980s. Many large machinery systems have slow time responses and use analog PI controllers. This modeling is good enough to describe the system characteristics. On the other hands, if the converters used to drive an induction motor, the stator circuit is a natural first-order filter with time constant $\tau = L/R$, which is usually lower than the power supply cycle $T = 1/f_O$. Carefully selecting the inverter carrier frequency f_C, the relation (4.33) is easily satisfied.

Since 1980s the digital processors were applied in research and industrial application, where the systems perform faster. The AC/AC converters cannot be considered only as a proportional element. Since its output voltage is out of control once the pulse width is applied, it should be looked as a sample and linear-varying element. Therefore, its mathematical modeling in per-unit digital control system should be a FOH in both the s- and z-domains:

$$G(s) = \frac{1}{1 + Ts} \tag{4.35}$$

$$G(z) = \mathbb{Z}[G(s)] = \mathbb{Z}\left[\frac{1}{1 + Ts}\right] = \frac{z/T}{z - 1/e} \tag{4.36}$$

where T is the sampling interval $T = 1/f_C$. It means that the DC/AC PWM inverter performs a first-order inertial element with time constant T in the s-domain and a linear rising/falling element in one step (T) in a digital control system.

FURTHER READING

1. Luo F. L., Jackson R. D. and Hill R. J., Digital controller for thyristor current source, *IEE-Proc Part B*, Vol. 132, No. 1, January 1985, pp. 46–52.
2. Luo F. L. and Hill R. J., Disturbance response techniques for digital control systems, *IEEE Trans IE*, Vol. 32, No. 3, August 1985, pp. 245–253.
3. Luo F. L. and Hill R. J., Minimisation of interference effects in thyristor converters by feedback feedforward control, *IEEE-Trans Meas Contr*, Vol. 7, No. 4, July–September 1985, pp. 175–182.
4. Luo F. L. and Hill R. J., Influence of feedback filter on system stability area in digitally-controlled thyristor converters, *IEEE-Trans Ind Appl*, Vol. 22, No. 1, January/February 1986, pp. 18–24.
5. Luo F. L. and Hill R. J., Fast response and optimum regulation in digitally-controlled thyristor converters, *IEEE-Trans Ind Appl*, Vol. 22, No. 1, January/February 1986, pp. 10–17.
6. Luo F. L. and Hill R. J., System analysis of digitally-controlled thyristor converters, *IEEE-Trans Meas and Contr*, Vol. 8, No. 1, January/March 1986, pp. 39–45.
7. Luo F. L. and Hill R. J., System optimisation – self-adaptive controller for digitally-controlled thyristor current controller, *IEEE-Trans IE*, Vol. 33, No. 3, August 1986, pp. 254–261.
8. Luo F. L. and Hill R. J., Stability analysis of thyristor current controllers, *IEEE-Trans Ind Appl*, Vol. 23, No. 1, January/February 1987, pp. 49–56.
9. Luo F. L. and Hill R. J., Current source optimisation in AC–DC GTO thyristor converters, *IEEE-Trans IE*, Vol. 34, No. 4, November 1987, pp. 475–482.
10. Luo F. L. and Hill R. J., Microprocessor-based control of steel rolling mill digital DC drives, *IEEE-Trans Power Electr*, Vol. 4, No. 2, April 1989, pp. 289–297.
11. Luo F. L. and Hill R. J., Microprocessor-controlled power converter using single-bridge rectifier and GTO current switch, *IEEE-Trans Meas Contr*, Vol. 12, No. 1, January–March 1990, pp. 2–8.
12. Muth E. J., *Transform Method with Applications to Engineering and Operation Research*, Prentice-Hall, Englewood Cliffs, New Jersey, 1977.
13. Oliver G., Stefanovic R. and Jamil A., Digitally controlled thyristor current source, *IEEE-Trans IECI*, Vol. 26, 1979, pp. 185–191.
14. Fallside F. and Jackson R. D., Direct digital control of thyristor amplifiers, *IEE-Proc, Part B*, Vol. 116, No. 5, 1969, pp. 873–878.

15. Arrillaga J., Galanos G. and Posner E. T., Direct digital control of HVDC converters, *IEEE-Trans PAS*, Vol. 89, 1970, pp. 2056–2065.
16. Daniels A. R. and Lipczyski R. T., Digital firing angle circuit for thyristor motor controllers, *IEE-Proc, Part B*, Vol. 125, No. 3, 1969, pp. 245–256.
17. Dewan S. B. and Dunford W. G., A microprocessor-based controller for a three-phase controlled bridge rectifiers, *IEEE-Trans IA*, Vol. 19, 1983, pp. 113–119.
18. Cheung W. N., The realisation of converter control using sampled-and-delay method, *IEE-Proc, Part B*, Vol. 127, No. 5, 1971, pp. 701–705.

Chapter 5

Digitally Controlled AC/DC Rectifiers

As described in Chapter 3, all AC/DC rectifiers are treated as a zero-order-hold (ZOH) element in digital control systems. We will discuss this model in various circuits in this Chapter.

5.1 INTRODUCTION

AC/DC rectifiers are the first group of the power switching circuits applied in industrial applications. In the 1940s, Mercury-arc rectifiers were very popular in DC power supply. In 1960s, semiconductor manufacture development brought power devices, such as power diode, thyristor (or silicon controlled rectifier, SCR), gate turn-off (GTO), Triac, bipolar transistor (BT), insulated gate bipolar transistors (IGBT) and metal-oxide semiconductor field effected transistor (MOSFET) and so on, into the DC power supply. The DC power supply equipment is totally changed. The corresponding control circuit is gradually changed from analog-to-digital control system since 1980s. The mathematical modeling for all AC/DC rectifiers is discussed widely in worldwide. Finally, a ZOH is generally accepted to be used to simulate the AC/DC rectifiers used.

In this Chapter we assume that the input voltage is a sinusoidal wave with the frequency $f = 50\,\mathrm{Hz}$. The transformer is used in the rectifiers with the following turn's ratio: 1:1 for Y/Y connection, $\sqrt{3}:1$ for Δ/Y connection, $1:\sqrt{3}$ for Y/Δ connection and 1:1 (or using $\sqrt{3}:\sqrt{3}$) for Δ/Δ connection in convenient analysis. The parameters used to describe the characteristics are listed below:

- $v(t)$: instantaneous phase voltage:

$$v(t) = \sqrt{2}V_{\mathrm{rms}}\sin(\omega t) \tag{5.1}$$

- V_{rms}: root-mean-square (rms) voltage
- V_m: maximum (amplitude) voltage:

$$V_m = \sqrt{2} V_{rms} \tag{5.2}$$

- V_O: average output voltage
- FF: form factor (FF):

$$FF = \frac{V_{rms}}{V_d} \tag{5.3}$$

- RF: ripple factor (RF):

$$RF = \frac{\sqrt{\sum_{n=1}^{\infty} V_n}}{V_d} \approx FF - 1 \tag{5.4}$$

- V_d: zeroth-order (DC) voltage component
- V_1: first-order (fundamental) harmonic voltage
- V_n: nth-order harmonic voltage
- PF: power factor (PF):

$$PF = \cos \phi = \frac{P}{S} \tag{5.5}$$

- P: real power
- Q: reactive power
- S: apparent power:

$$S = P + jQ \tag{5.6}$$

- THD: total harmonic distortion (THD):

$$THD = \frac{\sqrt{\sum_{n=2}^{\infty} V_n}}{V_1} \tag{5.7}$$

The generally used AC/DC diode rectifiers are introduced below:

1. Single-phase half-wave rectifiers
2. Single-phase full-wave rectifiers
3. Three-phase half-wave rectifiers
4. Three-phase full-wave rectifiers
5. Six-phase half-wave rectifiers
6. Six-phase full-wave rectifiers
7. Other circuits

Since the diode rectifiers have constant output voltages, their applications are limited. Replacing the diodes by thyristors or other switching devices, such as GTO, Triac and IGBT, we obtain the corresponding controlled AC/DC rectifiers. Our investigation objects are the controlled AC/DC rectifiers. The output voltage of the controlled AC/DC rectifiers relies on the firing angle of the firing pulse. When the firing angle $\alpha = 0$, we obtain the maximum output voltage of all controlled AC/DC rectifiers, which are equal to those of the diode AC/DC rectifiers. Well understanding the characteristics of the diode rectifiers is very helpful for our further discussion.

5.1.1 SINGLE-PHASE HALF-WAVE DIODE RECTIFIER

The simplest AC/DC diode rectifier is the single-phase half-wave diode rectifier shown in Figure 5.1. The only diode is conducted in the duration of 180° (or π rad) a cycle. Since the load is an R–L circuit with freewheel diode, the output voltage average value is:

$$V_O = \frac{1}{2\pi} \int_0^{2\pi} V_m \sin(\omega t)\,d(\omega t) = \frac{\sqrt{2}V_{rms}}{2\pi} \int_0^{\pi} \sin(\omega t)\,d(\omega t) = \frac{\sqrt{2}}{\pi} V_{rms} = 0.45 V_{rms}$$

(5.8)

$$FF = \frac{\pi}{2} = 1.57 \tag{5.9}$$

$$RF = 0.57 \tag{5.10}$$

$$PF = \frac{1}{\sqrt{2}} = 0.707 \tag{5.11}$$

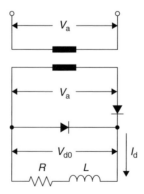

Figure 5.1   Single-phase half-wave diode rectifier with L–R load.

5.1.2 SINGLE-PHASE FULL-WAVE DIODE RECTIFIER

The single-phase full-wave diode rectifier has two configurations. The first circuit is called *center-tap (midpoint) circuit* shown in Figure 5.2(a) consisting of two diodes, and the second circuit is called *bridge (Graetz) circuit* shown in Figure 5.2(b) consisting of four diodes. Each diode is conducted in a cycle in 180°.

Parameters

Since the load is an *R–L* circuit, the output voltage average value is:

$$V_O = \frac{1}{2\pi} \int_0^{2\pi} V_m \sin(\omega t)\, d(\omega t) = \frac{2\sqrt{2}}{\pi} V_{rms} = 0.9 V_{rms} \tag{5.12}$$

$$FF = \frac{\pi}{2\sqrt{2}} = 1.11 \tag{5.13}$$

$$RF = 0.11 \tag{5.14}$$

Power Factor

The power factor (*PF*) of the center-tap (midpoint) circuit shown in Figure 5.2(a) is:

$$PF = \frac{1}{\sqrt{2}} = 0.707 \tag{5.15}$$

The power factor of the bridge (Graetz) circuit shown in Figure 5.2(b) is:

$$PF = 1 \tag{5.16}$$

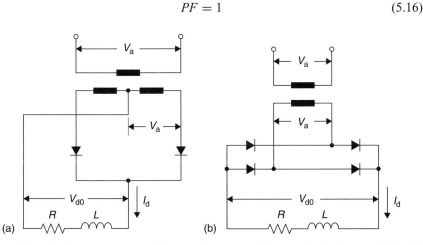

Figure 5.2 Single-phase full-wave diode rectifier with *L–R* load. (a) Center-tap (midpoint) circuit and (b) bridge (Graetz) circuit.

5.1.3 Three-Phase Half-Wave Diode Rectifier

The three-phase half-wave diode rectifier shown in Figure 5.3 has four configurations, all consisting of three diodes.

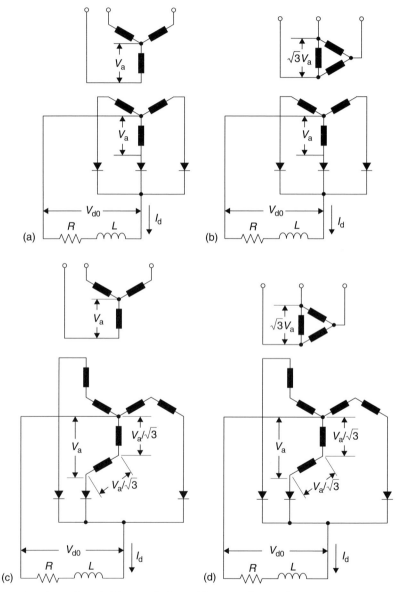

Figure 5.3 Three-phase half-wave diode rectifier. (a) Y/Y circuit, (b) Δ/Y circuit, (c) Y/Y bending circuit and (d) Δ/Y bending circuit.

The first circuit is called Y/Y circuit shown in Figure 5.3(a), the second circuit is called Δ/Y circuit shown in Figure 5.3(b), the third circuit is called Y/Y bending circuit shown in Figure 5.3(c) and the fourth circuit is called Δ/Y bending circuit shown in Figure 5.3(d). Each diode is conducted in 120°-cycle. Since the load is an *R–L* circuit, the output voltage average value is:

$$V_O = \frac{1}{2\pi/3} \int_{\pi/6}^{5\pi/6} V_m \sin(\omega t)\, d(\omega t) = \frac{3\sqrt{6}}{2\pi} V_{rms} = 1.17 V_{rms} \tag{5.17}$$

$$FF = 1.01615 \tag{5.18}$$

$$RF = 0.01615 \tag{5.19}$$

$$PF = 0.686 \tag{5.20}$$

5.1.4 THREE-PHASE FULL-WAVE DIODE RECTIFIER

The three-phase full-wave diode rectifier shown in Figure 5.4 has four configurations, all consisting of six diodes. The first circuit is called Y/Y circuit (Figure 5.6(a)), the second circuit is called Δ/Y circuit shown in Figure 5.4(b), the third circuit is called Y/Δ circuit shown in Figure 5.4(c) and the fourth circuit is called Δ/Δ circuit shown in Figure 5.4(d). Each diode is conducted in a cycle in 120°. Since the load is an *R–L* circuit, the output voltage average value is:

$$V_O = \frac{2}{2\pi/3} \int_{\pi/6}^{5\pi/6} V_m \sin(\omega t)\, d(\omega t) = \frac{3\sqrt{6}}{\pi} V_{rms} = 2.34 V_{rms} \tag{5.21}$$

$$FF = 1.00088 \tag{5.22}$$

$$RF = 0.00088 \tag{5.23}$$

$$PF = 0.956 \tag{5.24}$$

5.1.5 THREE-PHASE DOUBLE-ANTI-STAR WITH INTERPHASE-TRANSFORMER RECTIFIER

The three-phase double-anti-star with interphase-transformer diode rectifier shown in Figure 5.5 is three-phase half-wave diode rectifier, it has two configurations all

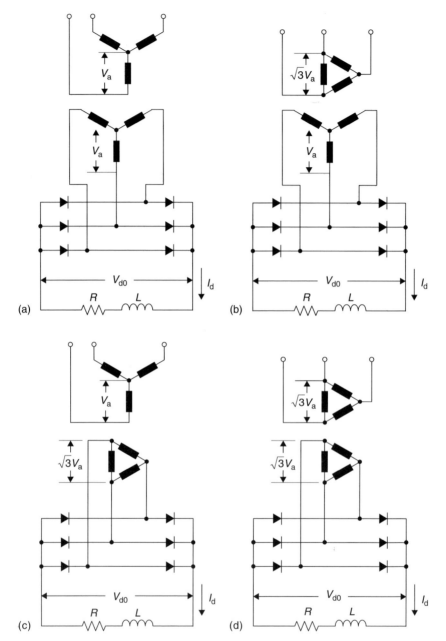

Figure 5.4   Three-phase full-wave diode rectifier. (a) Y/Y circuit, (b) Δ/Y circuit, (c) Y/Δ circuit and (d) Δ/Δ circuit.

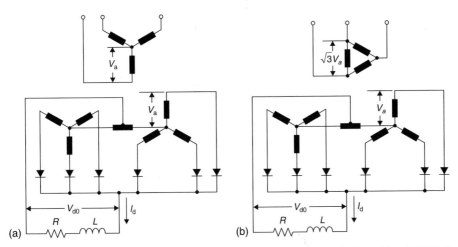

Figure 5.5   Three-phase double-anti-star with interphase-transformer diode rectifier. (a) Y/Y–Y circuit and (b) Δ/Y–Y circuit.

consisting of six diodes. The first circuit is called Y/Y–Y circuit shown in Figure 5.5(a) and the second circuit is called Δ/Y–Y circuit shown in Figure 5.5(b). Each diode is conducted in a cycle in 120°. Since the load is an *R–L* circuit, the output voltage average value is:

$$V_O = \frac{1}{2\pi/3} \int_{\pi/6}^{5\pi/6} V_m \sin(\omega t)\, d(\omega t) = \frac{3\sqrt{6}}{2\pi} V_{rms} = 1.17 V_{rms} \qquad (5.25)$$

$$FF = 1.01615 \qquad (5.26)$$

$$RF = 0.01615 \qquad (5.27)$$

$$PF = 0.686 \qquad (5.28)$$

5.1.6 SIX-PHASE HALF-WAVE DIODE RECTIFIER

The six-phase half-wave diode rectifier shown in Figure 5.6 has four configurations, all consisting of six diodes.

The first circuit is called Y/star circuit shown in Figure 5.6(a), the second circuit is called Δ/star circuit shown in Figure 5.6(b), the third circuit is called Y/star bending circuit shown in Figure 5.6(c) and the fourth circuit is called Δ/star bending circuit shown in Figure 5.6(d). Each diode is conducted in a cycle in 60°. Since the load is an

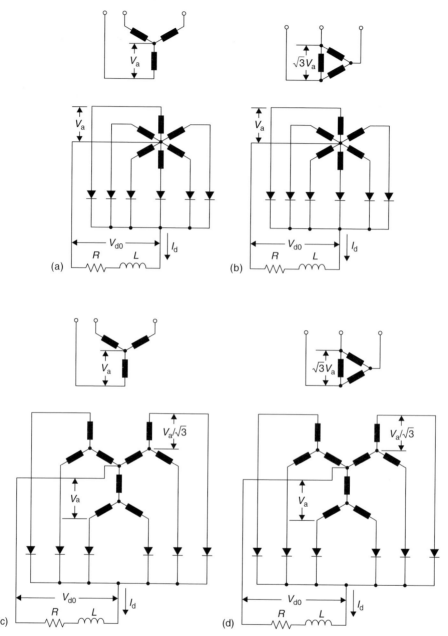

Figure 5.6 Six-phase half-wave diode rectifier. (a) Y/star circuit, (b) Δ/star circuit, (c) Y/star bending circuit and (d) Δ/star bending circuit.

R–L circuit, the output voltage average value is:

$$V_O = \frac{1}{\pi/3} \int\limits_{\pi/3}^{2\pi/3} V_m \sin(\omega t)\, d(\omega t) = \frac{3\sqrt{2}}{\pi} V_{rms} = 1.35 V_{rms} \qquad (5.29)$$

$$FF = 1.00088 \qquad (5.30)$$

$$RF = 0.00088 \qquad (5.31)$$

$$PF = 0.552 \qquad (5.32)$$

5.1.7 SIX-PHASE FULL-WAVE DIODE RECTIFIER

The six-phase full-wave diode rectifier shown in Figure 5.7 has two configurations, all consisting of 12 diodes. The first circuit is called six-phase bridge circuit shown in Figure 5.7(a) and the second circuit is called hexagon bridge circuit shown in Figure 5.7(b). Each diode is conducted in a cycle in 60°. Since the load is an R–L circuit, the output voltage average value is:

$$V_O = \frac{2}{\pi/3} \int\limits_{\pi/3}^{2\pi/3} V_m \sin(\omega t)\, d(\omega t) = \frac{6\sqrt{2}}{\pi} V_{rms} = 2.7 V_{rms} \qquad (5.33)$$

$$FF = 1.00088 \qquad (5.34)$$

$$RF = 0.00088 \qquad (5.35)$$

$$PF = 0.956 \qquad (5.36)$$

5.2 MATHEMATICAL MODELING FOR AC/DC RECTIFIERS

Replacing the diodes by any controlled devices, such as thyristors, GTO, Triac and IGBT, we obtain the corresponding controlled AC/DC rectifiers. The output voltage of the controlled AC/DC rectifiers relies on the firing angle α of the firing pulse. When the firing angle $\alpha = 0$, we obtain the maximum output voltage of all controlled AC/DC rectifiers, which are equal to those of the diode AC/DC rectifiers. In general condition the load is an R–L circuit with the time constant $\tau = L/R$, which is usually larger than the sampling interval T. Therefore, the output current is continuous and the output average voltage is generally equal to:

$$V_d = V_{d\text{-max}} \cos \alpha \qquad (5.37)$$

where V_d is the output average voltage and $V_{d\text{-max}}$ is the maximum output DC voltage corresponding to the firing angle $\alpha = 0$.

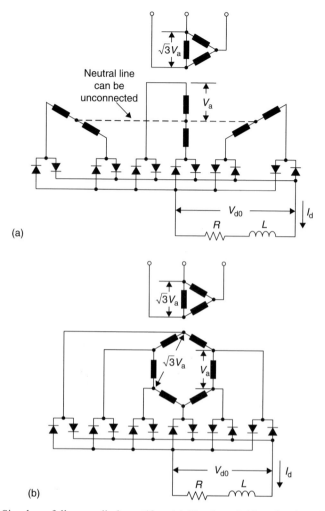

(a)

(b)

Figure 5.7 Six-phase full-wave diode rectifier. (a) Six-phase bridge circuit and (b) hexagon bridge circuit.

By per-unit system the voltage transfer gain is unity. The transfer function in the time domain is a constant value (unit-step function) $u(t)$, and it has the following form in the *s*-domain:

$$G(s) = \frac{1}{s} \tag{5.38}$$

In digital control system, all AC/DC rectifiers are treated as a ZOH, which has the transfer function:

$$G(z) = \frac{z}{z - 1} \tag{5.39}$$

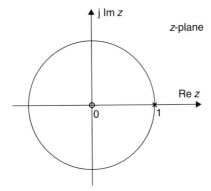

Figure 5.8   The zero and pole in the *z*-plane.

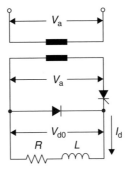

Figure 5.9   Single-phase half-wave controlled AC/DC rectifier with *L–R* load.

It means the rectifier is the element that possesses one zero at $z = 0$ and one pole at $z = 1$, which is located on the unit-cycle. The zero and pole in the *z*-plane are shown in Figure 5.8. Therefore, a rectifier is a critical stable element. In industrial applications, closed-loop control is required to increase the stability margin.

5.3 SINGLE-PHASE HALF-WAVE CONTROLLED AC/DC RECTIFIER

The single-phase half-wave controlled AC/DC rectifier is shown in Figure 5.9. The load is an *R–L* circuit with a freewheeling diode. The SCR is conducted in the period from α to π, i.e. the conduction angle is $\pi - \alpha$.

The open-loop control block diagram is shown in Figure 5.10. The sampling interval is $T = 1/f$, where f is the AC power supply source frequency. If $f = 50$ Hz, then $T = 20$ ms. This control can be implemented by a digital computer, which offers a firing pulse a cycle in 20 ms. The actuator is usually an *R–L* load. The final output parameter is the current I_O shown in Figure 5.10.

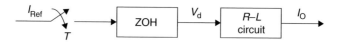

Figure 5.10 Open-loop control block diagram.

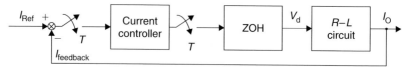

Figure 5.11 Closed-loop control block diagram.

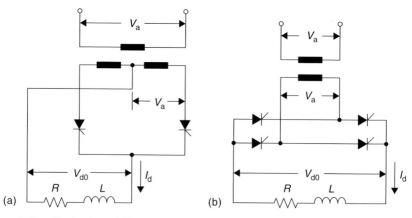

Figure 5.12 Single-phase full-wave controlled AC/DC rectifier with L–R load. (a) Center-tap (midpoint) circuit and (b) bridge (Graetz) circuit.

The closed-loop control block diagram is shown in Figure 5.11. The sampling interval is $T = 1/f$. A current controller is always requested in a closed-loop control system. It can be a proportional-plus-integral (PI) controller in digital form. This control can be implemented by a digital computer, which offers a firing pulse a cycle in 20 ms. The actuator is usually an R–L load. The final output parameter is the current I_O shown in Figure 5.11.

5.4 SINGLE-PHASE FULL-WAVE AC/DC RECTIFIER

The single-phase full-wave controlled AC/DC rectifier is shown in Figures 5.12(a) and (b). The load is an R–L circuit with continuous load current. Each SCR is conducted in the period of conduction angle π, from α to $(\pi + \alpha)$.

The open-loop control block diagram is still shown in Figure 5.10. The sampling interval is $T = 1/2f$, where f is the AC power supply source frequency. If $f = 50$ Hz, then $T = 10$ ms. This control can be implemented by a digital computer, which offers a firing pulse to each SCR a cycle in 20 ms. The actuator is usually an R–L load. The final output parameter is the current I_O shown in Figure 5.10.

The closed-loop control block diagram is shown in Figure 5.11. The sampling interval is $T = 1/2f$, where f is the AC power supply source frequency. If $f = 50$ Hz, then $T = 10$ ms. This control can be implemented by a digital computer, which offers a firing pulse to each SCR a cycle in 20 ms. The actuator is usually an R–L load. The final output parameter is the current I_O shown in Figure 5.11.

5.5 THREE-PHASE HALF-WAVE CONTROLLED AC/DC RECTIFIER

The three-phase half-wave controlled AC/DC rectifier is shown in Figures 5.13(a)–(d). The load is an R–L circuit with continuous load current. Each SCR is conducted in the period of conduction angle, $2\pi/3$, from α to $(2\pi/3 + \alpha)$.

The open-loop control block diagram is still shown in Figure 5.10. The sampling interval is $T = 1/3f$, where f is the AC power supply source frequency. If $f = 50$ Hz, then $T = 6.67$ ms. This control can be implemented by a digital computer, which offers a firing pulse to each SCR a cycle in 20 ms. The actuator is usually an R–L load. The final output parameter is the current I_O shown in Figure 5.10.

The closed-loop control block diagram is shown in Figure 5.11. The sampling interval is $T = 1/3f$, where f is the AC power supply source frequency. If $f = 50$ Hz, then $T = 6.67$ ms. This control can be implemented by a digital computer, which offers a firing pulse to each SCR a cycle in 20 ms. The actuator is usually an R–L load. The final output parameter is the current I_O shown in Figure 5.11.

5.6 THREE-PHASE FULL-WAVE CONTROLLED AC/DC RECTIFIER

The three-phase full-wave controlled AC/DC rectifier is shown in Figures 5.14(a)–(d). The load is an R–L circuit with continuous load current. Each SCR is conducted in the period of conduction angle $2\pi/3$, from α to $(2\pi/3 + \alpha)$.

The open-loop control block diagram is still shown in Figure 5.10. The sampling interval is $T = 1/3f$, where f is the AC power supply source frequency. If $f = 50$ Hz, then $T = 6.67$ ms. This control can be implemented by a digital computer, which offers a firing pulse to each SCR a cycle in 20 ms. The actuator is usually an R–L load. The final output parameter is the current I_O shown in Figure 5.10.

The closed-loop control block diagram is shown in Figure 5.11. The sampling interval is $T = 1/3f$, where f is the AC power supply source frequency. If $f = 50$ Hz, then $T = 6.67$ ms. This control can be implemented by a digital computer, which offers a firing pulse to each SCR a cycle in 20 ms. The actuator is usually an R–L load. The final output parameter is the current I_O shown in Figure 5.11.

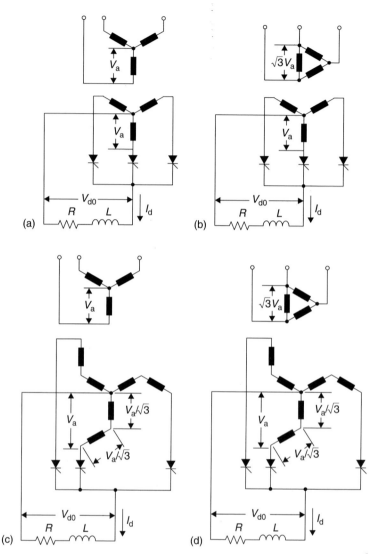

Figure 5.13 Thee-phase half-wave controlled AC/DC rectifier. (a) Y/Y circuit, (b) Δ/Y circuit, (c) Y/Y bending circuit and (d) Δ/Y bending circuit.

5.7 THREE-PHASE DOUBLE-ANTI-STAR WITH INTERPHASE-TRANSFORMER CONTROLLED AC/DC RECTIFIER

The three-phase double-anti-star with interphase-transformer controlled AC/DC rectifier is shown in Figure 5.15(a) and (b). The load is an R–L circuit with continuous

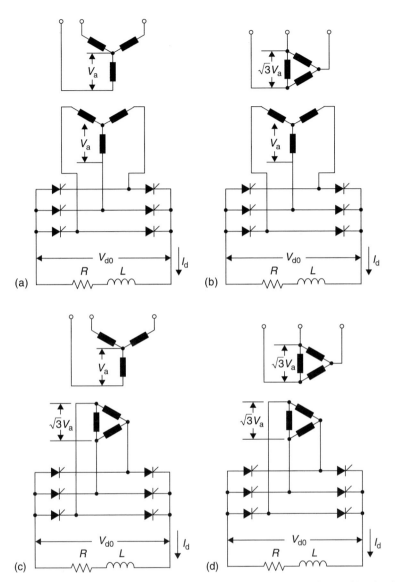

Figure 5.14   Three-phase full-wave controlled AC/DC rectifier. (a) Y/Y circuit, (b) Δ/Y circuit, (c) Y/Δ circuit and (d) Δ/Δ circuit.

load current. Each SCR is conducted in the period of conduction angle $2\pi/3$, from α to $(2\pi/3 + \alpha)$.

The open-loop control block diagram is still shown in Figure 5.10. The sampling interval is $T = 1/3f$, where f is the AC power supply source frequency. If $f = 50$ Hz, then $T = 6.67$ ms. This control can be implemented by a digital computer, which offers

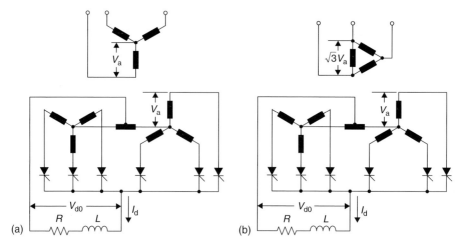

Figure 5.15 Three-phase double-anti-star with interphase-transformer controlled rectifier. (a) Y/Y–Y circuit and (b) Δ/Y–Y circuit.

a firing pulse to each SCR a cycle in 20 ms. The actuator is usually an R–L load. The final output parameter is the current I_O shown in Figure 5.10.

The closed-loop control block diagram is shown in Figure 5.11. The sampling interval is $T = 1/3f$, where f is the AC power supply source frequency. If $f = 50$ Hz, then $T = 6.67$ ms. This control can be implemented by a digital computer, which offers a firing pulse to each SCR a cycle in 20 ms. The actuator is usually an R–L load. The final output parameter is the current I_O shown in Figure 5.11.

5.8 SIX-PHASE HALF-WAVE CONTROLLED AC/DC RECTIFIER

The three-phase half-wave controlled AC/DC rectifier is shown in Figure 5.16(a) and (b). The load is an R–L circuit with continuous load current. Each SCR is conducted in the period of conduction angle $2\pi/3$, from α to $(2\pi/3 + \alpha)$.

The open-loop control block diagram is still shown in Figure 5.10. The sampling interval is $T = 1/3f$, where f is the AC power supply source frequency. If $f = 50$ Hz, then $T = 6.67$ ms. This control can be implemented by a digital computer, which offers a firing pulse to each SCR a cycle in 20 ms. The actuator is usually an R–L load. The final output parameter is the current I_O shown in Figure 5.10.

The closed-loop control block diagram is shown in Figure 5.11. The sampling interval is $T = 1/3f$, where f is the AC power supply source frequency. If $f = 50$ Hz, then $T = 6.67$ ms. This control can be implemented by a digital computer, which offers a firing pulse to each SCR a cycle in 20 ms. The actuator is usually an R–L load. The final output parameter is the current I_O shown in Figure 5.11.

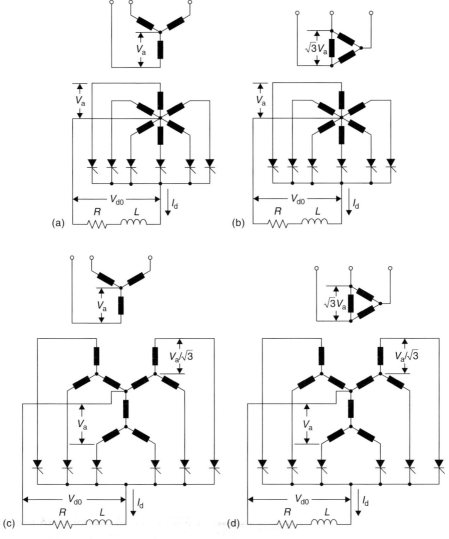

Figure 5.16 Six-phase half-wave controlled AC/DC rectifier. (a) Y/star circuit, (b) Δ/star circuit, (c) Y/star bending circuit and (d) Δ/star bending circuit.

5.9 SIX-PHASE FULL-WAVE CONTROLLED AC/DC RECTIFIER

The three-phase full-wave controlled AC/DC rectifier is shown in Figure 5.17(a) and (b). The load is an R–L circuit with continuous load current. Each SCR is conducted in the period of conduction angle $\pi/3$, from α to $(\pi/3 + \alpha)$.

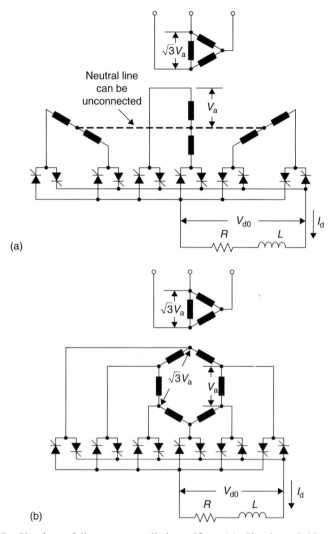

Figure 5.17 Six-phase full-wave controlled rectifier. (a) Six-phase bridge circuit and (b) hexagon bridge circuit.

The open-loop control block diagram is still shown in Figure 5.10. The sampling interval is $T = 1/6f$, where f is the AC power supply source frequency. If $f = 50$ Hz, then $T = 3.33$ ms. This control can be implemented by a digital computer, which offers a firing pulse to each SCR a cycle in 20 ms. The actuator is usually an R–L load. The final output parameter is the current I_O shown in Figure 5.10.

The closed-loop control block diagram is shown in Figure 5.11. The sampling interval is $T = 1/6f$, where f is the AC power supply source frequency. If $f = 50$ Hz, then $T = 3.33$ ms. This control can be implemented by a digital computer, which offers a firing pulse to each SCR a cycle in 20 ms. The actuator is usually an R–L load. The final output parameter is the current I_O shown in Figure 5.11.

FURTHER READING

1. Luo F. L. and Jackson R. D., *Digitally Controlled Thyristor Current Sources*, Cambridge University Technical Report CUED/B Elect/TR69, 1984.
2. Parrish E. A. and McVey E. S., A theoretical model for single-phase silicon-controlled-rectifier systems, *IEEE Trans AC*, Vol. 12, 1967, pp. 577–589.
3. McMurray W., The closed-loop stability of power converters with an integrating controller, *IEEE Trans IA*, Vol. 18, 1982, pp. 521–531.
4. Luo F. L., Jackson R. D. and Hill R. D., Digital controller for thyristor current sources, *IEE Proc Part-B*, Vol. 132, No. 1, January 1985, pp. 46–52.
5. Luo F. L. and Hill R. D., Disturbance response techniques for digital control systems, *IEEE Trans IE*, Vol. 32, No. 3, August 1985, pp. 245–253.
6. Oliver G., Stefannovic V. and Jamil A., Digitally controlled thyristor current sources, *IEEE Trans IECI*, Vol. 26, No. 3, August 1985, pp. 185–191.
7. Lee Y. S., *Computer-Aided Analysis and Design of Switch-Mode Power Supplies*, Marcel Dekker, Inc., New York, 1993.
8. Jacobina C., Beltrao M., Cabral E. and Nogueira A., Induction motor drive systems for low-power applications, *IEEE Trans IA*, Vol. 35, No. 1, August 1999, pp. 52–60.
9. Veas D. R., Dixon J. W. and Ooi B. T., A novel load current control method for a leading power factor voltage sources PEM rectifier, *IEEE Trans PEL*, Vol. 9, No. 2, 1994, pp. 153–159.
10. McMurray W., Modulation of the copping frequency in DC choppers and PWM inverters having current hysteresis controllers, *IEEE Trans IA*, Vol. 20, No. 4, 1984, pp. 763–768.
11. Dixon J. W. and Ooi B. T., Indirect current control of a unity power factor sinusoidal current boost type three-phase rectifier, *IEEE Trans IE*, Vol. 35, No. 4, 1988, pp. 508–515.
12. Moran L., Dixon J. W. and Wallace R., A three-phase active power filter operating with fixed switching frequency for reactive power and current harmonic compensation, *IEEE Trans IE*, Vol. 42, No. 4, 1995, pp. 402–408.
13. Ooi B. T., Dixon J. W., Kulkarni A. B. and Nishimoto M., An integrated AC drive system using a controlled-current PWM rectifier/inverter link, *IEEE Trans PEL*, Vol. 3, No. 1, 1988, pp. 64–71.
14. Dixon J. W. and Ooi B. T., Series and parallel operation of hysteresis current-controlled PWM rectifiers, *IEEE Trans IA*, Vol. 25, No. 4, 1989, pp. 644–651.
15. Boost M. A. and Ziogas P., State-of-the-art PWM techniques, a critical evaluation, *IEEE Trans IA*, Vol. 24, No. 2, 1988, pp. 271–280.
16. Moltgen G., Converter engineering, and introduction to operation and theory, John Wiley and Sons, New York, 1984.

Chapter 6

Digitally Controlled DC/AC Inverters

As described in Chapter 3, all DC/AC pulse-width-modulation (PWM) inverters are treated as a first-order-hold (FOH) element in digital control systems. We will discuss this model in various circuits in this Chapter.

6.1 INTRODUCTION

DC/AC inverters are a newly developed group of the power switching circuits applied in industrial applications in comparison with other power switching circuits. Although choppers were popular in DC/AC power supply long time ago, power DC/AC inverters were used in industrial application since later 1980s. Semiconductor manufacture development brought power devices, such as gate turn-off thyristor, Triac, bipolar transistor, insulated gate bipolar transistor and metal-oxide semiconductor field effected transistor (GTO, Triac, BT, IGBT, MOSFET, respectively) and so on, in higher switching frequency (say from thousands Hz upon few MHz) into the DC/AC power supply since 1980s. Due to the devices such as thyristor (silicon controlled rectifier, SCR) with low switching frequency, the corresponding equipment is low power rate.

Square-waveform DC/AC inverters were used in early ages before 1980s. In those equipment thyristors, GTOs and Triacs could be used in low-frequency switching operation. High-frequency/high-power devices such as power BTs and IGBTs were produced in the 1980s. The corresponding equipment implementing the PWM technique has large range of the output voltage and frequency, and low total harmonic distortion (THD). Nowadays, most DC/AC inverters are DC/AC PWM inverters in different prototypes.

DC/AC inverters are used for inverting DC power source into AC power applications. They are generally used in following applications:

1. Variable voltage/frequency AC supplies in adjustable speed drives (ASDs), such as induction motor drives, synchronous machine drives and so on.
2. Constant regulated voltage AC power supplies, such as uninterruptible power supplies (UPSs).
3. Static var compensations.
4. Active filters.
5. Flexible AC transmission systems (FACTSs).
6. Voltage compensations.

Adjustable speed induction motor drive systems are widely applied in industrial applications. These systems requested the DC/AC power supply with variable frequency usually from 0 to 400 Hz in fractional horsepower (HP) to hundreds of HP. A large number of the DC/AC inverters were in the world market. The typical block circuit is shown in Figure 6.1.

From this block diagram we can see that the power DC/AC inverter produces variable frequency and voltage to implement the ASD.

The power devices used for ASD can be thyristors, Triacs and GTOs in the 1970s and early 1980s. Power IGBT was popular in the 1990s, and greatly changed the manufacturing of DC/AC inverters. The DC/AC power supply equipment is totally changed. The corresponding control circuit is gradually changed from analog control to digital control system since late 1980s. The mathematical modeling for all AC/DC rectifiers is well discussed widely in worldwide. Finally, an FOH is generally accepted to be used for simulation of all DC/AC inverters.

The generally used DC/AC inverters are introduced below:

1. Single-phase half-bridge voltage source inverter (VSI)
2. Single-phase full-bridge VSI
3. Three-phase full-bridge VSI
4. Three-phase full-bridge current source inverter (CSI)
5. Multistage PWM inverters

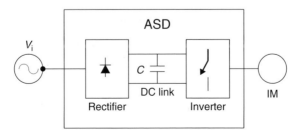

Figure 6.1 A standard ASD scheme.

6. Multilevel PWM inverters
7. Soft-switching inverters.

As mentioned in Chapter 3, we list some parameters as follows:

- The modulation index m_a (also known as the amplitude-modulation ratio in Chapter 3):

$$m_a = \frac{V_C}{V_\Delta} \tag{6.1}$$

where V_C is the amplitude of the control or the preliminary reference signal, and V_Δ is the amplitude of the triangle signal. Generally, linear-modulation operation is considered, so that m_a is usually smaller than unity (e.g. $m_a = 0.8$).
- The normalized carrier frequency index m_f (also known as the frequency-modulation ratio in Chapter 3):

$$m_f = \frac{f_\Delta}{f_C} \tag{6.2}$$

where f_Δ is the frequency of the triangle signal, and f_C is the frequency of the control signal or the preliminary reference signal. Generally, in order to obtain low THD, the m_f has usually taken large number (e.g. $m_f = 9$).

In order to well understand each inverter, we have shown some typical circuits below.

6.1.1 SINGLE-PHASE HALF-BRIDGE VSI

A single-phase half-bridge VSI is shown in Figure 6.2. The carrier-based PWM technique is applied in this single-phase half-bridge VSI. Two large capacitors are required to provide a neutral point N, therefore, each capacitor keep the half of the input DC voltage. Two switches S+ and S− are switched by the PWM signal.

Figure 6.3 shows the ideal waveforms associated with the half-bridge VSI. We can find out the output of the phase delayed between the output current and voltage.

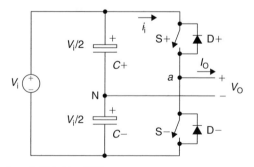

Figure 6.2   Single-phase half-bridge VSI.

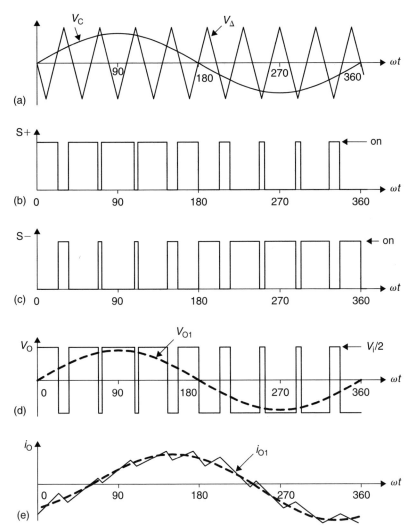

Figure 6.3 Ideal waveforms associated with the single-phase half-bridge VSI ($m_a = 0.8$, $m_f = 9$). (a) Carrier and modulating signals, (b) switch S+ state, (c) switch S− state, (d) AC output voltage and (e) AC output current.

6.1.2 SINGLE-PHASE FULL-BRIDGE VSI

A single-phase full-bridge VSI is shown in Figure 6.4.

The carrier-based PWM technique is applied in this single-phase full-bridge VSI. Two large capacitors may be used to provide a neutral point N, therefore, each capacitor keep the half of the input DC voltage. Four switches S_1+ and $S_1−$ plus S_2+ and $S_2−$ are applied and switched by the PWM signal. Figure 6.5 shows the ideal waveforms

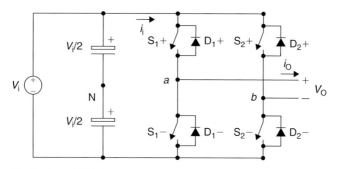

Figure 6.4   Single-phase full-bridge VSI.

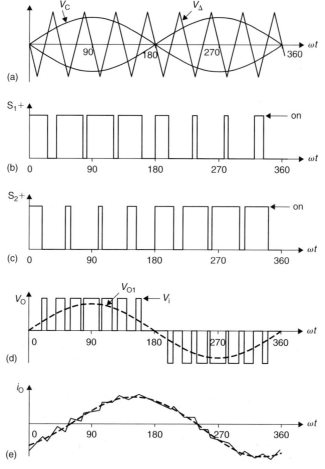

Figure 6.5   Ideal waveforms associated with the full-bridge VSI ($m_a = 0.8$, $m_f = 8$). (a) Carrier and modulating signals, (b) switch S_1+ and S_1- state, (c) switch S_2+ and S_2- state, (d) AC output voltage and (e) AC output current.

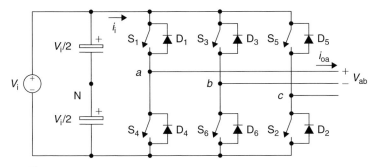

Figure 6.6 Three-phase full-bridge VSI.

associated with the full-bridge VSI. We can find out the output of the phase delayed between the output current and voltage.

6.1.3 THREE-PHASE FULL-BRIDGE VSI

A three-phase full-bridge VSI is shown in Figure 6.6.

The carrier-based PWM technique is applied in this single-phase full-bridge VSI. Two large capacitors may be used to provide a neutral point N, therefore, each capacitor keep the half of the input DC voltage. Six switches S_1–S_6 are applied and switched by the PWM signal. Figure 6.7 shows the ideal waveforms associated with the full-bridge VSI. We can find out the output of the phase delayed between the output current and voltage.

6.1.4 THREE-PHASE FULL-BRIDGE CSI

A three-phase full-bridge CSI is shown in Figure 6.8.

The carrier-based PWM technique is applied in this single-phase full-bridge CSI. The main objective of these static power converters is to produce AC output current waveforms from a DC current power supply. Six switches S_1–S_6 are applied and switched by the PWM signal. Figure 6.9 shows the ideal waveforms associated with the full-bridge CSI.

We can find out the output of the phase ahead between the output voltage and current.

6.1.5 MULTISTAGE PWM INVERTER

Multistage PWM inverter consists of many cells. Each cell can be a single- or three-phase input plus single-phase output VSI, which is shown in Figure 6.10. If the

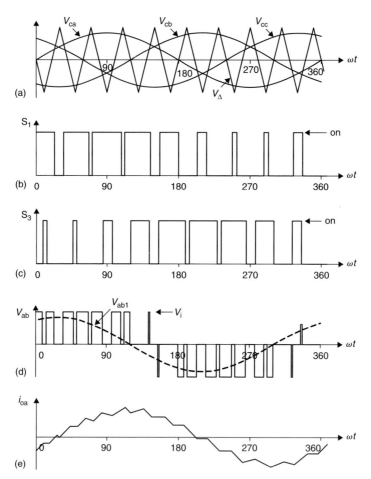

Figure 6.7 Ideal waveforms associated with the three-phase full-bridge VSI ($m_a = 0.8$, $m_f = 9$). (a) Carrier and modulating signals, (b) switch S_1+ state, (c) switch S_3 state, (d) AC output voltage and (e) AC output current.

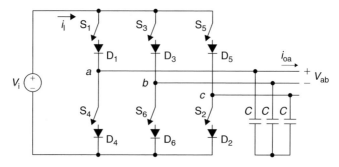

Figure 6.8 Three-phase full-bridge CSI.

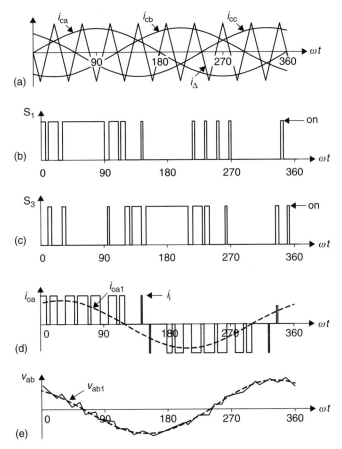

Figure 6.9 Ideal waveforms associated with the three-phase full-bridge CSI ($m_a = 0.8$, $m_f = 9$). (a) Carrier and modulating signals, (b) switch S_1+ state, (c) switch S_3 state, (d) AC output current and (e) AC output voltage.

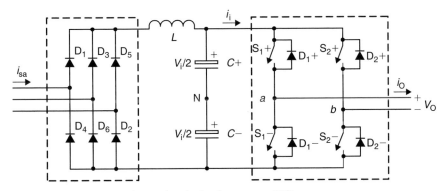

Figure 6.10 Three-phase input plus single-phase output VSI.

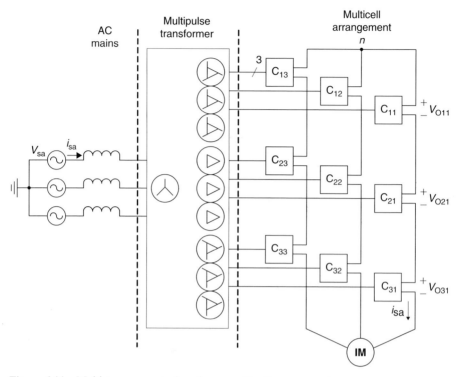

Figure 6.11   Multistage converter based on a multicell arrangement.

three-phase AC supply is a secondary winding of a main transformer, it is floating and isolated from other cells and common ground point. Therefore, all cells can be linked in series or parallel manner.

A three-stage PWM inverter is shown in Figure 6.11. Each phase consist of three cells with difference phase-angle shift by 20° each other.

The carrier-based PWM technique is applied in this three-phase multistage PWM inverter. Figure 6.12 shows the ideal waveforms associated with the full-bridge VSI. We can find out the output of the phase delayed between the output current and voltage.

6.1.6 MULTILEVEL PWM INVERTER

A three-level PWM inverter is shown in Figure 6.13. The carrier-based PWM technique is applied in this multilevel PWM inverter. Figure 6.14 shows the ideal waveforms associated with the multilevel PWM inverter. We can find out the output of the phase delayed between the output current and voltage.

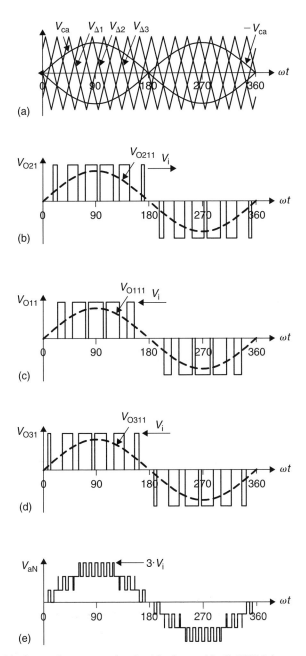

Figure 6.12 Ideal waveforms associated with the multicell PWM inverter (three stages, $m_a = 0.8$, $m_f = 6$). (a) Carrier and modulating signals, (b) cell c_{11} AC output voltage, (c) cell c_{21} AC output voltage, (d) cell c_{31} AC output voltage and (e) phase a load voltage.

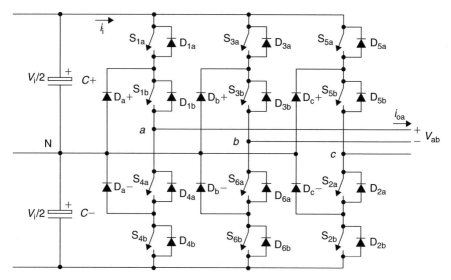

Figure 6.13   Three-phase three-level PWM VSI.

6.2 MATHEMATICAL MODELING FOR DC/AC PWM INVERTERS

By carefully investigating the PWM inversion process, we can see that in each pulse width $T = 1/f_\Delta$, and the modulation ratio m_a is proportional to the control signal $v_C(t)$. If the frequency ratio m_f is large enough, the value of the control signal $v_C(t)$ in a sampling period T can be considered a constant value. The output voltage value of a VSI is proportional to the input control signal. The corresponding output current value of a VSI is an increasing or decreasing wave. The corresponding waveforms have been shown in Figures 6.3 and 6.5. In general condition, the load is an R–L circuit with the time constant $\tau = L/R$, which is usually larger than the sampling interval T. Therefore, the output current is continuous and is generally accumulated interval by interval. The expression in per-unit system can be written as:

$$i_{O\text{-}k} = i_{O\text{-}(k-1)}(1 \pm e^{-t/T}) \tag{6.3}$$

where $i_{O\text{-}k}$ is the kth-step output current and $i_{O\text{-}(k-1)}$ is the previous step output current.

By per-unit system the voltage transfer gain is unity. The transfer function in the time domain is an exponential function, and it has the following form in the s-domain:

$$G(s) = \frac{1}{1 + sT} \tag{6.4}$$

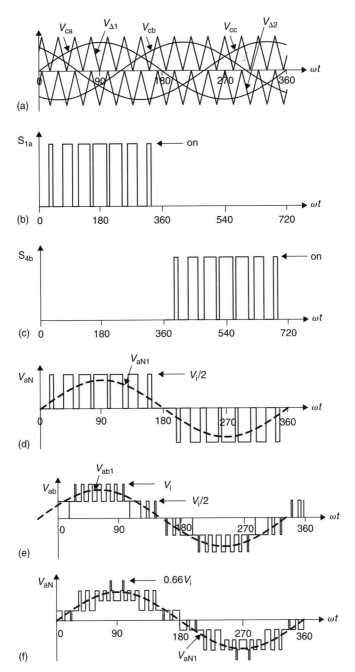

Figure 6.14 Ideal waveforms associated with three-phase three-level VSI (three levels, $m_a = 0.8$, $m_f = 15$). (a) Carrier and modulating signals, (b) switch S_{1a} status, (c) switch S_{4b} status, (d) inverter phase aN voltage, (e) AC output line voltage and (f) AC output phase voltage.

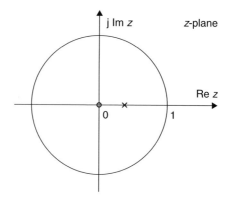

Figure 6.15   Zero and pole of the FOH in the z-plane.

In digital control system, all DC/AC PWM inverters are treated as an FOH has the transfer function in the z-domain:

$$G(z) = \frac{z}{z - 1/e} \qquad (6.5)$$

It means the DC/AC PWM inverter is the first-order-element that possesses one zero at $z = 0$ and one pole at $z = 1/e$, which is located on the unit-cycle. The zero and pole in the z-plane are shown in Figure 6.15. Therefore, a rectifier is a critical stable element. In industrial applications, closed-loop control is required to increase the stability margin.

6.3 SINGLE-PHASE HALF-WAVE VSI

The single-phase half-wave VSI is shown in Figure 6.2. The load is an R–L circuit. The open-loop control block diagram is shown in Figure 6.16. The sampling interval is $T = 1/f_\Delta$, where f_Δ is the triangle frequency. If $f = 400$ Hz, then $T = 2.5$ ms. This control can be implemented by a digital computer, which offers a pulse cycle in 2.5 ms. The actuator is usually an R–L load. The final output parameter is the current I_O shown in Figure 6.16.

The closed-loop control block diagram is shown in Figure 6.17. The sampling interval is $T = 1/f_\Delta$. A current controller is always requested in a closed-loop control system. It can be a proportional-plus-integral (PI) controller in digital form. This control can be implemented by a digital computer, which offers a firing pulse cycle in 2.5 ms. The actuator is usually an R–L load. The final output parameter is the current I_O shown in Figure 6.17.

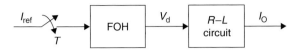

Figure 6.16 Open-loop control of the DC/AC PWM inverters.

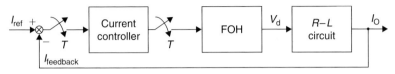

Figure 6.17 Closed-loop control of the DC/AC PWM inverters.

6.4 SINGLE-PHASE FULL-BRIDGE PWM VSI

The single-phase full-wave VSI is shown in Figure 6.4. The load is an R–L circuit. The open-loop control block diagram is shown in Figure 6.16. The sampling interval is $T = 1/f_\Delta$, where f_Δ is the triangle frequency. If $f = 400\,\text{Hz}$, then $T = 2.5\,\text{ms}$. This control can be implemented by a digital computer, which offers a pulse cycle in 2.5 ms. The actuator is usually an R–L load. The final output parameter is the current I_O shown in Figure 6.16.

The closed-loop control block diagram is shown in Figure 6.17. The sampling interval is $T = 1/f_\Delta$. A current controller is always requested in a closed-loop control system. It can be a PI controller in digital form. This control can be implemented by a digital computer, which offers a firing pulse cycle in 2.5 ms. The actuator is usually an R–L load. The final output parameter is the current I_O shown in Figure 6.17.

6.5 THREE-PHASE FULL-BRIDGE PWM VSI

The three-phase full-bridge PWM VSI is shown in Figure 6.6. The open-loop control block diagram is still shown in Figure 6.16. The sampling interval is $T = 1/f_\Delta$, where f_Δ is the triangle frequency. If $f = 400\,\text{Hz}$, then $T = 2.5\,\text{ms}$. This control can be implemented by a digital computer, which offers a pulse cycle in 2.5 ms. The actuator is usually an R–L load. The final output parameter is the current I_O shown in Figure 6.16.

The closed-loop control block diagram is shown in Figure 6.17. The sampling interval is $T = 1/f_\Delta$. A current controller is always requested in a closed-loop control system. It can be a PI controller in digital form. This control can be implemented by a digital computer, which offers a firing pulse cycle in 2.5 ms. The actuator is usually an R–L load. The final output parameter is the current I_O shown in Figure 6.17.

6.6 THREE-PHASE FULL-BRIDGE PWM CSI

The three-phase full-bridge PWM CSI is shown in Figure 6.8. The open-loop control block diagram is still shown in Figure 6.16. The sampling interval is $T = 1/f_\Delta$, where f_Δ is the triangle frequency. If $f = 400$ Hz, then $T = 2.5$ ms. This control can be implemented by a digital computer, which offers a pulse cycle in 2.5 ms. The actuator is usually an R–L load. The final output parameter is the current I_O shown in Figure 6.16.

The closed-loop control block diagram is shown in Figure 6.17. The sampling interval is $T = 1/f_\Delta$. A current controller is always requested in a closed-loop control system. It can be a PI controller in digital form. This control can be implemented by a digital computer, which offers a firing pulse cycle in 2.5 ms. The actuator is usually an R–L load. The final output parameter is the current I_O shown in Figure 6.17.

6.7 MULTISTAGE PWM INVERTER

The multistage PWM inverter based on a multicell arrangement is shown in Figure 6.11. The open-loop control block diagram is still shown in Figure 6.16. The sampling interval is $T = 1/f_\Delta$, where f_Δ is the triangle frequency. If $f = 400$ Hz, then $T = 2.5$ ms. This control can be implemented by a digital computer, which offers a pulse cycle in 2.5 ms. The actuator is usually an R–L load. The final output parameter is the current I_O shown in Figure 6.16.

The closed-loop control block diagram is shown in Figure 6.17. The sampling interval is $T = 1/f_\Delta$. A current controller is always requested in a closed-loop control system. It can be a PI controller in digital form. This control can be implemented by a digital computer, which offers a firing pulse cycle in 2.5 ms. The actuator is usually an R–L load. The final output parameter is the current I_O shown in Figure 6.17.

6.8 MULTILEVEL PWM INVERTER

The multistage PWM inverter based on a multicell arrangement is shown in Figure 6.13. The open-loop control block diagram is still shown in Figure 6.16. The sampling interval is $T = 1/f_\Delta$, where f_Δ is the triangle frequency. If $f = 400$ Hz, then $T = 2.5$ ms. This control can be implemented by a digital computer, which offers a pulse cycle in 2.5 ms. The actuator is usually an R–L load. The final output parameter is the current I_O shown in Figure 6.16.

The closed-loop control block diagram is shown in Figure 6.17. The sampling interval is $T = 1/f_\Delta$. A current controller is always requested in a closed-loop control system. It can be a PI controller in digital form. This control can be implemented by a digital computer, which offers a firing pulse cycle in 2.5 ms. The actuator is usually an R–L load. The final output parameter is the current I_O shown in Figure 6.17.

FURTHER READING

1. Pan Z. Y. and Luo F. L., Novel soft-switching inverter for brushless DC motor variable speed drive system, *IEEE Trans PEL*, Vol. 19, No. 2, March 2004, pp. 280–288.
2. Pan Z. Y. and Luo F. L., Novel resonant pole inverter for brushless DC motor drive system, *IEEE Trans PEL*, Vol. 20, No. 1, January 2005, pp. 173–181.
3. Pan Z. Y. and Luo F. L., Transformer based resonant DC link inverter for brushless DC motor drive system, *IEEE Trans PEL*, Vol. 20, No. 4, July 2005, pp. 865–873.
4. Luo F. L. and Pan Z. Y., Novel soft switching inverter for brushless DC motor variable speed drives systems, *Proceedings of IEE International Conference IPEC '03*, Singapore, 27–29 November 2003, pp. 1045–1050.
5. Pan Z. Y. and Luo F. L., Transformer based resonant DC link inverter for brushless DC motor drive system, *Proceedings of the IEEE International Conference PESC '04*, Aachen, Germany, 20–25 June 2004, pp. 3866–3872.
6. Joos G. and Espinoza J., Three-phase series var compensation based on a voltage controlled current source inverter with supplemental modulation index control, *IEEE Trans PEL*, Vol. 14, No. 3, May 1999, pp. 587–598.
7. Jain P., Espinoza J. and Jin H., Performance of a single-stage UPS system for single-phase trapezoidal-shaped AC voltage supplies, *IEEE Trans PEL*, Vol. 13, No. 5, 1998, pp. 912–923.
8. Kamran F. and Habetler T., A novel on-line UPS with universal filtering capabilities, *IEEE Trans PEL*, Vol. 13, No. 3, 1998, pp. 410–418.
9. Wu B., Dewan S. and Slemon G., PWM-CSI inverter for induction motor drives, *IEEE Trans IA*, Vol. 28, No. 1, 1992, pp. 64–71.
10. Antunes F., Braga H. and Barbi I., Application of a generalized current multilevel cell to current-source inverters, *IEEE Trans IE*, Vol. 46, No. 1, 1999, pp. 31–38.
11. Pande M., Jin H. and Joos G., Modulated inverter control technique for compensating switch delays and nonideal DC buses in voltage-source inverters, *IEEE Trans IE*, Vol. 44, No. 2, 1997, pp. 182–190.
12. Espinoza J. and Joos G., DSP implementation of output voltage reconstruction in CSI based converters, *IEEE Trans IE*, Vol. 45, No. 6, 1998, pp. 895–904.
13. Kazmierkowski M. and Malesani L., Current control techniques for three-phase voltage-source PWM converters, a survey, *IEEE Trans IE*, Vol. 45, No. 5, 1998, pp. 691–703.
14. Aoki N., Satoh K. and Nabae A., Damping circuit to suppress motor terminal overvoltage and ringing in PWM inverter-fed AC motor drive system with long motor leads, *IEEE Trans IA*, Vol. 35, No. 5, 1999, pp. 1014–1020.
15. Rendusara D. and Enjeti P., An improved inverter output filter configuration reduces common and differential modes dv/dt at the motor terminals in PWM drive systems, *IEEE Trans PEL*, Vol. 13, No. 6, 1998, pp. 1135–1143.
16. Chen S. and Lipo T., Bearing current and shaft voltages of an induction motor under hard- and soft-switching inverter excitation, *IEEE Trans IA*, Vol. 34, No. 5, 1998, pp. 1042–1048.

Chapter 7

Digitally Controlled DC/DC Converters

As described in Chapter 3, all power DC/DC converters are treated as a second-order-hold (SOH) element in digital control systems. We will discuss this model in various circuits in this chapter.

7.1 INTRODUCTION

Power DC/DC converters have plenty of topologies, and the corresponding conversion technique is a big research topic. By an uncompleted statistics, there are more than 500 topologies of power DC/DC converters existing. Dr. F. L. Luo and Dr. H. Ye have firstly categorized all existing prototypes of the power DC/DC converters into six generations theoretically and evolutionarily since 2001. Their work is an outstanding contribution in the development of DC/DC conversion technology, and has been recognized and assessed by experts worldwide.

- First-generation (classical/traditional) converters
- Second-generation (multi-quadrant) converters
- Third-generation (switched-component, **SI/SC**) converters
- Fourth-generation (soft-switching: **ZCS/ZVS/ZT**) converters
- Fifth-generation (synchronous rectifier, **SR**) converters
- Sixth-generation (multiple energy-storage elements resonant, **MER**) converters

7.1.1 THE FIRST-GENERATION CONVERTERS

The first-generation converters perform in a single-quadrant mode and in low-power range (up to around 100 W). Since its development lasts a long time, it has briefly five categories:

- Fundamental converters
- Transformer-type converters
- Developed converters
- Voltage-lift converters
- Super-lift converters

Fundamental converters

Three types of fundamental DC/DC topologies were constructed, which are **Buck** converter, **Boost** converter and **Buck–Boost** converter. They can be derived from single-quadrant operation choppers. For example, buck converter was derived from A-type chopper. These converters have two main problems: linkage between input and output, and very large output voltage ripple.

Buck converter

Buck converter is a step-down DC/DC converter. It works in the first-quadrant operation. It can be derived from Quadrant I chopper. Its circuit diagram, and switch-on and switch-off equivalent circuit are shown in Figure 7.1. The output voltage is calculated by the formula:

$$V_O = \frac{t_{on}}{T} V_{in} = k V_{in} \tag{7.1}$$

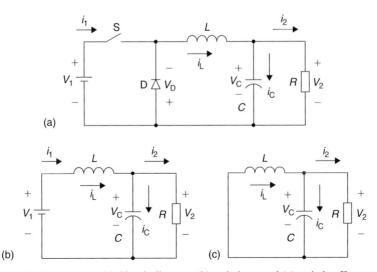

Figure 7.1   Buck converter. (a) Circuit diagram, (b) switch-on and (c) switch-off.

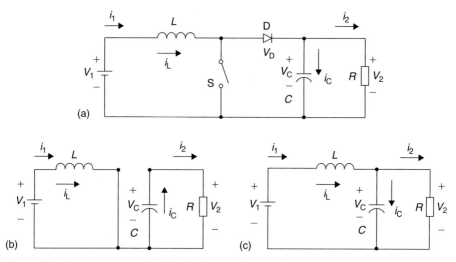

Figure 7.2 Boost converter. (a) Circuit diagram, (b) switch-on and (c) switch-off.

where T is the repeating period ($T = 1/f$), in which f is the chopping frequency; t_{on} is the switch-on time and k is the conduction duty cycle ($k = t_{on}/T$).

Boost converter

Boost converter is a step-up DC/DC converter. It works in the second-quadrant operation. It can be derived from Quadrant II chopper. Its circuit diagram, and switch-on and switch-off equivalent circuit are shown in Figure 7.2. The output voltage is calculated by the formula:

$$V_O = \frac{T}{T - t_{on}} V_{in} = \frac{1}{1 - k} V_{in} \tag{7.2}$$

where T is the repeating period ($T = 1/f$), in which f is the chopping frequency; t_{on} is the switch-on time and k is the conduction duty cycle ($k = t_{on}/T$).

Buck–Boost converter

Buck–boost converter is a step-down/up DC/DC converter. It works in the third-quadrant operation. Its circuit diagram and switch-on and switch-off equivalent circuit, and waveforms are shown in Figure 7.3. The output voltage is calculated by the formula:

$$V_O = \frac{t_{on}}{T - t_{on}} V_{in} = \frac{k}{1 - k} V_{in} \tag{7.3}$$

where T is the repeating period ($T = 1/f$), in which f is the chopping frequency; t_{on} is the switch-on time and k is the conduction duty cycle ($k = t_{on}/T$).

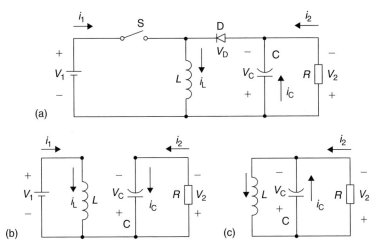

Figure 7.3 Buck–boost converter. (a) Circuit diagram, (b) switch-on and (c) switch-off.

Using this converter is easy to obtain the random output voltage, which can be a higher or lower input voltage. It provides very great convenience for industrial applications.

Transformer-Type Converters

Since all fundamental DC/DC converters keep the linkage from input side to output side and the voltage transfer gain is low, transformer-type converters were developed during 1960s to 1980s. These are a large number of converters such as **Forward** converter, **Push–Pull** converter, **Fly-back** converter, **Half-Bridge** converter, **Bridge** converter and **zeta** (or **ZETA**) converter. Usually, these converters have high transfer voltage gain and high insulation between both sides. Their gain usually depends on the transformer's turns ratio N, which can be several thousand times.

Forward converter

Forward converter is a transformer-type buck converter with the turns ratio, N. It works in the first-quadrant operation. Its circuit diagram is shown in Figure 7.4. The output voltage is calculated by the formula:

$$V_O = kNV_{in} \qquad (7.4)$$

where N is the transformer turns ratio and k is the conduction duty cycle ($k = t_{on}/T$).

In order to exploit the transformer iron core magnetic ability a tertiary winding can be employed in the transformer. Its corresponding circuit diagram is shown in Figure 7.5.

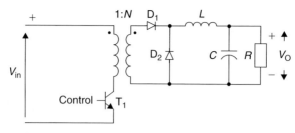

Figure 7.4 Forward converter.

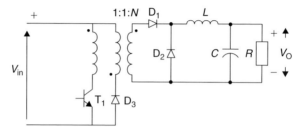

Figure 7.5 Forward converter with tertiary winding.

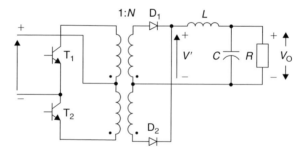

Figure 7.6 Push–pull converter.

Push–Pull Converter

Boost converter works in push–pull state, which effectively avoids the iron core saturation. Its circuit diagram is shown in Figure 7.6. Since there are two switches, which work alternatively, the output voltage is doubled. The output voltage is calculated by the formula:

$$V_O = 2kNV_{in} \tag{7.5}$$

where N is the transformer turns ratio and k is the conduction duty cycle ($k = t_{on}/T$).

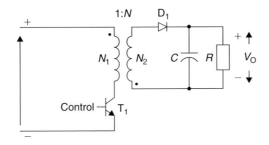

Figure 7.7 Fly-back converter.

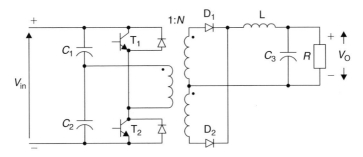

Figure 7.8 Half-bridge converter.

Fly-back Converter

Fly-back converter is a transformer-type converter using the demagnetizing effect. Its circuit diagram is shown in Figure 7.7. The output voltage is calculated by the formula:

$$V_O = \frac{k}{1-k} N V_{in} \tag{7.6}$$

where N is the transformer turns ratio and k is the conduction duty cycle ($k = t_{on}/T$).

Half-bridge converter

In order to reduce the primary side in one winding, half-bridge converter was constructed. Its circuit diagram is shown in Figure 7.8. The output voltage is calculated by the formula:

$$V_O = k N V_{in} \tag{7.7}$$

where N is the transformer turns ratio and k is the conduction duty cycle ($k = t_{on}/T$).

Bridge Converter

Bridge converter employs more switches and gains double output (D/O) voltage. Its circuit diagram is shown in Figure 7.9. The output voltage is calculated by the formula:

$$V_O = 2kNV_{in} \qquad (7.8)$$

where N is the transformer turns ratio and k is the conduction duty cycle ($k = t_{on}/T$).

Zeta Converter

Zeta converter is a transformer-type converter with a low-pass filter. Its output voltage ripple is small. Its circuit diagram is shown in Figure 7.10. The output voltage is calculated by the formula:

$$V_O = \frac{k}{1-k} NV_{in} \qquad (7.9)$$

where N is the transformer turns ratio and k is the conduction duty cycle ($k = t_{on}/T$).

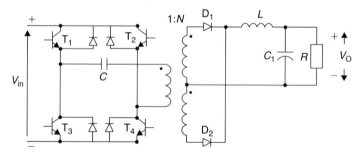

Figure 7.9   Bridge converter.

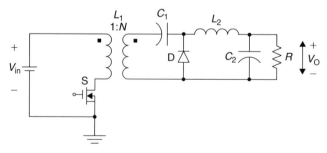

Figure 7.10   Zeta converter.

Forward Converter with Tertiary Winding and Multiple Outputs

Some industrial applications require multiple outputs. This requirement is easily realized by constructing multiple secondary windings and the corresponding conversion circuit. For example, a forward converter with tertiary winding and three outputs is shown in Figure 7.11. The output voltage is calculated by the formula:

$$V_O = kN_i V_{in} \qquad (7.10)$$

where N_i is the transformer turns ratio to the secondary winding, $i = 1$, 2 and 3, respectively, and k is the conduction duty cycle ($k = t_{on}/T$). In principle, this structure is available for all transformer-type DC/DC converters for multiple outputs applications.

Developed Converters

Developed-type converters overcome the second fault of the fundamental DC/DC converters. They are derived from fundamental converters with adding a low-pass filter. Preliminary idea was published in a conference paper in 1977. The author formed three types of converters that derived from fundamental DC/DC converters plus a low-pass filter. This conversion technique was very popular during 1970s to 1990s. Typical prototypes converters are positive output (P/O) **Luo**-converter, negative output (N/O) **Luo**-converter, double output (D/O) **Luo**-converter, **Cúk**-converter, single-ended primary inductance converter (**SEPIC**) and Watkins–Johnson converters. The output voltage ripple of all developed-type converters is usually small, which can be lower than 2%.

In order to obtain the random output voltage, which can be higher or lower input voltage, all developed converters provide very great convenience for industrial applications. Therefore, the output voltage gain of all developed converters is:

$$V_O = \frac{k}{1-k} V_{in} \qquad (7.11)$$

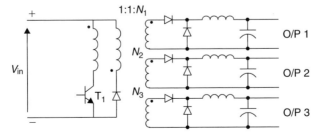

Figure 7.11 Forward converter with tertiary winding and three outputs.

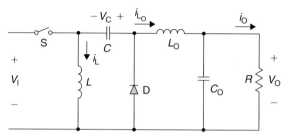

Figure 7.12   P/O Luo-converter.

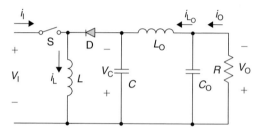

Figure 7.13   N/O Luo-converter.

P/O Luo-Converter

P/O **Luo**-converter is the elementary circuit of the series "P/O **Luo**-converters". It can be derived from buck–boost converter. Its circuit diagram is shown in Figure 7.12. The output voltage is calculated by the formula (7.11).

N/O Luo-Converter

N/O **Luo**-converter is the elementary circuit of the series "N/O **Luo**-converters". It can also be derived from buck–boost converter. Its circuit diagram is shown in Figure 7.13. The output voltage is calculated by the formula (7.11).

D/O Luo-Converter

In order to obtain mirror symmetrical P/O and N/O voltage D/O **Luo**-converter was constructed. D/O **Luo**-converter is the elementary circuit of the series "D/O **Luo**-converters". It can also be derived from buck–boost converter. Its circuit diagram is shown in Figure 7.14. The output voltage is calculated by the formula (7.11).

Cúk-Converter

Cúk-converter is derived from boost converter. Its circuit diagram is shown in Figure 7.15. The output voltage is calculated by the formula (7.11).

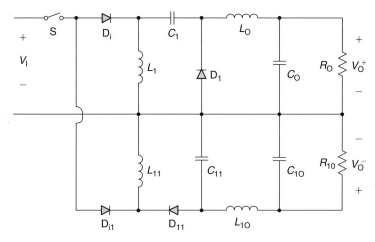

Figure 7.14 D/O Luo-converter.

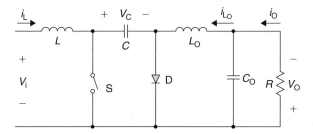

Figure 7.15 Cúk-converter.

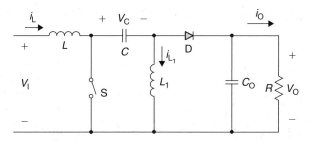

Figure 7.16 SEPIC.

SEPIC

The **SEPIC** is derived from boost converter. Its circuit diagram is shown in Figure 7.16. The output voltage is calculated by the formula (7.11).

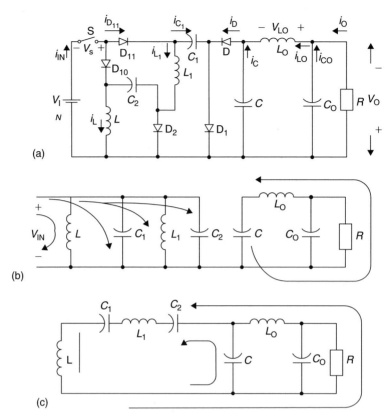

Figure 7.17 Re-lift circuit of the N/O Luo-converter. (a) Circuit diagram, (b) switch-on and (c) switch-off.

Voltage-Lift Converters

Voltage-lift (VL) technique is a good method to lift the output voltage in high level. This technique is widely applied in electronic circuit design. After long-term industrial application and research this method had been successfully used in DC/DC conversion technique. Using this method the output voltage can be easily lifted by tens to hundreds times. VL converters can be sorted in **self**-lift, **re**-lift, **triple**-lift, **quadruple**-lift and **high-stage**-lift converters. The main contributors in this area are Dr. F. L. Luo and Dr. H. Ye.

Figure 7.17 shows the re-lift circuit of the N/O Luo-converters. The voltage transfer gain of the re-lift circuit of the N/O Luo-converter is:

$$G = \frac{V_O}{V_{in}} = \frac{2}{1-k} \qquad (7.12)$$

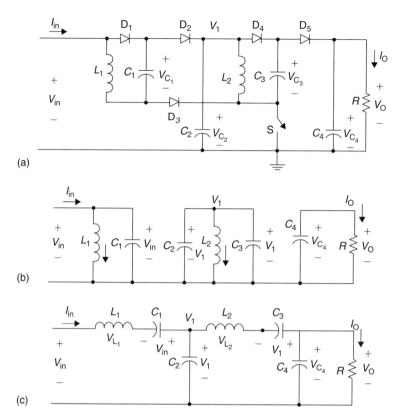

Figure 7.18 Re-lift circuit of P/O SL Luo-converter. (a) Circuit diagram, (b) equivalent circuit during switching-on and (c) equivalent circuit during switching-off.

Super-Lift Converters

VL technique is a popular method widely used in electronic circuit design. It has been successfully employed in DC/DC converter applications in recent years, and opened a way to design high-voltage gain converters. Three series Luo-converters [1–9] are the examples of VL technique implementations. However, the output voltage increases in stage by stage just along the arithmetic progression [10]. A novel approach – super-lift (SL) technique has been developed, which implements the output voltage increasing in stage by stage along the geometric progression. It effectively enhances the voltage transfer gain in power-law. The typical circuits are sorted into four series: **P/O SL Luo-converters**, **N/O SL Luo-converters**, **P/O cascade boost** converters and **N/O cascade boost** converters. All series have many sub-series such as main series, additional series and so on. Each sub-series have many circuits such as elementary circuit, re-lift circuits and so on.

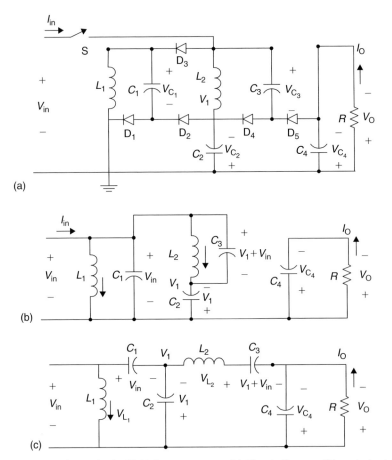

Figure 7.19   Re-lift circuit of N/O SL Luo-converters. (a) Circuit diagram, (b) equivalent circuit during switching-on and (c) equivalent circuit during switching-off.

Figure 7.18 shows the re-lift circuit of the main series of the P/O SL Luo-converters. The voltage transfer gain of the re-lift circuit of the P/O SL Luo-converter is:

$$G = \frac{V_O}{V_{in}} = \left(\frac{2-k}{1-k}\right)^2 \qquad (7.13)$$

Figure 7.19 shows the re-lift circuit of the main series of N/O SL Luo-converters. The voltage transfer gain of the re-lift circuit of the N/O SL Luo-converter is:

$$G = \frac{V_O}{V_{in}} = \left(\frac{2-k}{1-k}\right)^2 - 1 \qquad (7.14)$$

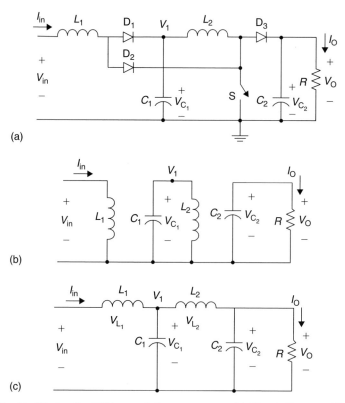

Figure 7.20 Re-lift circuit of P/O cascade boost converter. (a) Circuit diagram, (b) equivalent circuit during switching-on and (c) equivalent circuit during switching-off.

Figure 7.20 shows the re-lift circuit of the main series of P/O cascade boost converter. The voltage transfer gain of the re-lift circuit of the P/O cascade boost converter is:

$$G = \frac{V_O}{V_{in}} = \left(\frac{1}{1-k}\right)^2 \tag{7.15}$$

Figure 7.21 shows the re-lift circuit of the main series of N/O cascade boost converter. The voltage transfer gain of the re-lift circuit of the N/O cascade boost converter is:

$$G = \frac{V_O}{V_{in}} = \left(\frac{1}{1-k}\right)^2 - 1 \tag{7.16}$$

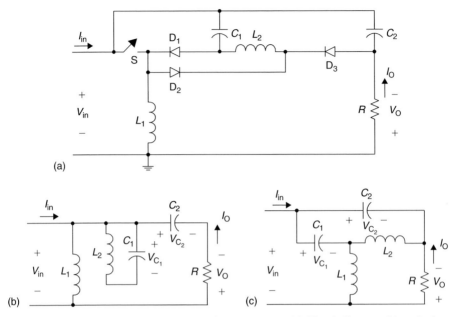

Figure 7.21 Re-lift circuit of N/O cascade boost converter. (a) Circuit diagram, (b) equivalent circuit during switching-on and (c) equivalent circuit during switching-off.

7.1.2 THE SECOND-GENERATION CONVERTERS

The second-generation converters are called multiple-quadrant operation converters. These converters perform in two- and four-quadrant operation with medium output power range (say 100 W or higher). The topologies can be sorted into two main categories: The first ones are the converters derived from the multiple-quadrant choppers and/or from the first-generation converters. The second ones are constructed with transformers. Usually, one-quadrant operation requires at least one switch. Therefore, a two-quadrant operation converter has at least two switches, and a four-quadrant operation converter has at least four switches.

Multiple-quadrant choppers were employed in industrial applications for a long time. It can be used to implement the DC motor multiple-quadrant operation. As the choppers titles, there are:

- Class-A converter (one-quadrant operation)
- Class-B converter (two-quadrant operation)
- Class-C converter
- Class-D converter
- Class-E (four-quadrant operation) converter

These converters are derived from multi-quadrant choppers, such as Class-B converter is derived from B-type chopper and Class-E converter is derived from E-type chopper.

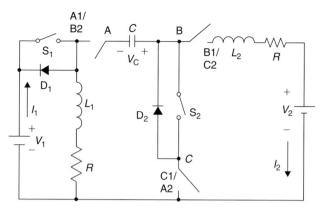

Figure 7.22   Multi-quadrant operation Luo-converter.

Class-A converter works in Quadrant I, corresponding to the forward-motoring operation of a DC motor drive. Class-B converter works in Quadrants I and II operation, corresponding to the forward-running motoring and regenerative braking operation of a DC motor drive. Class-C converter works in Quadrant I and VI operation. Class-D converter works in Quadrants III and VI operation, corresponding to the reverse-running motoring and regenerative braking operation of a DC motor drive. Class-E converter works in four-quadrant operation, corresponding to the four-quadrant operation of a DC motor drive. In recent years many papers investigate the Class-E converters for industrial applications.

Multi-quadrant operation converters can be derived from the first-generation converters. For example, multi-quadrant operation Luo-converters are derived from P/O and N/O Luo-converter. Figure 7.22 shows the multi-quadrant operation Luo-converter. The energy can be converted from the source V_1 to the load $+V_2$ (Quadrant I operation) and from the source V_1 to the load $-V_2$ (Quadrant III operation). Vice versa, the energy can be converted from the load $+V_2$ to the source V_1 (Quadrant II operation) and from the load $-V_2$ to the source V_1 (Quadrant IV operation).

The transformer-type multi-quadrant converters easily change the current direction by transformer polarity and diode rectifier. The main types of such converters can be derived from **Forward** converter, **Half-Bridge** converter and **Bridge** converter.

7.1.3 THE THIRD-GENERATION CONVERTERS

The third-generation converters are called switched component converters, and made of either inductor or capacitors, which are so-called switched-inductor and switched-capacitors. They can perform in two- or four-quadrant operation with high-output power range (say 1000 W). Since they are made of only inductor or capacitors, they are small. Consequently, the power density and efficiency are high.

Switched-Capacitor Converters

Switched-capacitor DC/DC converters consist of only capacitors. As there is no inductor in the circuit, their size is small. They have outstanding advantages such as low-power losses, low electromagnetic interference (EMI). Since its electromagnetic radiation is low, switched-capacitor DC/DC converters are specially required in certain equipment. Switched-capacitor can be integrated into integrated-chip. Hence, its size is largely reduced. Once switched-capacitor converter was developed, it has drawn much attention. Hundreds of papers published to discuss its characteristics and advantages. However, most of these converters in the literature perform single-quadrant operation. Some of them work in the push–pull status. In addition, their control circuit and topologies are very complex, especially, for the large difference between input and output voltages.

Figure 7.23 shows a two-quadrant operation switched-capacitor DC/DC converter. This circuit performs both VL technique and current-amplification technique. The input source voltage is 48 V and output load voltage is 14 V. The size of this converter is the volume in 24 in.^{3} and the transfer power can be 400–515 W. Therefore, the power density can be transferred is up to 16.7–21 W/in.^{3}

Switched-Inductor Converters

Switched-capacitors have many advantages, but its circuit is not simple. If the difference between the input and output voltages is large, many capacitors must be required. Switched-inductor has the outstanding advantage that only one inductor is required for any one switched-inductor converter no matter how large the difference between the input and output voltages. Eventually, only one inductor is required for any one switched-inductor converter working in multiple-quadrant operation. These characteristics are very important for large power conversion. In the present time, large power conversion equipment almost uses switched-inductor converter. For example, the MIT

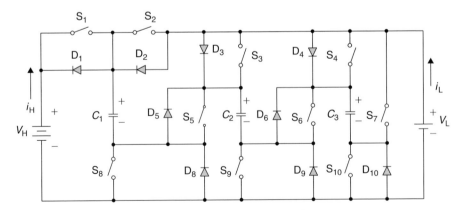

Figure 7.23   Two-quadrant operation switched-capacitor DC/DC converter.

DC/DC converter designed by Prof. John G. Kassakian for his new system in the 2005 automobiles is a two-quadrant switched-inductor DC/DC converter.

Figure 7.24 shows a four-quadrant operation switched-inductor DC/DC converter. The operation state and switch/diode status are shown in Table 7.1.

The input source voltage is 14 V and output load voltage is 42 V. The size of this converter is the volume in 275 in.[3] and the transfer power can be 4.5–7.2 kW. Therefore, the power density that can be transferred is up to 16.3–26.2 W/in.[3]

7.1.4 THE FOURTH-GENERATION CONVERTERS

The fourth-generation DC/DC converters are called soft-switching converters. There are four types of soft-switching methods:

- Resonant-switch converters
- Load-resonant converters
- Resonant-DC-link converters
- High-frequency-link integral-half-cycle converters

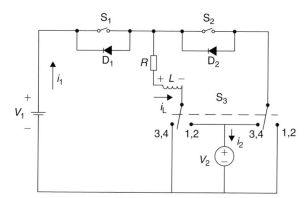

Figure 7.24 Four-quadrant operation switched-inductor DC/DC converter.

Table 7.1

Switch's status (not mentioned switches are off)

Quadrant number	State	S_1	D_1	S_2	D_2	S_3	Source	Load
Q_I, Mode A	ON	ON				ON 1/2	V_1+I_1+	V_2+I_2+
Forward motoring	OFF				ON	ON 1/2		
Q_{II}, Mode B	ON			ON		ON 1/2	V_1+I_1-	V_2+I_2-
Forward regenerative	OFF		ON			ON 1/2		
Q_{III}, Mode C	ON	ON				ON 3/4	V_1+I_1+	V_2-I_2-
Reverse motoring	OFF				ON	ON 3/4		
Q_{IV}, Mode D	ON			ON		ON 3/4	V_1+I_1-	V_2-I_2+
Reverse regenerative	OFF		ON			ON 3/4		

Till now only resonant-switch conversion method has been paid more attention. This resonance method is available for working independently to load. There are three main categories, which are zero-current-switching (ZCS), zero-voltage-switching (ZVS) and zero-transition (ZT) converters. Most topologies usually perform in single-quadrant operation in the literature. Actually, these converters can perform in two- and four-quadrant operation with high-output power range (say several thousand watts (w)).

According to the transferred power becomes large, the power losses increase largely. Main power losses are produced during the switch-on and switch-off period. How to reduce the power losses across the switches is the clue to increase the power transfer efficiency. Soft-switching technique successfully solved this problem. Prof. Fred C. Lee is the pioneer of the soft-switching technique. He established a research center and manufacturing base to realize the ZCS and ZVS DC/DC converters. His first paper induced his outstanding research fruits in 1984. ZCS and ZVS converters have three resonant states: over resonance (completed resonance); optimum resonance (critical resonance) and quasi-resonance (sub-resonance). Only quasi-resonance state has two clear cross-zero points as a repeating period. Many papers (since 1984) have been published to develop the ZCS quasi-resonant-converters (QRCs) and ZVS-QRCs.

Zero-Current-Switching Quasi-Resonant Converters

ZCS-QRC equips resonant circuit in the switch side to keep the switch-on and switch-off at zero-current condition. There are two states: full- and half-wave state. Most of the engineers enjoy the half-wave state. This technique has half-wave current resonance waveform with two zero-cross points.

Figure 7.25 shows a ZCS quasi-resonant (QR) DC/DC converter. The input source voltage $V_1 = 50$ V and output load voltage $V_2 = 30$ V. The load $R = 3\ \Omega$ and load current $I_2 = 10$ A. The circuit diagram is shown in Figure 7.25(a).

To simplify the analysis and calculation, the load current is assumed as a constant value. Therefore, the equivalent circuit is shown in Figure 2.25(b), and the corresponding waveforms of the resonant current $i_{L_r}(t)$ and resonant voltage $v_{C_r}(t)$ are shown in Figure 2.25(c). Since the power losses are very low, the energy transfer efficiency (η) can be very high. The transferred power is 300 W, and the power density is about 15 W/in.^{3} with the converter's volume (size: $2 \times 2.5 \times 4$ in.^{3}) to be 20 in.^{3}

Zero-Voltage-Switching Quasi-Resonant Converters

ZVS-QRC equips resonant circuit in the switch side to keep the switch-on and switch-off at zero-voltage condition. There are two states: full- and half-wave state. Most of engineers enjoy the half-wave state. This technique has half-wave voltage resonance waveform with two zero-cross points.

Figure 7.26 shows a two-quadrant operation ZVS QR DC/DC converter. The input source voltage is 14 V and output load voltage is 42 V. The size of this converter has the volume in 40 in.^{3} and the transfer power is 700 W. Therefore, the power density is up to 17.6 W/in.^{3}

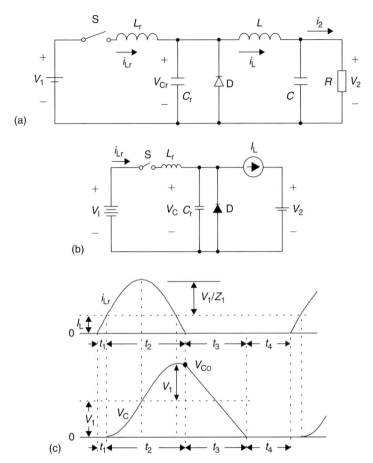

Figure 7.25 A ZCS QR DC/DC converter. (a) Circuit diagram, (b) equivalent circuit diagram and (c) current and voltage waveforms.

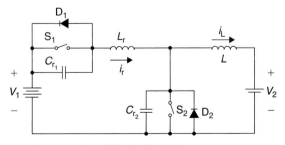

Figure 7.26 Two-quadrant operation ZVS QR DC/DC converter.

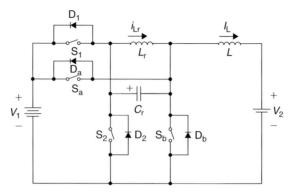

Figure 7.27   Two-quadrant operation ZT DC/DC converter.

Table 7.2

Switches (S) and diodes' (D) status (the blank status means off).

S&D	Mode A (Q_I)								Mode B (Q_{II})							
	Δt_1	Δt_2	Δt_3	Δt_4	Δt_5	Δt_6	Δt_7	Δt_8	Δt_1	Δt_2	Δt_3	Δt_4	Δt_5	Δt_6	Δt_7	Δt_8
S_1				ON	ON	ON										
D_1									ON							ON
S_a	ON	ON	ON													
D_a															ON	ON
S_2												ON	ON	ON		
D_2	ON							ON								
S_b									ON	ON	ON					
D_b						ON	ON									

Zero-Transition Converters

Using ZCS-QRC and ZVS-QRC largely reduce the power losses across the switches. Consequently, the switch device power rates become lower and converter power efficiency is increased. However, ZCS-QRC and ZVS-QRC have large current and voltage stresses. Therefore the device's current and voltage peak rates usually are 3–5 times higher than the working current and voltage. It is not only costly, but also ineffectively. ZT technique overcomes this fault. It implements zero-voltage plus ZCS (ZV-ZCS) technique without significant current and voltage stresses.

Figure 7.27 shows a two-quadrant operation ZT DC/DC converter. The input source voltage is 14 V and output load voltage is 42 V. The operation state and switch/diode status are shown in Table 7.2.

The waveforms of Quadrant I operation of the ZT DC/DC converter are shown in Figure 7.28 and the waveforms of Quadrant II operation of the ZT DC/DC converter are shown in Figure 7.29. The size of this converter is the volume in 40 in.[3] and the transfer power is 700 W. Therefore, the power density is up to 17.6 W/in.[3]

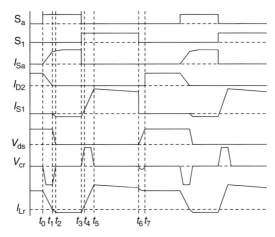

Figure 7.28 Waveforms of Quadrant I operation of the ZT DC/DC converter.

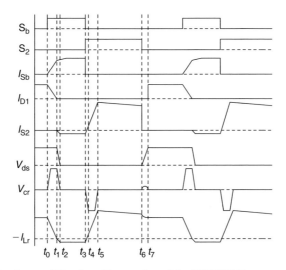

Figure 7.29 Waveforms of Quadrant II operation of the ZT DC/DC converter.

7.1.5 THE FIFTH-GENERATION CONVERTERS

The fifth-generation converters are called SR DC/DC converters. This type converter was required by the development of computing technology progress. Corresponding to the development of the micro-power consumption technique and high-density IC manufacture, the power supplies with low output voltage and strong current are widely required in communications, computer equipment and other industrial applications. Intel company developed Zelog-type computers governed the world market for a long time. Inter-80 computers used 5 V power supply. In order to increase the memory size

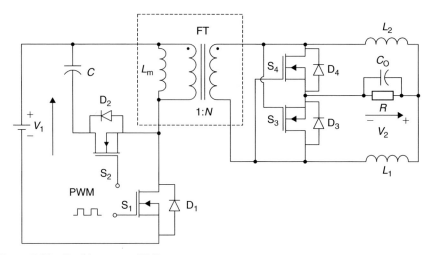

Figure 7.30 Double-current SR Luo-converter.

and operation speed, LSIC technique has been quickly developed. Along the density of IC manufacture increased, the gaps between the layers become narrow and narrow. Correspondingly the micro-power-consumption technique was completed. Therefore the new type computers Pentium I, II, III and IV use the power supply of 3.3 V. The future computers have larger memory and require lower power supply voltage, e.g. 2.5, 1.8, and 1.5 V, even if 1.1 V. Such low power supply voltage cannot be obtained by the traditional diode rectifier bridge because the diode voltage drop is too large. Since this requirement, new types of metal-oxide semiconductor field effect transistor (MOSFET) were developed. They have very low conduction resistance (6–8 mΩ) and forward voltage drop (0.05–0.2 V).

Many papers have been published since 1990s and many prototypes were developed. The fundamental topology is derived from the forward converter. Active-clamped circuit, flat-transformers, double current circuit, soft-switching methods and multiple current methods can be used in SR DC/DC converters.

Figure 7.30 shows double-current SR Luo-converter. The flash transformer turn's ratio is $N = 1{:}12$. The input source voltage is 30 V, and output load voltage is 1.8 V and output current can be 30–36 A. Power transfer efficiency can be 82–90%.

7.1.6 THE SIXTH-GENERATION CONVERTERS

The sixth-generation converters are called multiple energy-storage (MER) elements resonant power converters (RPC). Current source resonant inverters (CSRIs) are the heart of many systems and equipment, e.g. uninterruptible power supply (UPS) and high-frequency annealing (HFA) apparatus. Many topologies shown in open literature are the series resonant converters (SRC) and parallel resonant converters (PRC) that consist of two, three or four energy-storage elements. However, they have a lot of the

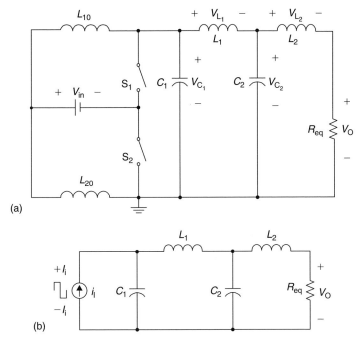

Figure 7.31 Cascade double Γ–CL current source resonant inverter. (a) Circuit diagram and (b) equivalent circuit.

limitations. These limitations of two-, three- or/and four-element resonant topologies can be overcome by special design. These converters have sorted into three main categories:

- Two energy-storage elements resonant DC/AC and DC/AC/DC converters;
- Three energy-storage elements resonant DC/AC and DC/AC/DC converters;
- Four energy-storage elements (2L–2C) resonant DC/AC and DC/AC/DC converters.

By mathematical calculation there are **8** prototypes of 2-element converters, **38** prototypes of 3-element converters and **98** prototypes of 4-element (2L–2C) converters. Carefully analyzing these prototypes we can find out that not so many circuits can be realized. If we keep the output in low-pass bandwidth, the series components must be inductors and shunt components must be capacitors. Furthermore analysis, the first component of the resonant-filter network can be an inductor in series, or a capacitor in shunt. In the first case, only alternative (square wave) voltage source can be applied to the network. In the second case, only alternative (square wave) current source can be applied to the network.

Figure 7.31(a) shows a cascade double Γ–CL current source resonant inverter. It consists of four energy-storage elements, the double Γ–CL: C_1–L_1 and C_2–L_2. Its

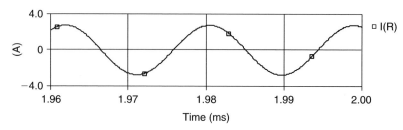

Figure 7.32   Output current waveform of the cascade double Γ–CL CSRI.

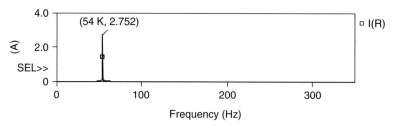

Figure 7.33   The FFT spectrum of the output current of the cascade double Γ–CL CSRI.

equivalent diagram is shown in Figure 7.31(b). The energy source is a DC voltage V_{in} chopped by two main switches S_1 and S_2 to construct a bipolar current source, $i_i = \pm I_i$. The pump inductors L_{10} and L_{20} are equal to each other, and are large enough to keep the source current nearly constant during operation. The real load absorbs the delivered energy, its equivalent load should be proposed resistive, R_{eq}. The input source voltage is $V_{in} = 30\,\text{V}$, $L_{10} = L_{20} = 20\,\text{mH}$, $I_{in} = \pm 1\,\text{A}$, $R_{eq} = 10\,\Omega$, $C_1 = C_2 = C = 0.22\,\mu\text{F}$, and $L_1 = L_2 = L = 100\,\mu\text{H}$. The output current waveform is nearly pure sinusoidal function shown in Figure 7.32. The corresponding fast Fourier transformation (FFT) spectrum is shown in Figure 7.33 and the total harmonic distortion (THD) is mostly a 0.

7.1.7 ALL PROTOTYPES AND DC/DC CONVERTER FAMILY TREE

As we know that there are more than 500 topologies of DC/DC converters existing. In order to manage and sort them it is urgently necessary to categorize all prototypes. From all accumulated knowledge we can build all DC/DC converters altogether. The family tree is shown in Figure 7.34.

7.2 MATHEMATICAL MODELING FOR POWER DC/DC CONVERTERS

Since the output voltage of a power DC/DC converter is out of control in a period T once the duty cycle k is applied, therefore, it is the element to keep the output voltage in a period $T = 1/f$. By per-unit system, the voltage transfer gain is unity (1) in a

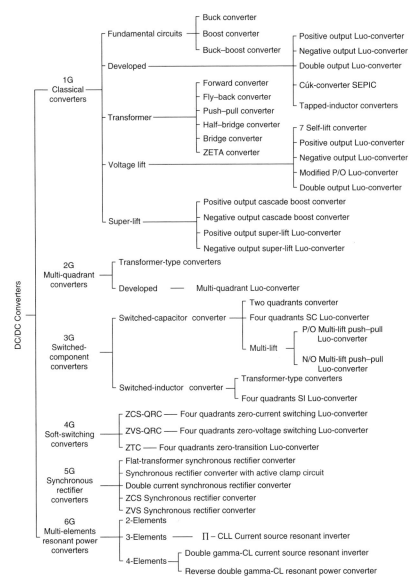

Figure 7.34 DC/DC converter family tree.

sampling interval. As discussed in Chapter 2, a power DC/DC converter is a second-order element, and its transfer function is:

$$G(s) = \frac{V_O}{V_I}\bigg|_{\text{per-unit}} = \frac{1}{1 + s\tau + s^2 \tau \tau_d} \qquad (7.17)$$

where τ is the time constant and τ_d is the damping time constant.

The mathematical modeling for the power DC/DC converters was proposed by Dr. F. L. Luo and Dr. H. Ye in the period 2001–2004. This new methodology was carefully described in Chapter 2. As the second-order transfer function simulated all power DC/DC converters, the mathematical modeling in per-unit digital control system should be a second-order transfer function in the s-domain:

$$G(s) = \frac{1}{1 + s\tau + s^2 \tau \tau_d} \tag{7.18}$$

where τ is the time constant (2.23) and τ_d is the damping time constant (2.25). In general situation, τ_d is smaller than the critical value $\tau/4$, there are two real poles $-\sigma_1$ and $-\sigma_2$ located in the left-hand half-plane in the s-plane. As discussed in Chapter 2, Equation (7.18) is rewritten as:

$$G(s) = \frac{1}{1 + s\tau + s^2 \tau \tau_d} = \frac{1/\tau\tau_d}{(s + \sigma_1)(s + \sigma_2)} \tag{7.19}$$

where

$$\sigma_1 = \frac{\tau + \sqrt{\tau^2 - 4\tau\tau_d}}{2\tau\tau_d} \quad \text{and} \quad \sigma_2 = \frac{\tau - \sqrt{\tau^2 - 4\tau\tau_d}}{2\tau\tau_d}$$

Correspondingly, a power DC/DC converter is an single-order hold (SOH) in the z-domain.

$$G(z) = \mathbb{Z}[G(s)] = \mathbb{Z}\left[\frac{1/\tau\tau_d}{(s + \sigma_1)(s + \sigma_2)}\right] = \frac{1}{\tau\tau_d(\sigma_2 - \sigma_1)}\left(\frac{z}{z - e^{-T\sigma_1}} - \frac{z}{z - e^{-T\sigma_2}}\right) \tag{7.20}$$

Expanding and simplifying Equation (7.20),

$$\begin{aligned}
G(z) &= \frac{M}{\sqrt{\tau^2 - 4\tau\tau_d}}\left[\frac{z}{z - e^{-\sigma_1 T}} - \frac{z}{z - e^{-\sigma_2 T}}\right] \\[2mm]
&= \frac{Mz}{\sqrt{\tau^2 - 4\tau\tau_d}}\left[\frac{e^{-\sigma_1 T} - e^{-\sigma_2 T}}{(z - e^{-\sigma_1 T})(z - e^{-\sigma_2 T})}\right] \\[2mm]
&= \frac{Mz\left(e^{-\left(\frac{\tau - \sqrt{\tau^2 - 4\tau\tau_d}}{2\tau\tau_d}\right)T} - e^{-\left(\frac{\tau + \sqrt{\tau^2 - 4\tau\tau_d}}{2\tau\tau_d}\right)T}\right)}{\sqrt{\tau^2 - 4\tau\tau_d}\left(z - e^{-\left(\frac{\tau - \sqrt{\tau^2 - 4\tau\tau_d}}{2\tau\tau_d}\right)T}\right)\left(z - e^{-\left(\frac{\tau + \sqrt{\tau^2 - 4\tau\tau_d}}{2\tau\tau_d}\right)T}\right)}
\end{aligned} \tag{7.21}$$

As

$$\sigma_1 - \sigma_2 = \frac{\sqrt{\tau^2 - 4\tau\tau_d}}{\tau\tau_d}$$

It means that the DC/DC converter performs a second-order response without oscillation.

In some applications, τ_d is greater than the critical value $\tau/4$, there are a couple of conjugating poles $-s_1$ and $-s_2$ located in the left-hand half-plane in the s-plane. As discussed in Chapter 2, Equation (7.18) is rewritten as:

$$G(s) = \frac{1}{1 + s\tau + s^2\tau\tau_d} = \frac{1/\tau\tau_d}{(s + s_1)(s + s_2)} \tag{7.22}$$

where $s_1 = \sigma + j\omega$ and $s_2 = \sigma - j\omega$

$$\sigma = \frac{1}{2\tau_d} \quad \text{and} \quad \omega = \frac{\sqrt{4\tau\tau_d - \tau^2}}{2\tau\tau_d}$$

Correspondingly, the power DC/DC converter is an SOH in the s-domain and is rewritten as:

$$G(s) = \frac{M/\tau\tau_d}{(s + \sigma + j\omega)(s + \sigma - j\omega)} = \frac{M/\tau\tau_d}{(s + \sigma)^2 + \omega^2} = \frac{M}{\tau\tau_d\omega} \times \frac{\omega}{(s + \sigma)^2 + \omega^2}$$

$$= \frac{2M}{\sqrt{4\tau\tau_d - \tau^2}} \times \frac{\omega}{(s + \sigma)^2 + \omega^2} \tag{7.23}$$

Applying the transformation, the mathematical modeling for the SOH for large damping time constant is:

$$G(z) = \frac{2M}{\sqrt{4\tau\tau_d - \tau^2}} \frac{ze^{-aT}\sin\omega T}{z^2 - 2ze^{-aT}\cos\omega T + e^{-2aT}} \tag{7.24}$$

where $\sigma = a = 1/2\tau_d$. Expanding and simplifying Equation (7.24):

$$G(z) = \frac{2Mze^{-aT}\sin\omega T}{\sqrt{4\tau\tau_d - \tau^2}(z^2 - 2ze^{-aT}\cos\omega T + e^{-2aT})}$$

$$= \frac{2Mze^{-T/2\tau_d}\sin\left(\frac{\sqrt{4\tau\tau_d - \tau^2}}{2\tau\tau_d} \times T\right)}{\sqrt{4\tau\tau_d - \tau^2}\left(z^2 - 2ze^{-T/2\tau_d}\cos\left(\frac{\sqrt{4\tau\tau_d - \tau^2}}{2\tau\tau_d}\right) + e^{-T/\tau_d}\right)} \tag{7.25}$$

It means that the DC/DC converter performs a second-order response with oscillation in one-step delay (T) in a digital control system.

7.3 FUNDAMENTAL DC/DC CONVERTER

Fundamental DC/DC converters such as buck, boost and buck–boost converters consist of one capacitor and one inductor; therefore, the modeling is simple. The treatment process was introduced in Chapter 2 in detail.

In most industrial applications, power DC/DC converters work in the continuous conduction mode (CCM), and perform in the case with small damping time constant. For example, the inductor L is in mH and capacitor C is in μF. Usually, the time constant τ is large enough, and damping time constant τ_d is smaller than the critical value $\tau/4$. Refer to the buck converter shown in Figure 7.1. Assume the following conditions.

$V_1 = 40$ V, $L = 1$ mH, $C = 40\,\mu$F, $f = 20$ kHz $(T = 50\,\mu$s$)$, $k = 0.4$ and $R = 1\,\Omega$ power losses are ignored. We get $V_2 = 16$ V, $I_2 = I_L = 16$ A, $I_1 = 6.4$ A. The mathematical model components are:

$$PE = V_1 I_1 T = 40 \times 6.4 \times 50\mu = 12.8\,\text{mJ}$$

$$W_L = \frac{1}{2}LI_L^2 = \frac{1}{2}1m \times 16^2 = 128\,\text{mJ}$$

$$W_C = \frac{1}{2}CV_C^2 = \frac{1}{2}40\mu \times 16^2 = 5.12\,\text{mJ}$$

$$SE = W_L + W_C = 128 + 5.12 = 133.12\,\text{mJ}$$

$$EF = \frac{SE}{PE} = \frac{133.12}{12.8} = 10.4$$

$$CIR = \frac{W_C}{W_L} = \frac{5.12}{128} = 0.04$$

since $EL = 0$, efficiency $(\eta) = 1$.

$$\tau = \frac{2T \times EF}{1 + CIR} = \frac{100\mu \times 10.4}{1.04} = 1\,\text{ms}$$

$$\tau_d = \frac{2T \times EF}{1 + CIR}CIR = \frac{100\mu \times 10.4 \times 0.04}{1.04} = 40\,\mu\text{s}$$

$$\xi = \frac{\tau_d}{\tau} = CIR = 0.04 < 0.25$$

The corresponding transfer function in per-unit system is:

$$G(s) = \frac{1}{1 + s\tau + s^2\tau\tau_d} = \frac{1/\tau\tau_d}{(s + \sigma_1)(s + \sigma_2)} \quad (7.26)$$

with

$$\sigma_1 = \frac{\tau + \sqrt{\tau^2 - 4\tau\tau_d}}{2\tau\tau_d} = \frac{1m + \sqrt{1\mu - 160n}}{80n} = \frac{1 + 0.9165}{80\mu}$$

$$= \frac{1}{0.0000417} = 23.96\,\text{kHz}$$

$$\sigma_2 = \frac{\tau - \sqrt{\tau^2 - 4\tau\tau_d}}{2\tau\tau_d} = \frac{1m - \sqrt{1\mu - 160n}}{80n} = \frac{1 - 0.9165}{80\mu}$$

$$= \frac{1}{0.000958} = 1.044\,\text{kHz}$$

We know that σ_1 is much larger than σ_2, so that the term $e^{-\sigma_1 t}$ is much smaller than the term $e^{-\sigma_2 t}$. It is reasonable to ignore the expression involving the term $e^{-\sigma_1 t}$.

The unit-step response is:

$$v_2(t) = 16(1 + K_1 e^{-\sigma_1 t} + K_2 e^{-\sigma_2 t}) = 16(1 + 0.0455 e^{-\frac{t}{0.0000417}} - 1.0455 e^{-\frac{t}{0.000958}}) \tag{7.27}$$

where:

$$K_1 = -\frac{1}{2} + \frac{\tau}{2\sqrt{\tau^2 - 4\tau\tau_d}} = -0.5 + \frac{1}{2\sqrt{1^2 - 0.16}} = -0.5 + 0.5455 = 0.0455$$

$$K_1 = -\frac{1}{2} + \frac{\tau}{2\sqrt{\tau^2 - 4\tau\tau_d}} = -0.5 - \frac{1}{2\sqrt{1^2 - 0.16}} = -0.5 - 0.5455 = -1.0455$$

or

$$v_2(t) \approx 16(1 - e^{-\frac{t}{0.000958}}) \tag{7.28}$$

The impulse response is:

$$\Delta v_2(t) = \frac{U}{\sqrt{1 - 4\tau_d/\tau}}(e^{-\sigma_2 t} - e^{-\sigma_1 t}) = 1.0911 U(e^{-\frac{t}{0.000958}} - e^{-\frac{t}{0.0000417}}) \tag{7.29}$$

or

$$\Delta v_2(t) \approx U e^{-\frac{t}{0.000958}} \tag{7.30}$$

where U is the interference. From the above analysis and calculation, the corresponding transfer function in per-unit system in the s-domain can be approximately written as:

$$G(s) \approx \frac{1}{1 + s\tau_e} \tag{7.31}$$

where τ_e is the equivalent time constant $\tau_e = 0.000958\,\text{s} \approx 1\,\text{ms} = \tau$.

The corresponding transfer function in per-unit system in the z-domain can be approximately written as:

$$G(z) \approx \frac{z}{z - e^{-T/\tau_e}} \tag{7.32}$$

where T is the sampling interval ($T = 50\,\mu\text{s}$). Since T/τ_e is very small nearly 0.05, so that:

$$G(z) \approx \frac{z}{z - e^{-T/\tau_e}} \approx \frac{z}{z - 1 + T/\tau_e} \approx \frac{z}{z - 0.95} \tag{7.33}$$

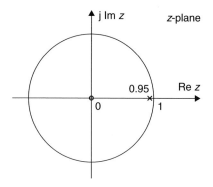

Figure 7.35   The zero's and pole's locations of a buck converter.

The transfer function has one pole and one zero. The pole is at $z = 0.95$ inside the unity-cycle, the zero is at $z = 0$ at the original point. Therefore, this converter is stable element. The zero's and pole's locations of this buck converter are shown in Figure 7.35.

7.4 DEVELOPED DC/DC CONVERTERS

Most power DC/DC converters consist of multiple (more than two) passive energy-stored components. In traditional method, they have higher-order transfer function. For example, the developed DC/DC converters such as positive/negative output Luo-converters in Figures 7.12 and 7.13, Cúk converter and SEPIC in Figures 7.15 and 7.16, consist of two capacitors and two inductors. In most industrial applications, power DC/DC converters work in the CCM, and perform in the case with small damping time constant. For example, the inductor, L, is in mH and capacitor, C, is in μF. Usually, the time constant τ is large enough, and damping time constant τ_d is smaller than the critical value $\tau/4$.

Refer to the N/O Luo-converter shown in Figure 7.13. Assume the following conditions.

$V_1 = 40\,\text{V}$, $L = L_O = 5\,\text{mH}$, $C = C_O = 20\,\mu\text{F}$, $f = 20\,\text{kHz}$ ($T = 50\,\mu\text{s}$), $k = 0.5$ and $R = 2\,\Omega$ power losses are ignored. We get $V_O = 40\,\text{V}$, $V_C = V_{CO} = 40\,\text{V}$, $I_O = I_{LO} = 20\,\text{A}$, $I_1 = I_L = 20\,\text{A}$. The mathematical model components are:

$$PE = V_1 I_1 T = 40 \times 20 \times 50\mu = 40\,\text{mJ}$$

$$W_L = \frac{1}{2}LI_L^2 + \frac{1}{2}L_O I_{LO}^2 = 5m \times 20^2 = 2\,\text{J}$$

$$W_C = \frac{1}{2}C_O V_{CO}^2 + \frac{1}{2}CV_C^2 = 20\mu \times 40^2 = 32\,\text{mJ}$$

$$SE = W_L + W_C = 2000 + 32 = 2032\,\text{mJ}$$

$$EF = \frac{SE}{PE} = \frac{2032}{40} = 50.8$$

$$CIR = \frac{W_C}{W_L} = \frac{32}{2000} = 0.016$$

since $EL = 0$, efficiency $(\eta) = 1$.

$$\tau = \frac{2T \times EF}{1 + CIR} = \frac{100\mu \times 50.8}{1.016} = 5\,\text{ms}$$

$$\tau_d = \frac{2T \times EF}{1 + CIR} CIR = \frac{100\mu \times 50.8 \times 0.016}{1.016} = 80\,\mu\text{s}$$

$$\xi = \frac{\tau_d}{\tau} = CIR = 0.016 \ll 0.25$$

Since the damping time constant τ_d is much smaller than time constant τ, the corresponding transfer function in per-unit system is considered as:

$$G(s) = \frac{1}{1 + s\tau + s^2 \tau \tau_d} \approx \frac{1}{1 + s\tau} = \frac{1}{1 + 0.005s} \qquad (7.34)$$

The unit-step response is:

$$v_O(t) = 40(1 - e^{-\frac{t}{\tau}}) = 40(1 - e^{-\frac{t}{0.005}}) \qquad (7.35)$$

The impulse response is:

$$\Delta v_O(t) = Ue^{-\frac{t}{\tau}} = Ue^{-\frac{t}{0.005}} \qquad (7.36)$$

where U is the interference. The corresponding transfer function in per-unit system in the z-domain can be approximately written as:

$$G(z) \approx \frac{z}{z - e^{-T/\tau}} = \frac{z}{z - e^{-50\mu/5m}} = \frac{z}{z - e^{-0.01}} = \frac{z}{z - 0.99} \qquad (7.37)$$

where T is the sampling interval ($T = 50\,\mu\text{s}$) and τ is the time constant ($\tau = 5\,\text{ms}$), so that T/τ is very small to be equal to 0.01. The transfer function has one pole and one zero. The pole is at $z = 0.99$ inside the unity-cycle; therefore, this converter is stable circuit. The zero's and pole's locations of this Cúk converter are shown in Figure 7.36.

7.5 SOFT-SWITCHING CONVERTERS

In order to reduce the power losses soft-switching technique has been applied in both research and industrial application since 1984. Soft-switching converters are sorted

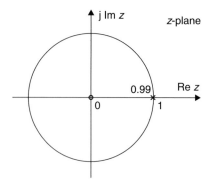

Figure 7.36   The zero's and pole's locations of a N/O Luo-converter.

in three categories:

- ZCS converters,
- ZVS converters,
- ZT converters.

All soft-switching converters have resonant circuit to implement the soft-switching process: ZCS or/and ZVS, and ZT operations. Usually, the resonant circuit consists of the components with small values of inductance and capacitance. The stored energy in the resonant circuit is very small. An L_r–C_r resonant circuit is shown in Figure 7.25. For high-accuracy consideration we can estimate the energy stored in a resonant circuit is:

$$E_{res} = W_{L_r} + W_{C_r} = \frac{1}{2} \int_{t_{res}} \left[L_r i_{L_r}^2(t) + C_r v_{C_r}^2(t) \right] dt \qquad (7.38)$$

where t_{res} is the resonant process time-length, $i_{L_r}(t)$ is the resonant inductor instantaneous current in the resent process and $v_{C_r}(t)$ is the resonant capacitor instantaneous voltage in the resent process. For a pure resonance period the functions $i_{L_r}(t)$ and $v_{C_r}(t)$ should be sinusoidal function. The energy stored in the inductor and capacitor is fully transferred to each other in a quarter cycles:

$$i_{L_r}(t) = I_O \sin \omega_r t \qquad (7.39)$$

$$v_{C_r}(t) = V \sin\left(\omega_r t - \frac{\pi}{2}\right) \qquad (7.40)$$

with

$$I_O = \frac{V}{\sqrt{L_r/C_r}} \quad \text{and} \quad \omega_r = \frac{1}{\sqrt{L_r C_r}}$$

The average energy stored in the inductor is:

$$W_{Lr} = \frac{1}{2} \int_{t_{res}} L_r i_{L_r}^2(t) dt = \frac{1}{2} L_r \left(\frac{I_O}{\sqrt{2}} \right)^2 = \frac{C_r}{4} V^2 \tag{7.41}$$

The average energy stored in the capacitor is:

$$W_{Cr} = \frac{1}{2} \int_{t_{res}} C_r v_{C_r}^2(t) dt = \frac{1}{2} C_r \left(\frac{V}{\sqrt{2}} \right)^2 = \frac{C_r}{4} V^2 \tag{7.42}$$

Therefore the average energy stored in whole resonant circuit is:

$$E_r = W_{L_r} + W_{C_r} = \frac{1}{2} C_r V^2 \tag{7.43}$$

The resonant inductor L_r is usually in μH that is much lower than the main inductor, and the capacitor C_r is usually in μF that maybe lower than the main capacitor. Therefore the stored energy in the resonant circuit can be ignored in brief calculation.

For example, a ZCS buck converter is shown in Figure 7.25. The components are: $V_1 = 50\,V$, $I_1 = 6\,A$, $L = 10\,mH$, $C = 60\,\mu F$, $L_r = 4\,\mu H$, $C_r = 1\,\mu F$, output voltage $V_2 = 30\,V$ the load $R = 3\,\Omega$ and $I_2 = 10\,A$. Therefore, we have:

$$\omega = \frac{1}{\sqrt{L_r C_r}} = \frac{1}{\sqrt{4\mu \times 1\mu}} = 500000\,\text{rad} \tag{7.44}$$

$$Z = \sqrt{\frac{L_r}{C_r}} = \sqrt{\frac{4\mu}{1\mu}} = 2\,\Omega \tag{7.45}$$

$$\alpha = \sin^{-1} \frac{ZI_2}{V_1} = \sin^{-1} \frac{2 \times 10}{50} = \sin^{-1} 0.4 = 0.41\,\text{rad} \tag{7.46}$$

$$t_1 = \frac{L_r I_2}{V_1} = \frac{4 \times 10}{50} = 0.8\,\mu s \tag{7.47}$$

$$t_2 = \frac{\pi + \alpha}{\omega} = \frac{3.552}{500000} = 7.1\,\mu s \tag{7.48}$$

$$t_3 = \frac{V_1(1 + \cos \alpha)C_r}{I_2} = \frac{50 \times 1.9165 \times 1\mu}{10} = 4.58\,\mu s \tag{7.49}$$

$$t_4 = \frac{V_1(t_1 + t_2)}{V_2 I_2} \left(I_2 + \frac{V_1}{Z} \frac{2 \cos \alpha}{\alpha + \pi/2} \right) - (t_1 + t_2 + t_3) = 5.05\,\mu s \tag{7.50}$$

$$T = t_1 + t_2 + t_3 + t_4 = 17.53\,\mu s \tag{7.51}$$

$$f = 1/T = \frac{1}{17.528} = 57\,\text{kHz} \tag{7.52}$$

$$k = \frac{t_1 + t_2}{T} = \frac{7.9}{17.53} = 0.45\,\mu\text{s} \tag{7.53}$$

The resonant stored energy is:

$$E_r = W_{L_r} + W_{C_r} = \frac{1}{2}C_r V_1^2 = 0.5\mu \times 50^2 = 1.25\,\text{mJ} \tag{7.54}$$

Other parameters are:

$$PE = V_1 I_1 T = 50 \times 6 \times 17.53\mu = 5.26\,\text{mJ}$$

$$W_L = \frac{1}{2}L I_L^2 = 5m \times 10^2 = 500\,\text{mJ}$$

$$W_C = \frac{1}{2}C V_C^2 = 30\mu \times 30^2 = 27\,\text{mJ}$$

$$SE = W_L + W_C + E_r = 500 + 27 + 1.25 = 528.25\,\text{mJ}$$

$$EF = \frac{SE}{PE} = \frac{528.25}{5.26} = 100.43$$

$$CIR = \frac{W_C}{W_L} = \frac{27}{500} = 0.054$$

since $EL = 0$, efficiency $\eta = 1$.

$$\tau = \frac{2T \times EF}{1 + CIR} = \frac{2 \times 17.53\mu \times 100.43}{1.054} = 3.34\,\text{ms}$$

$$\tau_d = \frac{2T \times EF}{1 + CIR}CIR = \frac{2 \times 17.53\mu \times 100.43 \times 0.054}{1.054} = 180\,\mu\text{s}$$

$$\xi = \frac{\tau_d}{\tau} = CIR = 0.054 \ll 0.25$$

Since the damping time constant τ_d is much smaller than time constant τ, the corresponding transfer function in per-unit system is considered as:

$$G(s) = \frac{1}{1 + s\tau + s^2\tau\tau_d} \approx \frac{1}{1 + s\tau} = \frac{1}{1 + 0.00334s} \tag{7.55}$$

The unit-step response is:

$$v_O(t) = 30(1 - e^{-\frac{t}{\tau}}) = 30(1 - e^{-\frac{t}{0.00334}}) \tag{7.56}$$

The impulse response is:

$$\Delta v_O(t) = U e^{-\frac{t}{\tau}} = U e^{-\frac{t}{0.00334}} \tag{7.57}$$

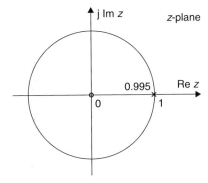

Figure 7.37   The zero's and pole's locations of a ZCS QR DC/DC converter.

where U is the interference. The corresponding transfer function in per-unit system in the z-domain can be approximately written as:

$$G(z) \approx \frac{z}{z - e^{-T/\tau}} = \frac{z}{z - e^{-17.53\mu/3.34m}} = \frac{z}{z - e^{-0.00525}} = \frac{z}{z - 0.995} \quad (7.58)$$

where T is the sampling interval ($T = 17.53\ \mu$s) and τ is the time constant ($\tau = 3.34$ ms), so that T/τ is very small to be equal to 0.00525. The transfer function has one pole and one zero. The pole is at $z = 0.995$ inside the unity-cycle; therefore, this converter is stable circuit. The zero's and pole's locations of this ZCS QR DC/DC converter are shown in Figure 7.37.

7.6 MULTI-ELEMENT RESONANT POWER CONVERTERS

The sixth-generation converters work in the resonant state to reduce the power losses. This technique was created in 1980s and used to be popular in 1990s. Depending on the number of the passive components there are few categories:

- Two-element RPC;
- Three-element RPC;
- Four-element (2L–2C) RPC.

All RPCs work in forced resonant state. It means that the applied frequency may not be equal to the natural circuit resonant frequency. The stored energy in the resonant circuit is AC form. A cascade double Γ–LC current source resonant inverter is shown in Figure 7.31. From the equivalent circuit, the input current $i_i(t)$ is:

$$i_i(t) = \begin{cases} 1\ \text{A} & nT \le t < (n+0.5)T \\ -1\ \text{A} & (n+0.5)T \le t < (n+1)T \end{cases} \quad (7.59)$$

The input impedance is given by:

$$Z(\omega) = \frac{R_{eq}(1 - \omega^2 L_1 C_2) + j\omega(L_1 + L_2 - \omega^2 L_1 L_2 C_2)}{1 - \omega^2(L_1 C_1 + L_2 C_1 + L_2 C_2) + \omega^4 L_1 L_2 C_1 C_2 + j\omega R_{eq}(C_1 + C_2 - \omega^2 L_1 C_1 C_2)} \tag{7.60}$$

or

$$Z(\omega) = \frac{R_{eq}(1 - \omega^2 L_1 C_2) + j\omega(L_1 + L_2 - \omega^2 L_1 L_2 C_2)}{B(\omega)} \tag{7.61}$$

where

$$B(\omega) = 1 - \omega^2(L_1 C_1 + L_2 C_1 + L_2 C_2) + \omega^4 L_1 L_2 C_1 C_2 + j\omega R_{eq}(C_1 + C_2 - \omega^2 L_1 C_1 C_2) \tag{7.62}$$

Voltage and current on capacitor C_1:

$$\frac{V_{C_1}(\omega)}{I_i(\omega)} = \frac{R_{eq}(1 - \omega^2 L_1 C_2) + j\omega(L_1 + L_2 - \omega^2 L_1 L_2 C_2)}{B(\omega)} \tag{7.63}$$

$$\frac{I_{C_1}(\omega)}{I_i(\omega)} = \frac{R_{eq}(1 - \omega^2 L_1 C_2) + j\omega(L_1 + L_2 - \omega^2 L_1 L_2 C_2)}{B(\omega)/j\omega C_1} \tag{7.64}$$

Voltage and current on inductor L_1:

$$\frac{V_{L_1}(\omega)}{I_i(\omega)} = \frac{-R_{eq}\omega^2 L_1 C_2 + j\omega L_1(1 - \omega^2 L_2 C_2)}{B(\omega)} \tag{7.65}$$

$$\frac{I_{L_1}(\omega)}{I_i(\omega)} = \frac{(1 - \omega^2 L_2 C_2) + j R_{eq}\omega C_2}{B(\omega)} \tag{7.66}$$

Voltage and current on capacitor C_2:

$$\frac{V_{C_2}(\omega)}{I_i(\omega)} = \frac{R_{eq} + j\omega L_2}{B(\omega)} \tag{7.67}$$

$$\frac{I_{C_2}(\omega)}{I_i(\omega)} = \frac{-\omega^2 L_2 C_2 + j R_{eq}\omega C_2}{B(\omega)} \tag{7.68}$$

Voltage and current on inductor L_2:

$$\frac{V_{L_2}(\omega)}{I_i(\omega)} = \frac{j\omega L_2}{B(\omega)} \tag{7.69}$$

$$\frac{I_{L_2}(\omega)}{I_i(\omega)} = \frac{1}{B(\omega)} \tag{7.70}$$

The output voltage and current on the resistor R_{eq}:

$$\frac{V_O(\omega)}{I_i(\omega)} = \frac{R_{eq}}{B(\omega)} \tag{7.71}$$

The current transfer gain is given by:

$$g(\omega) = \frac{I_O(\omega)}{I_i(\omega)} = \frac{1}{B(\omega)} \tag{7.72}$$

Usually, the input impedance and output current gain draw more attention rather than all transfer functions listed in previous section. To simplify the operation, select:

$$L_1 = L_2 = L; \quad C_1 = C_2 = C; \quad \omega_0 = \frac{1}{\sqrt{LC}}$$

$$Z_0 = \sqrt{\frac{L}{C}}; \quad Q = \frac{Z_0}{R_{eq}} = \frac{\omega_0 L}{R_{eq}} = \frac{1}{\omega_0 C R_{eq}}; \quad \beta = \frac{\omega}{\omega_0}$$

Obtain

$$B(\beta) = 1 - 3\beta^2 + \beta^4 + j\frac{2 - \beta^2}{Q}\beta \tag{7.73}$$

Therefore:

$$Z(\beta) = \frac{(1 - \beta^2) + jQ(2 - \beta^2)}{1 - 3\beta^2 + \beta^4 + j\frac{2-\beta^2}{Q}\beta} R_{eq} = |Z|\angle\phi \tag{7.74}$$

where

$$|Z| = \frac{\sqrt{(1 - \beta^2)^2 + Q^2(2 - \beta^2)^2}}{\sqrt{(1 - 3\beta^2 + \beta^4)^2 + \beta^2\left(\frac{2-\beta^2}{Q}\right)^2}} R_{eq} \quad \text{and}$$

$$\phi = \tan^{-1}\frac{2 - \beta^2}{1 - \beta^2}Q - \tan^{-1}\frac{(2 - \beta^2)\beta}{(1 - 3\beta^2 + \beta^4)Q}$$

The current transfer gain becomes:

$$g(\beta) = \frac{1}{1 - 3\beta^2 + \beta^4 + j\frac{2-\beta^2}{Q}\beta} = |g|\angle\theta \tag{7.75}$$

where

$$|g| = \frac{1}{\sqrt{(1 - 3\beta^2 + \beta^4)^2 + \beta^2\left(\frac{2-\beta^2}{Q}\right)^2}} \tag{7.76}$$

and

$$\theta = -\tan^{-1}\frac{(2 - \beta^2)\beta}{(1 - 3\beta^2 + \beta^4)Q}$$

Therefore, the voltage and current on capacitor C_1:

$$\frac{V_{C_1}(\beta)}{I_i(\beta)} = \frac{(1-\beta^2) + j\beta Q(2-\beta^2)}{B(\beta)} R_{eq} \qquad (7.77)$$

$$\frac{I_{C_1}(\beta)}{I_i(\beta)} = j\frac{(1-\beta^2) + j\beta Q(2-\beta^2)}{B(\beta)Q} \qquad (7.78)$$

Voltage and current on inductor L_1:

$$\frac{V_{L_1}(\beta)}{I_i(\beta)} = \frac{-\beta^2 + j\beta Q(1-\beta^2)}{B(\beta)} R_{eq} \qquad (7.79)$$

$$\frac{I_{L_1}(\beta)}{I_i(\beta)} = \frac{(1-\beta^2) + j\beta/Q}{B(\beta)} \qquad (7.80)$$

Voltage and current on capacitor C_2:

$$\frac{V_{C_2}(\beta)}{I_i(\beta)} = \frac{1 + j\beta Q}{B(\beta)} R_{eq} \qquad (7.81)$$

$$\frac{I_{C_2}(\beta)}{I_i(\beta)} = \frac{-\beta^2 + j\beta/Q}{B(\beta)} \qquad (7.82)$$

Voltage and current on inductor L_2:

$$\frac{V_{L_2}(\beta)}{I_i(\beta)} = \frac{j\beta Q}{B(\beta)} R_{eq} \qquad (7.83)$$

$$\frac{I_{L_2}(\omega)}{I_i(\omega)} = \frac{1}{B(\omega)} \qquad (7.84)$$

The output voltage and current on the resistor R_{eq}:

$$\frac{V_o(\omega)}{I_i(\omega)} = \frac{R_{eq}}{B(\omega)} \qquad (7.85)$$

The current transfer gain is given by:

$$g(\beta) = \frac{I_o(\beta)}{I_i(\beta)} = \frac{1}{B(\beta)} \qquad (7.86)$$

We can calculate the energy stored in this resonant circuit as:

$$E_{res} = W_{Lr} + W_{Cr} = \frac{1}{2}\int_{t_{res}} \left[L_1 i_{L_1}^2(t) + L_2 i_{L_2}^2(t) + C_1 v_{C_2}^2(t) + C_2 v_{C_2}^2(t)\right]dt \qquad (7.87)$$

where t_{res} is the forced resonant process time-length $T = 1/f$, $i_{L_1}(t)$ and $i_{L_2}(t)$ are the resonant inductor instantaneous currents in the resent process, and $v_{C_1}(t)$ and $v_{C_2}(t)$ are the resonant capacitor instantaneous voltages in the resent process. The energy stored in the inductor and capacitor is fully transferred to each other in a quarter cycles:

$$i_L(t) = I_O \sin \omega t \tag{7.88}$$

$$v_C(t) = V \sin\left(\omega t - \frac{\pi}{2}\right) \tag{7.89}$$

where the natural resonant angular frequency is:

$$\omega_O = \frac{1}{\sqrt{LC}}$$

The applied frequency is $f = \omega/2\pi$. Usually, the applied frequency ω is not equal to the natural resonant angular frequency ω_O. The relevant frequency β is selected as $\beta = 1.59$ in this example. $V_{in} = 30\,V$, $L_{10} = L_{20} = 20\,mH$, $I_{in} = \pm 1\,A$, $R_{eq} = 10\,\Omega$, $C_1 = C_2 = C = 0.22\,\mu F$ and $L_1 = L_2 = L = 100\,\mu H$. Therefore, $Z_0 = 21.32\,\Omega$, $Q = Z_0/R_{eq} = 2.132$. The output current is nearly pure sinusoidal function shown in Figure 7.32. The applied frequency $f = 54\,kHz$, i.e. $T = 18.5\,\mu s$. Therefore, we have the parameters below:

$$B(\beta) = 1 - 3\beta^2 + \beta^4 + j\frac{2 - \beta^2}{Q}\beta = 1 - 3 \times 1.59^2 + 1.59^4 + j\frac{2 - 1.59^2}{2.132}1.59 \tag{7.90}$$

$$|B(\beta)| = 0.4386 \quad \text{or} \quad |g| = \frac{1}{|B(\beta)|} = 2.28$$

$$V_{C_1} = \frac{(1 - \beta^2) + j\beta Q(2 - \beta^2)}{B(\beta)}R_{eq} \tag{7.91}$$

$$= \frac{(1 - 1.59^2) + j \times 1.59 \times 2.132(2 - 1.59^2)}{0.4386} \times 10 = 75.66\,V \tag{7.92}$$

$$I_{L_1} = \frac{(1 - \beta^2) + j\beta/Q}{B(\beta)} = \frac{(1 - 1.59^2) + j(1.59/2.132)}{0.4386} = 3.88\,A \tag{7.93}$$

$$V_{C_2} = \frac{1 + j\beta Q}{B(\beta)}R_{eq} = \frac{1 + j \times 1.59 \times 2.132}{0.4386} \times 10 = 80.6\,V \tag{7.94}$$

$$I_{L_2} = \frac{1}{B(\beta)} = 2.28\,A \tag{7.95}$$

The stored energy as:

$$W_L = \frac{1}{2}LI_{L_1}^2 + \frac{1}{2}LI_{L_2}^2 = 50\mu \times (3.88^2 + 2.28^2) = 1012.6\,\mu J \tag{7.96}$$

$$W_C = \frac{1}{2}CV_{C_1}^2 + \frac{1}{2}CV_{C_2}^2 = 0.11\mu \times (75.66^2 + 80.6^2) = 1344.3\,\mu J \tag{7.97}$$

Other parameters are:

$$PE = V_1 I_1 T = 30 \times 1 \times 18.5\mu = 555\,\mu J$$

$$SE = W_L + W_C = 1012.6 + 1344.3 = 2356.9\,\mu J$$

$$EF = \frac{SE}{PE} = \frac{2356.9}{555} = 4.25$$

$$CIR = \frac{W_C}{W_L} = \frac{1344.3}{1012.6} = 1.33$$

since $EL = 0$, efficiency $\eta = 1$.

$$\tau = \frac{2T \times EF}{1 + CIR} = \frac{2 \times 18.5\mu \times 4.25}{2.33} = 67.5\,\mu s$$

$$\tau_d = \frac{2T \times EF}{1 + CIR}CIR = \frac{2 \times 18.5\mu \times 4.25 \times 1.33}{2.33} = 90\,\mu s$$

$$\xi = \frac{\tau_d}{\tau} = CIR = 1.33 > 0.25$$

Since the damping time constant τ_d is greater than time constant τ, the corresponding transfer function in per-unit system is considered as:

$$G(s) = \frac{1}{1 + s\tau + s^2\tau\tau_d} = \frac{1/\tau\tau_d}{(s + s_1)(s + s_2)} \tag{7.98}$$

where

$$s_1 = \sigma + j\omega \quad \text{and} \quad s_2 = \sigma - j\omega$$

$$\sigma = \frac{1}{2\tau_d} = \frac{1}{0.00018}\,\text{Hz} \quad \text{and} \quad \omega = \frac{\sqrt{4\tau\tau_d - \tau^2}}{2\tau\tau_d} = \frac{\sqrt{24.3n - 4.556n}}{12.15n}$$

$$= \frac{140.5\mu}{12.15n} = 11,565\,\text{rad/s}$$

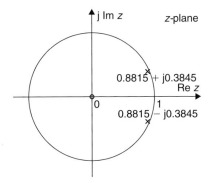

Figure 7.38 The zero's and pole's locations of a cascade double Γ–CL CSRI.

The transfer function is rewritten as:

$$G(s) = \frac{1/\tau\tau_d}{(s+\sigma)^2 + \omega^2} = \frac{2}{\sqrt{4\tau_d/\tau - 1}} \frac{\omega}{(s+\sigma)^2 + \omega^2} \tag{7.99}$$

The unit-step response is:

$$v_O(t) = 21.32\,[1 - e^{-\frac{t}{0.00018}}(\cos 11565t - 0.48 \sin 11565t)] \text{ V}$$

The impulse response is:

$$\Delta v_O(t) = 0.96 U e^{-\frac{t}{0.00018}} \sin 11565t \tag{7.100}$$

where U is the interference. The corresponding transfer function in per-unit system in the z-domain can be approximately written as:

$$G(z) = \frac{2}{\sqrt{4\tau_d/\tau - 1}} \frac{ze^{-aT} \sin \omega T}{z^2 - 2ze^{-aT} \cos \omega T + e^{-2aT}} \tag{7.101}$$

where $a = \sigma = \dfrac{1}{2\tau_d} = \dfrac{1}{180\ \mu s}$, $\omega = 11{,}565$ rad/s and $T = 18.5\ \mu s$

$$G(z) = \frac{2}{\sqrt{4\tau_d/\tau - 1}} \frac{ze^{-aT} \sin \omega T}{z^2 - 2ze^{-aT} \cos \omega T + e^{-2aT}}$$

$$= \frac{0.96 \times ze^{-0.103} \sin 0.214}{z^2 - 2ze^{-0.103} \cos 0.214 + e^{-0.206}}$$

$$\text{or } G(z) = \frac{0.96 \times z \times 0.902 \times 0.2124}{z^2 - 2z \times 0.902 \times 0.9772 + 0.814} = \frac{0.039065z}{z^2 - 1.763z + 0.814}$$

The transfer function has one zero and two poles. The poles are at $z = 0.8815 \pm j\,0.3845$ inside the unity-cycle; therefore, this converter is stable circuit. The zero's and pole's locations of this cascade double Γ–LC CSRI are shown in Figure 7.38.

FURTHER READING

1. Luo F. L., Re-Lift converter: design, test, simulation and stability analysis, *IEE-EPA Proc*, Vol. 145, No. 4, July 1998, pp. 315–325.
2. Luo F. L., Negative output Luo-converters: voltage lift technique, *IEE-EPA Proc*, Vol. 146, No. 2, March 1999, pp. 208–224.
3. Luo F. L., Positive output Luo-converters: voltage lift technique, *IEE-EPA Proc*, Vol. 146, No. 4, July 1999, pp. 415–432.
4. Luo F. L., Double output Luo-converters: advanced voltage lift technique, *IEE-EPA Proc*, Vol. 147, No. 6, November 2000, pp. 469–485.
5. Luo F. L., Six self-lift DC/DC converters: voltage lift technique, *IEEE Trans Ind Electron*, Vol. 48, No. 6, December 2001, pp. 1268–1272.
6. Luo F. L., Ye H. and Rashid M. H., Multiple quadrant operation Luo-converters, *IEE-EPA Proc*, Vol. 149, No. 1, January 2002, pp. 9–18.
7. Luo F. L. and Ye H., Investigation and verification of a cascade double Γ-CL current source resonant inverter, *IEE-EPA Proc*, Vol. 149, No. 5, September 2002, pp. 369–378.
8. Luo F. L. and Ye H., Positive output super-lift Luo-converters, *Proc IEEE Int Conf PESC'2002*, Cairns, Australia, 23–27 June 2002, pp. 425–430.
9. Luo F. L. and Ye H., Positive output super-lift converters, *IEEE Trans Power Electron*, Vol. 18, No. 1, January 2003, pp. 105–113.
10. Luo F. L. and Ye H., Negative output super-lift Luo-converters, *Proc IEEE Int Conf PESC '03*, Acapulco, Mexico, June 15–19, 2003, pp. 1361–1366.
11. Luo F. L. and Ye H., Negative output super-lift converters, *IEEE Trans Power Electron*, Vol. 18, No. 5, September 2003, pp. 1113–1121.
12. Luo F. L. and Ye H., Positive output multiple-lift push–pull switched-capacitor Luo-converters, *IEEE Trans Ind Electron*, Vol. 51, No. 3, June 2004, pp. 594–602.
13. Erickson R., *Fundamentals of Power Electronics*, Chapman and Hall, New York, 1997.
14. Hart D. W., *Introduction to Power Electronics*, Prentice-Hall, Englewood Cliffs, New Jersey, 1997.
15. Chung H. S. H. and Hui S. Y. R., Reduction of EMI in power converters using fully soft-switching technique, *IEEE Trans Electromagnetics*, Vol. 40, No. 3, August 1998, pp. 282–287.
16. Hui S. Y. R., Gogani E. S. and Zhang J., Analysis of a quasi-resonant circuit for soft-switched inverters, *IEEE Trans PEL*, Vol. 11, No. 1, January 1999, pp. 106–114.

Chapter 8

Digitally Controlled AC/AC Converters

As described in Chapter 3, all AC/AC (AC/DC/AC) converters are treated as a first-order-hold (FOH) element in digital control systems. We will discuss this model in various circuits in this Chapter.

8.1 INTRODUCTION

AC/AC and/or AC/DC/AC converters are newly developed group of the power switching circuits applied in industrial applications in comparison with other power switching circuits. Although choppers were popular in AC/DC/AC power supply long time ago, power AC/DC/AC converters were used in industrial application since later 1980s. Semiconductor manufacture development brought Power devices such as gate turn-off thyristors (GTO), Triac, bipolar transistors (BT), insulated gate bipolar transistors (IGBT) and power MOS field effected transistors (MOSFET) and so on into the DC/AC power supply since 1980s. Due to the low switching frequency devices the equipment power rate is not very high.

The DC power supply equipment was totally changed since 1960s because of the thyristor (SCR) produced. The corresponding control circuit is gradually changed from analog control to digital control system since 1980s. The mathematical modeling for all AC/DC/AC converters is discussed widely in worldwide. Finally, a FOH is generally accepted to be used for simulate the AC/AC and AC/DC/AC converters.

AC/AC converters are used for converting one AC power source into another AC power application. They are generally used in following applications:

1. single-phase AC/AC voltage controllers;
2. three-phase AC/AC voltage controllers;

3. single-phase input single-phase output (SISO) cycloconverters;
4. three-phase input single-phase output (TISO) cycloconverters;
5. three-phase input three-phase output (TITO) cycloconverters;
6. AC/DC/AC pulse width modulation (PWM) converters;
7. matrix converters.

All AC/AC voltage converters convert the voltage from an AC source with high voltage and frequency to the lower output voltage and frequency with little phase angle delayed.

All AC/AC cycloconverters convert the voltage from an AC source with high voltage and frequency to the lower output voltage and frequency with little phase angle delayed.

All AC/DC/AC converters convert the voltage from an AC source via DC link, then invert to the output load with lower voltage and variable (higher or lower) frequency.

All AC/AC matrix converters directly convert the voltage from an AC source to the output load with lower voltage and variable (higher or lower) frequency.

We discuss some typical AC/DC converters in this Chapter.

8.1.1 SINGLE-PHASE AC/AC VOLTAGE CONTROLLER

Single-phase AC/AC voltage controllers have three typical control methods:

- phase angle control,
- on/off control,
- PWM AC chopper control.

All control methodologies have lower output voltage with same or lower frequency respecting to the input voltage and frequency.

Phase Angle Control

The basic power circuit of a single-phase AC–AC voltage controllers with phase-angle control as shown in Figure 8.1(a), is composed of a pair of SCRs connected back-to-back (also known as inverse-parallel or antiparallel) between the AC supply and the load. This connection provides a *bidirectional full-wave symmetrical* control and the SCR pair can be replaced by a Triac in Figure 8.1(b) for low-power applications. Alternate arrangements are as shown in Figure 8.1(c) with two diodes and two SCRs to provide a common cathode connection for simplifying the gating circuit without needing isolation, and in Figure 8.1(d) with one SCR and four diodes to reduce the device cost but with increased device conduction loss. An SCR and diode combination, known as a ***thyrode controller***, as shown in Figure 8.1(e), provides a *unidirectional half-wave asymmetrical voltage* control with device economy but introduces a DC component and more harmonics, and thus is not very practical to use except for a very low power heating load.

With *phase control*, the switches conduct the load current for a chosen period of each input cycle of voltage and with *on/off* control the switches connect the load either

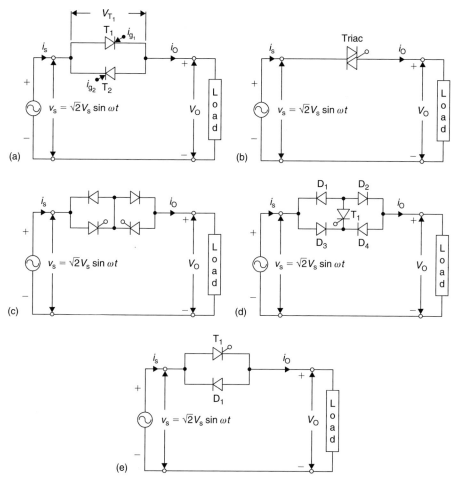

Figure 8.1 Single-phase AC/AC voltage controllers: (a) full-wave, two SCRs in inverse parallel; (b) full-wave with Triac; (c) full wave with two SCRs and two diodes; (d) full wave with four diodes and one SCR and (e) half wave with one SCR and one diode in antiparallel.

for a few cycles of input voltage and disconnect it for the next few cycles (*integral cycle control*) or the switches are turned on and off several times within alternate half-cycles of input voltage (*AC chopper or PWM AC voltage controller*).

For a full-wave, symmetrical phase control, the SCRs T_1 and T_2 in Figure 8.1(a) are gated at α and $\pi + \alpha$, respectively, from the zero crossing of the input voltage and by varying α, the power flow to the load is controlled through voltage control in alternate half-cycles. As long as one SCR is carrying current, the other SCR remains reverse-biased by the voltage drop across the conducting SCR. The principle of operation in each half-cycle is similar to that of the controlled half-wave rectifier and

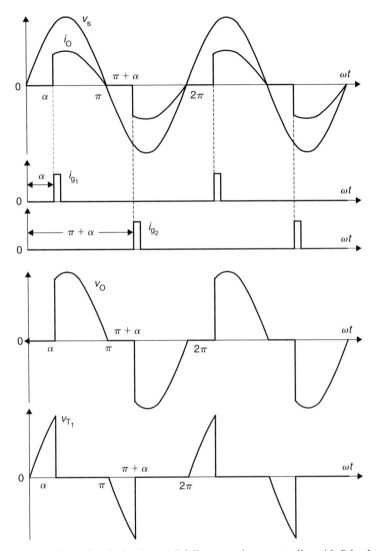

Figure 8.2   Waveforms for single-phase AC full-wave voltage controller with *R*-load.

one can use the same approach for analysis of the circuit. Figure 8.2 shows the typical voltage and current waveforms for the single-phase bidirectional phase controlled AC voltage controller of Figure 8.1(a) with resistive load. The output voltage and current waveforms have half-wave symmetry and thus no DC component.

Figure 8.3 shows the voltage and current waveforms for the controller in Figure 8.1(a) with *R–L* load. Due to the inductance, the current carried by the SCR T_1 may not fall to zero at $\omega t = \pi$ when the input voltage goes negative and may continue until $\omega t = \beta$, the extinction angle, as shown.

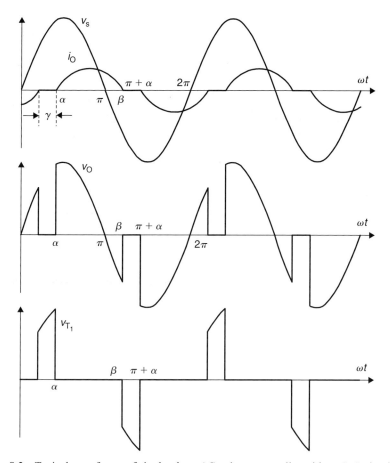

Figure 8.3 Typical waveforms of single-phase AC voltage controller with an *L–R* circuit.

On/Off Control

As an alternative to the phase control, the method of integral cycle, control or burst-firing is used for heating loads. Here, the switch is turned on for a time t_n with n integral cycles and turned off for a time t_m with m integral cycles shown in Figure 8.4. As the SCRs or Triacs used here are turned on at the zero-crossing of the input voltage and turn-off occurs at zero current, supply harmonics and radio frequency interference are very low.

However, sub-harmonic frequency components may be generated that are undesirable as they may set up sub-harmonic resonance in the power supply system, cause lamp flicker, and may interfere with the natural frequencies of motor loads causing shaft oscillations.

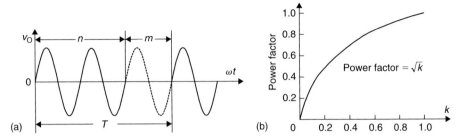

Figure 8.4 Single-phase AC/AC voltage controllers with on/off control: (a) typical load-voltage waveforms and (b) power factor with the duty cycle k.

For sinusoidal input voltage:

$$v = \sqrt{2}V_s \sin \omega t \tag{8.1}$$

and the rms output voltage is:

$$V_O = V_s\sqrt{k} \tag{8.2}$$

where $k = n/(n+m) =$ duty cycle and $V_s =$ rms phase voltage. The power factor is:

$$PF = \sqrt{k} \tag{8.3}$$

which is poorer for lower values of the duty cycle k.

PWM AC Chopper Control

As in the case of controlled rectifier, the performance of AC voltage controllers can be improved in terms of harmonics, quality of output current and input power factor by PWM control in PWM AC choppers. The circuit configuration of one such single-phase unit is shown in Figure 8.5.

Here, fully controlled switches S_1 and S_2 connected in antiparallel are turned on and off many times during the positive and negative half-cycles of the input voltage, respectively; S'_1 and S'_2 provide the freewheeling paths for the load current when S_1 and S_2 are off. An input capacitor filter may be provided to attenuate the high switching frequency current drawn from the supply and also to improve the input power factor. Figure 8.6 shows the typical output voltage and load-current waveform for a single-phase PWM AC chopper. It can be shown that the control characteristics of an AC chopper depend on the *modulation index m*, which theoretically varies from zero to unity.

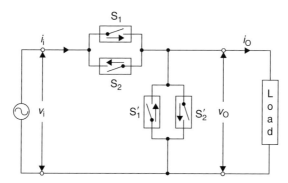

Figure 8.5 Single-phase PWM as chopper circuit.

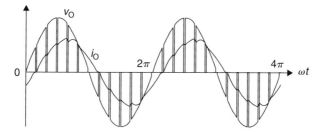

Figure 8.6 Typical output voltage and current waveforms of a single-phase PWM AC chopper.

8.1.2 THREE-PHASE AC/AC VOLTAGE CONTROLLER

In generally, all methods applied in single-phase AC/AC voltage controllers are available to apply in three-phase AC/AC voltage controllers.

Phase Angle Control

Several possible circuit configurations for three-phase phase controlled AC regulators with star- or delta-connected loads are shown in Figures 8.7(a)–(h). The configurations in Figures 8.7(a) and (b) can be realized by three single-phase AC regulators operating independently of each other and they are easy to analyze. In Figure 8.7(a), the SCRs are to be rated to carry line currents and withstand phase voltages, whereas in Figure 8.7(b) they should be capable of carrying phase currents and withstand the line voltages. Also, in Figure 8.7(b) the line currents are free from triplen harmonics while these are present in the closed delta. The power factor in Figure 8.7(b) is slightly higher. The firing angle control range for both these circuits is 0–180° for *R*-load.

The circuits in Figure 8.7(c) and (d) are three-phase three-wire circuits and are difficult to analyze. In both these circuits, at least two SCRs – one in each phase – must be gated simultaneously to get the controller started by establishing a current path between the supply lines. This necessitates two firing pulses spaced at 60° apart

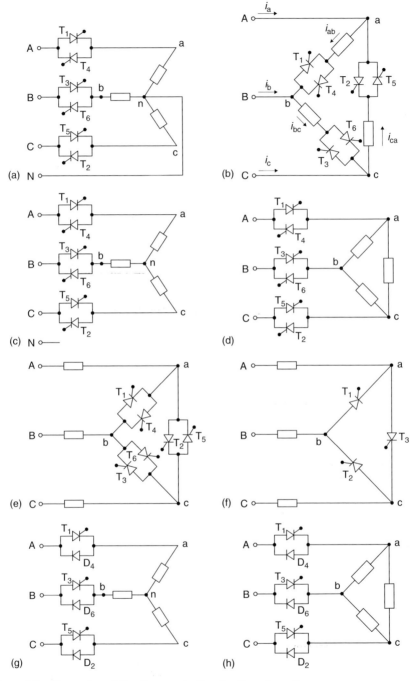

Figure 8.7   Three-phase AC voltage-controller circuit configurations.

per cycle for firing each SCR. The operation modes are defined by the number of SCRs conducting in these modes. The firing control range is $0-150°$. The triplen harmonics are absent in both these configurations.

Another configuration is shown in Figure 8.7(e) when the controllers are delta connected and the load is connected between the supply and the converter. Here, current can flow between two lines even if one SCR is conducting, so each SCR requires one firing pulse per cycle. The voltage and current ratings of SCRs are nearly the same as those of the circuit in Figure 8.7(b). It is also possible to reduce the number of devices to three SCRs in delta as shown in Figure 8.7(f) connecting one source terminal directly to one load circuit terminal. Each SCR is provided with gate pulses in each cycle spaced $120°$ apart. In both Figures 8.7(e) and (f) each end of each phase must be accessible. The number of devices in Figure 8.7(f) is fewer but their current ratings must be higher.

As in the case of the single-phase phase controlled voltage regulator, the total regulator cost can be reduced by replacing six SCRs by three SCRs and three diodes, resulting in three-phase *half-wave controlled* unidirectional AC regulators as shown in Figure 8.7(g) and (h) for star- and delta-connected loads. The main drawback of these circuits is the large harmonic content in the output voltage, particularly the second harmonic because of the asymmetry. However, the DC components are absent in the line. The maximum firing angle in the half-wave controlled regulator is $210°$.

On/Off Control

Similarly, there is on/off control method applied in three-phase AC/AC voltage controllers. Each phase has n_c cycles conducted with interrupted cycle. All interrupted cycle in three phases can be or cannot be synchronized.

PWM AC/AC Control

Similarly, there is PWM control method applied in three-phase AC/AC voltage controllers. Each phase may have same or different modulation rules.

8.1.3 SISO CYCLOCONVERTERS

In contrast to the AC voltage controllers operating at constant frequency discussed so far, a cycloconverter operates as a direct AC/AC frequency changer with an inherent voltage control feature. The basic principle of this converter to construct an alternating voltage wave of lower frequency from successive segments of voltage waves of higher frequency AC supply by a switching arrangement was conceived and patented in the 1920s. Grid controlled mercury-arc rectifiers were used in these converters installed in Germany in the 1930s to obtain $16\frac{2}{3}$ Hz single-phase supply for AC series traction motors from a three-phase 50 Hz system while at the same time a cycloconverter using 18 thyratrons supplying a 400-hp synchronous motor was in operation for some years as a power station auxiliary drive in the USA. However, the practical and commercial utilization of these schemes waited until the SCRs became available in the 1960s.

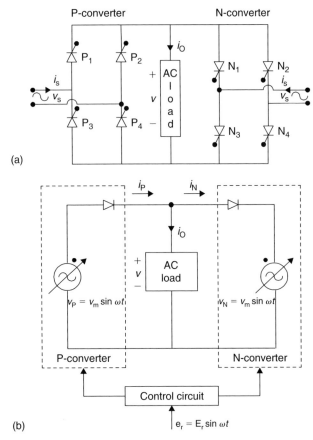

Figure 8.8 Single-phase AC/AC cycloconverter circuit configuration: (a) power circuit for a single-phase bridge cycloconverter and (b) simplified equivalent circuit of a cycloconverter.

With the development of large power SCRs and microprocessor-based control, the cycloconverter today is a matured practical converter for application in large-power low-speed variable-voltage variable-frequency (VVVF) AC drives in cement and steel rolling mills as well as in variable-speed constant-frequency (VSCF) systems in aircraft and naval ships.

A cycloconverter is a naturally commuted converter with the inherent capability of bidirectional power flow and there is no real limitation on its size unlike an SCR inverter with commutation elements. Here, the switching losses are considerably low, the regenerative operation at full power over complete speed range is inherent, and it delivers a nearly sinusoidal waveform resulting in minimum torque pulsation and harmonic heating effects. It is capable of operating even with the blowing out of an individual SCR fuse (unlike the inverter), and the requirements regarding turn-off time, current rise time and dv/dt sensitivity of SCRs are low. The main limitations of a naturally commutated cycloconverter are: (i) limited frequency range for

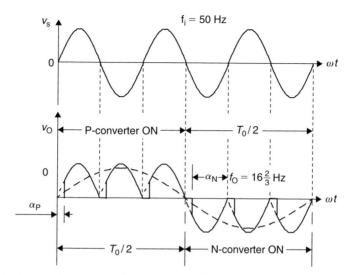

Figure 8.9 Input and output waveforms of a 50–$16\frac{2}{3}$ Hz cycloconverter with R-load.

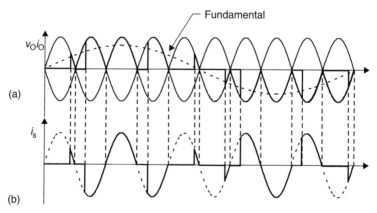

Figure 8.10 Waveforms of a single-phase/single-phase cycloconverter (50–10 Hz) with R-load: (a) load voltage and load current and (b) input supply current.

sub-harmonic-free and efficient operation; and (ii) poor input displacement/power factor, particularly at low output voltages.

Though rarely used, the operation of an SISO cycloconverter is useful to demonstrate the basic principle involved. Figure 8.8(a) shows the power circuit of a single-phase bridge-type cycloconverter, which is the same arrangement as that of the dual converter. Figure 8.8(b) shows the simplified control scheme.

Figure 8.9 shows the typical waveforms for a 50–$16\frac{2}{3}$ Hz single-phase supply. The output voltage has the frequency to be one-third of the input voltage frequency. Figure 8.10 shows the typical waveforms for a 50–10 Hz single-phase supply. The output voltage has the frequency to be one-fifth of the input voltage frequency.

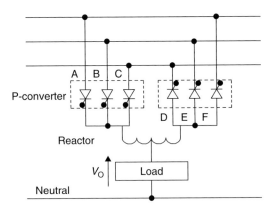

Figure 8.11   Three-phase half-wave (three-pulse) cycloconverter supplying a single-phase load.

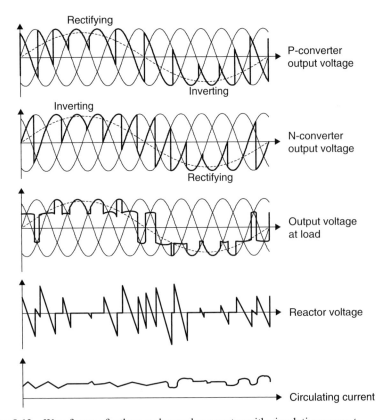

Figure 8.12   Waveforms of a three-pulse cycloconverter with circulating current.

8.1.4 TISO CYCLOCONVERTERS

Figure 8.11 shows a schematic diagram of a three-phase half-wave (three-pulse) cyclo-converter feeding a single-phase load. The control principle is same to the single-phase control.

Figure 8.12 shows typical waveforms of a three-pulse cycloconverter operating with circulating current. Each converter conducts continuously with rectifying and inverting modes as shown and the load is supplied with an average voltage of two converters reducing some of the ripple in the process, with the inter-group reactor behaving as a potential divider. The reactor limits the circulating current, with the value of its inductance to the flow of load current being one-fourth of its value to the flow of circulating current, as the inductance is proportional to the square of the number of turns. The fundamental waves produced by both the converters are the same. The reactor voltage is the instantaneous difference between the converter voltages, and the time integral of this voltage divided by the inductance (assuming negligible circuit resistance) is the circulating current. For a three-pulse cycloconverter, it can be observed that this current reaches its peak when $\alpha_P = 60°$ and $\alpha_N = 120°$.

8.1.5 TITO CYCLOCONVERTERS

Figure 8.13 shows the configuration of a three-phase half-wave (three-pulse) cyclocon-verter feeding a three-phase load. The basic process of a three-phase cycloconversion is illustrated in Figure 8.14 at 15 Hz, 0.6 power factor lagging load from a 50-Hz sup-ply. As the firing angle α is cycled from 0 at "a" to 180° at "j" half a cycle of output frequency is produced (the gating circuit is to be suitably designed to introduce this oscillation of the firing angle). For this load, it can be seen that, although the mean output voltage reverses at X, the mean output current (assumed sinusoidal) remains positive until Y. During XY, the SCRs A, B and C in the P-converter are "inverting". A similar period exists at the end of the negative half-cycle of the output voltage when D, E and F, SCRs in the N-converter are "inverting". Thus, the operation of the con-verter follows in the order of "rectification" and "inversion" in a cyclic manner, with the relative durations being dependent on the load power factor. The output frequency is that of the firing angle oscillation about a quiescent point of 90° (condition when the mean output voltage, given by $V_o = V_{do} \cos \alpha$ is zero).

For obtaining the positive half-cycle of the voltage, firing angle α is varied from 90° to 0° and then to 90°, and for the negative half-cycle, from 90° to 180° and back to 90°. Variation of α within the limits of 180° automatically provides for "natural" line commutation of the SCRs. It is shown that a complete cycle of low-frequency out-put voltage is fabricated from the segments of the three-phase input voltage by using the phase controlled converters. The P- or N-converter SCRs receive firing pulses that are timed such that each converter delivers the same mean output voltage. This is achieved, as in the case of the single-phase cycloconverter or the dual converter, by maintaining the firing angle constraints of the two groups as $\alpha_P = (180° - \alpha_N)$.

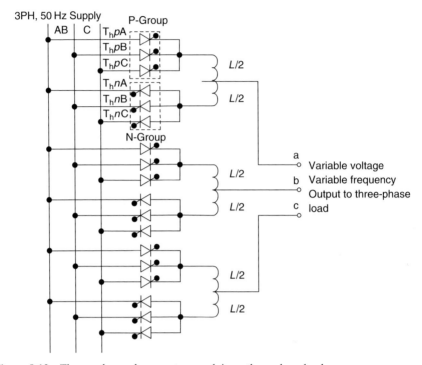

Figure 8.13   Three-pulse cycloconverter supplying a three-phase load.

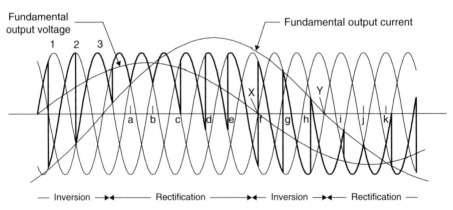

Figure 8.14   Output voltage waveform for one phase of a three-pulse cycloconverter operating at 15 Hz from a 50-Hz supply and 0.6 power factor lagging load.

However, the instantaneous voltages of two converters are not identical and a large circulating current may result unless limited by an inter-group reactor as shown (*circulating-current cycloconverter*) or completely suppressed by removing the gate pulses from the non-conducting converter by an inter-group blanking logic (*circulating-current-free cycloconverter*).

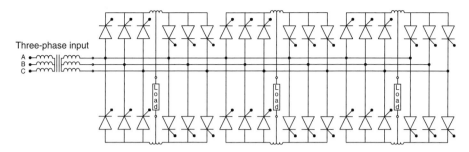

Figure 8.15   Three-phase six-pulse cycloconverter with isolated loads.

A six-pulse cycloconverter circuit configuration is shown in Figure 8.15.

Typical load-voltage waveforms for six-pulse (with 36 SCRs) are shown in Figure 8.16.

It is possible applying a 12-pulse converter which is obtained by connecting two 6-pulse configurations in series and appropriate transformer connections for the required phase-shifted. It may be seen that the higher pulse numbers will generate waveforms closer to the desired sinusoidal form and thus permit higher frequency output. The phase loads may be isolated from each other as shown or interconnected with suitable secondary winding connections.

8.1.6 AC/DC/AC CONVERTERS

AC/DC/AC converters are likely an adjustable speed drive (ASD) in Figure 6.1. The technology is based on the AC/DC rectifier and DC/AC inversion technique, we do not sped long time to repeat its control. Since DC/AC inverters have no restriction on the frequency, therefore, AC/DC/AC converters can convert an AC source into an AC load with lower voltage and variable frequency.

8.1.7 MATRIX CONVERTERS

The matrix converter shown in Figure 8.17 is a development of the force-commutated cycloconverter based on bidirectional fully controlled switches, incorporating PWM voltage control, as mentioned earlier. With the initial progress reported, it has received considerable attention as it provides a good alterative to the double-sided PWM voltage-source rectifier-inverters having the advantages of being a single-stage converter with only nine switches for three-phase to three-phase conversion and inherent bidirectional power flow, sinusoidal input/output waveforms with moderate switching frequency, the possibility of compact design due to the absence of DC link reactive components and controllable input power factor independent of the output load current. The main disadvantages of the matrix converters developed so far are the inherent restriction of the *voltage transfer ratio* (0.866), a more complex control and protection strategy, and

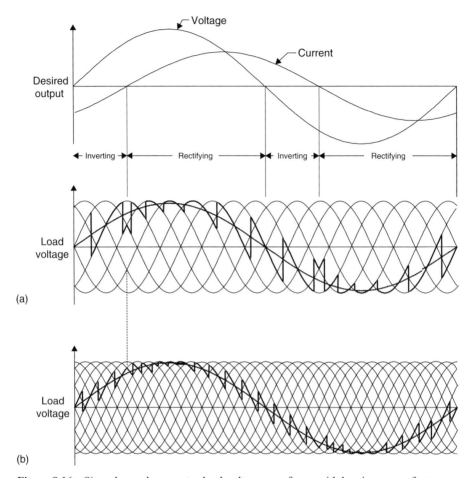

Figure 8.16   Six-pulse cycloconverter load-voltage waveforms with lagging power factor.

above all the non-availability of a fully controlled bidirectional high-frequency switch integrated in a silicon chip (Triac, though bilateral, cannot be fully controlled).

The power circuit diagram of the most practical three-phase to three-phase (3ϕ–3ϕ) matrix converter is shown in Figure 8.17(a), which uses nine bidirectional switches so arranged that any of three input phases can be connected to any output phase as shown in the switching matrix symbol in Figure 8.17(b). Thus, the voltage at any input terminal may be made to appear at any output terminal or terminals while the current in any phase of the load may be drawn from any phase or phases of the input supply. For the switches, the inverse-parallel combination of reverse-blocking self-controlled devices such as power MOSFETs or IGBTs or transistor-embedded diode bridge as shown, have been used so far. The circuit is called a matrix converter as it provides exactly one switch for each of the possible connections between the input

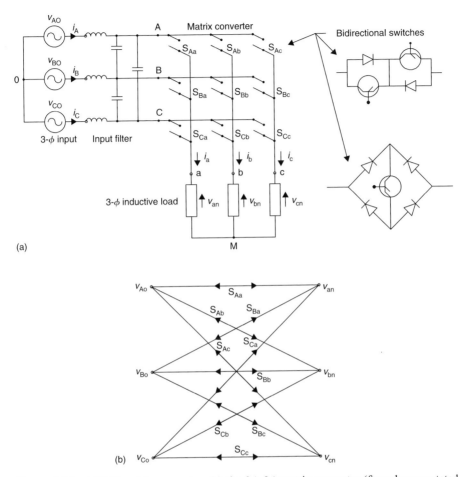

Figure 8.17 AC/AC matrix converter: (a) the 3ϕ–3ϕ matrix converter (forced-commutated cycloconverter) circuit with input filter and (b) switching matrix symbol for converter.

and the output. The switches should be controlled in such a way that, at any time, one and only one of the three switches connected to an output phase must be closed to prevent "short-circuiting" of the supply lines or interrupting the load-current flow in an inductive load. With these constraints, it can be visualized that from the possible $512 (=2^9)$ states of the converter, only 27 switch combinations are allowed as given in Table 8.1, which includes the resulting output line voltages and input phase currents. These combinations are divided into three groups. Group I consists of six combinations when each output phase is connected to a different input phase. In Group II, there are three subgroups, each having six combinations with two output phases short-circuited (connected to the same input phase). Group III includes three combinations with all output phases short-circuited.

Table 8.1

Three-phase/three-phase matrix converter switching combinations

Group	A	B	C	v_{ab}	v_{bc}	v_{ca}	i_A	i_B	i_C	S_{Aa}	S_{Ab}	S_{Ac}	S_{Ba}	S_{Bb}	S_{Bc}	S_{Ca}	S_{Cb}	S_{Cc}
	A	B	C	v_{AB}	v_{BC}	v_{CA}	i_a	i_b	i_c	1	0	0	0	1	0	0	0	1
	A	C	B	$-v_{CA}$	$-v_{BC}$	$-v_{AB}$	i_a	i_c	i_b	1	0	0	0	0	1	0	1	0
	B	A	C	$-v_{AB}$	$-v_{CA}$	$-v_{BC}$	i_b	i_a	i_c	0	1	0	1	0	0	0	0	1
I	B	C	A	v_{BC}	nu_{CA}	v_{AB}	i_c	i_a	i_b	0	1	0	0	0	1	0	1	0
	C	A	B	v_{CA}	v_{AB}	v_{BC}	i_b	i_c	i_a	0	0	1	1	0	0	0	1	0
	C	B	A	$-v_{BC}$	$-v_{AB}$	$-v_{CA}$	i_c	i_b	i_a	0	0	1	0	1	0	1	0	0
	A	C	C	$-v_{CA}$	0	v_{CA}	i_a	0	$-i_a$	1	0	0	0	0	1	0	0	1
	B	C	C	v_{BC}	0	$-v_{BC}$	0	i_a	$-i_a$	0	1	0	0	0	1	0	0	1
	B	A	A	$-v_{AB}$	0	$-v_{AB}$	$-i_a$	i_a	0	0	0	1	0	1	0	0	1	0
II-A	C	A	A	v_{CA}	0	$-v_{CA}$	$-i_a$	0	i_a	0	0	1	1	0	0	0	1	0
	C	B	B	$-v_{BC}$	0	v_{BC}	0	$-i_a$	i_a	0	0	1	0	1	0	0	1	0
	A	B	B	v_{AB}	0	$-v_{AB}$	i_a	$-i_a$	0	0	1	0	0	0	1	0	1	0
	C	A	C	$-v_{CA}$	$-v_{CA}$	0	i_b	0	$-i_b$	0	0	1	1	0	0	0	0	1
	C	B	C	$-v_{BC}$	v_{BC}	0	0	i_b	$-i_b$	0	0	1	0	1	0	0	0	1
	A	B	A	v_{AB}	$-v_{AB}$	0	$-i_b$	i_b	0	1	0	0	0	1	0	1	0	0
II-B	A	C	A	$-v_{CA}$	v_{CA}	0	$-i_b$	0	i_b	1	0	0	0	0	1	1	0	0
	B	C	B	v_{BC}	$-v_{BC}$	0	0	i_b	$-i_b$	0	1	0	0	0	1	0	1	0
	B	A	B	$-v_{AB}$	v_{AB}	0	i_b	$-i_b$	0	0	1	0	1	0	0	0	1	0
	C	C	A	0	v_{CA}	$-v_{CA}$	i_c	0	$-i_c$	0	0	1	0	0	1	1	0	0
	C	C	B	0	$-v_{BC}$	v_{BC}	0	i_c	$-i_c$	0	0	1	0	0	1	0	1	0
	A	A	B	0	v_{AB}	$-v_{AB}$	$-i_c$	i_c	0	1	0	1	0	1	0	0	1	0
II-C	A	A	C	0	$-v_{CA}$	v_{CA}	$-i_c$	0	i_c	1	0	1	0	0	0	0	0	1
	B	B	C	0	v_{BC}	$-v_{BC}$	0	$-i_c$	i_c	0	1	0	0	1	0	0	0	1
	B	B	A	0	$-v_{AB}$	v_{AB}	i_c	$-i_c$	0	0	1	0	0	1	0	1	0	0
	A	A	A	0	0	0	0	0	0	0	0	0	1	0	0	1	0	0
III	B	B	B	0	0	0	0	0	0	0	0	0	0	1	0	0	1	0
	C	C	C	0	0	0	0	0	0	0	0	0	0	0	1	0	0	1

With a given set of input three-phase voltages, any desired set of three-phase output voltages can be synthesized by adopting a suitable switching strategy. However, it has been shown that regardless of the switching strategy there are physical limits on the achievable output voltage with these converters as the maximum peak-to-peak output voltage cannot be greater than the minimum voltage difference between two phases of the input.

To have complete control of the synthesized output voltage, the envelope of the three-phase reference or target voltages must be fully contained within the continuous envelope of the three-phase input voltages. Initial strategy with the output frequency voltages as references reported the limit as 0.5 of the input as shown in Figure 8.18(a). This value can be increased to 0.866 by adding a third harmonic voltage of input frequency ($V_i/4)\cos 3\omega_i t$, to all target output voltages and subtracting from them a third harmonic voltage of output frequency ($V_O/6)\cos 3\omega_O t$ as shown in Figure 8.18(b). However, this process involves a considerable amount of additional computations in synthesizing the output voltages. The other alternative is to use the space vector

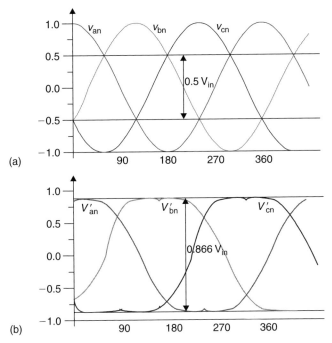

Figure 8.18 Output voltage limits for three-phase AC/AC matrix converter: (a) basic converter input voltages and (b) maximum attainable with inclusion of third harmonic voltages of input and outout frequency to the target voltages.

modulation (SVM) strategy as used in PWM inverters without adding third harmonic components but it also yields the maximum voltage transfer ratio as 0.866.

An AC input LC filter is used to eliminate the switching ripples generated in the converter and the load is assumed to be sufficiently inductive to maintain continuity of the output currents.

The converter in Figure 8.17 connects any input phase (A, B and C) to any output phase (a, b and c) at any instant. When connected, the voltages v_{an}, v_{bn}, v_{cn} at the output terminals are related to the input voltages v_{Ao}, v_{Bo}, v_{Co} as follows:

$$\begin{bmatrix} v_{an} \\ v_{bn} \\ v_{cn} \end{bmatrix} = \begin{bmatrix} S_{Aa} & S_{Ba} & S_{Ca} \\ S_{Ab} & S_{Bb} & S_{Cb} \\ S_{Ac} & S_{Bc} & S_{Cc} \end{bmatrix} \begin{bmatrix} v_{AO} \\ v_{BO} \\ v_{CO} \end{bmatrix} \tag{8.4}$$

where S_{Aa} through S_{Cc} are the switching variables of the corresponding switches shown in Figure 8.17. For a balanced linear star-connected load at the output terminals, the input phase currents are related to the output phase current phase currents by:

$$\begin{bmatrix} i_A \\ i_B \\ i_C \end{bmatrix} = \begin{bmatrix} S_{Aa} & S_{Ab} & S_{Ac} \\ S_{Ba} & S_{Bb} & S_{Bc} \\ S_{Ca} & S_{Cb} & S_{Cc} \end{bmatrix} \begin{bmatrix} i_a \\ i_b \\ i_c \end{bmatrix} \tag{8.5}$$

Note that the matrix of the switching variables in Equation (8.5) is a transpose of the respective matrix in Equation (8.4). The matrix converter should be controlled using a specific and appropriately timed sequence of the values of the switching variables, which will result in balanced output voltages having the desired frequency and amplitude, while the input currents are balanced and in phase (for unity IDF) or at an arbitrary angle (for controllable IDF) with respect to the input voltages. As the matrix converter, in theory, can operate at any frequency, at the output or input, including zero, it can be employed as a three-phase AC/DC converter, DC/three-phase AC converter, or even a buck/boost DC chopper and thus as a *universal power converter*.

The switches should be controlled in such a way that, at any time, one and only one of the three switches connected to an output phase must be closed to prevent "short-circuiting" of the supply lines or interrupting the load-current flow in an inductive load. With these constraints, it can be visualized that from the possible 512 ($=2^9$) states of the converter, only 27 switch combinations are allowed as given in Table 8.1.

These combinations are divided into three groups. Group I consists of six combinations when each output phase is connected to a different input phase. In Group II, there are three subgroups, each having six combinations with two output phases short-circuited (connected to the same input phase). Group III includes three combinations with all output phases short-circuited.

To have complete control of the synthesized output voltage, the envelope of the three-phase reference or target voltages must be fully contained within the continuous envelope of the three-phase input voltages. Initial strategy with the output frequency voltages as references reported the limit as 0.5 of the input as shown in Figure 8.18(a). This value can be increased to 0.866 by adding a third harmonic voltage of input frequency $(V_i/4)\cos 3\omega_i t$, to all target output voltages and subtracting from them a third harmonic voltage of output frequency $(V_O/6)\cos 3\omega_O t$ as shown in Figure 8.18(b). However, this process involves a considerable amount of additional computations in synthesizing the output voltages. The other alternative is to use the SVM strategy as used in PWM inverters without adding third harmonic components, but it also yields the maximum voltage transfer ratio as 0.866.

The control methods adopted so far for the matrix converter are quite complex and are subjects of continuing research. On the methods proposed for independent control of the output voltages and input currents, two methods are of wide use and will be reviewed briefly here: (i) the *Venturini* method based on a mathematical approach of transfer function analysis; and (ii) the space vector modulation (SVM) approach (as has been standardized now in the case of PWM control of the DC link inverter).

Venturini Method

Given a set of three-phase input voltages with constant amplitude V_i and frequency $f_i = \omega_i/2\pi$, this method calculates a switching function involving the duty cycles of each of the nine bidirectional switches and generates the three-phase output voltages by sequential piecewise sampling of the input waveforms. These output voltages follow a predetermined set of reference or target voltage waveforms and with a three-phase load connected, a set of input currents I_i, and angular frequency ω_i, should be in phase for unity IDF or at a specific angle for controlled IDF.

A transfer function approach is employed to achieve the previously mentioned features by relating the input and output voltages and the output and input currents as:

$$\begin{bmatrix} V_{O1}(t) \\ V_{O2}(t) \\ V_{O3}(t) \end{bmatrix} = \begin{bmatrix} m_{11}(t) & m_{12}(t) & m_{13}(t) \\ m_{21}(t) & m_{22}(t) & m_{23}(t) \\ m_{31}(t) & m_{32}(t) & m_{33}(t) \end{bmatrix} \begin{bmatrix} V_{i1}(t) \\ V_{i2}(t) \\ V_{i3}(t) \end{bmatrix} \qquad (8.6)$$

$$\begin{bmatrix} I_{i1}(t) \\ I_{i2}(t) \\ I_{i3}(t) \end{bmatrix} = \begin{bmatrix} m_{11}(t) & m_{21}(t) & m_{31}(t) \\ m_{12}(t) & m_{22}(t) & m_{32}(t) \\ m_{13}(t) & m_{23}(t) & m_{33}(t) \end{bmatrix} \begin{bmatrix} I_{O1}(t) \\ I_{O2}(t) \\ I_{O3}(t) \end{bmatrix} \qquad (8.7)$$

where the elements of the modulation matrix $m_{ij}(t)$ (i, $j = 1, 2, 3$) represent the duty cycles of a switch connecting output phase i to input phase j within a sample switching interval. The elements of $m_{ij}(t)$ are limited by the constraints:

$$0 \le m_{ij}(t) \le 1 \quad \text{and} \quad \sum_{j=1}^{3} m_{ij}(t) = 1 \quad (i = 1, 2, 3) \qquad (8.8)$$

The set of three-phase target or reference voltages to achieve the maximum voltage transfer ratio for unity IDF is:

$$\begin{bmatrix} V_{O1}(t) \\ V_{O2}(t) \\ V_{O3}(t) \end{bmatrix} = V_{Om} \begin{bmatrix} \cos \omega_O t \\ \cos(\omega_O t - 120°) \\ \cos(\omega_O t - 240°) \end{bmatrix} + \frac{V_{im}}{4} \begin{bmatrix} \cos 3\omega_i t \\ \cos 3\omega_i t \\ \cos 3\omega_i t \end{bmatrix} - \frac{V_{Om}}{6} \begin{bmatrix} \cos 3\omega_O t \\ \cos 3\omega_O t \\ \cos 3\omega_O t \end{bmatrix} \qquad (8.9)$$

where V_{Om} and V_{im} are the magnitudes of output and input fundamental voltages of angular frequencies ω_O and ω_i, respectively. With $V_{Om} \le 0.866\, V_{im}$, a general formula for the duty cycles $m_{ij}(t)$ is derived. For unity IDF condition, a simplified formula is:

$$\begin{aligned} m_{ij} = \frac{1}{3} &\left\{ 1 + 2q\cos(\omega_i t - 2(j-1)60°) \right. \\ &\times \left[\cos(\omega_O t - 2(i-1)60°) + \frac{1}{2\sqrt{3}}\cos(3\omega_i t) - \frac{1}{6}\cos(3\omega_O t) \right] \\ &\left. - \frac{2q}{3\sqrt{3}} \left[\cos(4\omega_i t - 2(j-1)60°) - \cos(2\omega_i t - 2(1-j)60°) \right] \right\} \qquad (8.10) \end{aligned}$$

where $i, j = 1, 2, 3$ and $q = V_{Om}/V_{im}$.

The method developed as in the preceding is based on a *Direct Transfer Function (DTF)* approach using a single modulation matrix for the matrix converter, employing the switching combinations of all three groups in Table 8.1. Another approach called *Indirect Transfer Function (ITF)* approach considers the matrix converter as a combination of PWM voltage source rectifier-PWM voltage source inverter (VSR-VSI) and employs the already well-established VSR and VSI PWM techniques for MC control utilizing the switching combinations of Groups II and III only of Table 8.1. The drawback of this approach is that the IDF is limited to unity and the method also generates higher and fractional harmonic components in the input and the output waveforms.

SVM Method

The SVM is now a well-documented inverter PWM control technique that yields high voltage gain and less harmonic distortion compared to the other modulation techniques as discussed. Here, the three-phase input currents and output voltages are represented as space vectors and SVM is applied simultaneously to the output voltage and input current space vectors. Applications of the SVM algorithm to control of matrix converters have appeared in the literature and shown to have inherent capability to achieve full control of the instantaneous output voltage vector and the instantaneous current displacement angle even under supply voltage disturbances. The algorithm is based on the concept that the MC output line voltages for each switching combination can be represented as a voltage space vector denned by:

$$V_O = \frac{2}{3}[v_{ab} + v_{bc} \exp(j120°) + v_{ca} \exp(-j120°)] \tag{8.11}$$

Of the three groups in Table 8.1, only the switching combinations of Groups II and III are employed for the SVM method. Group II consists of switching state voltage vectors having constant angular positions and are called *active* or *stationary* vectors. Each subgroup of Group II determines the position of the resulting output voltage space vector, and the six-state space voltage vectors form a six-sextant hexagon used to synthesize the desired output voltage vector. Group III comprises the *zero* vectors positioned at the center of the output voltage hexagon and these are suitably combined with the active vectors for the output voltage synthesis.

The modulation method involves selection of the vectors and their on-time computation. At each sampling period T_s, the algorithm selects four active vectors related to any possible combinations of output voltage and input current vectors in addition to the zero vector to construct a desired reference voltage. The amplitude and the phase angle of the reference voltage vector are calculated and the desired phase angle of the input current vector is determined in advance. For computation of the on-time periods of the chosen vectors, these are combined into two sets leading to two new vectors adjacent to the reference voltage vector in the sextant and having the same direction as the reference voltage vector. Applying the standard SVM theory, the general formulas derived for the vector on-times, which satisfy, at the same time, the reference output voltage and input current displacement angle are as follows:

$$t_1 = \frac{2qT_s}{\sqrt{3}\cos\phi_i}\sin(60° - \theta_O)\sin(60° - \theta_i)$$

$$t_2 = \frac{2qT_s}{\sqrt{3}\cos\phi_i}\sin(60° - \theta_O)\sin\theta_i$$

$$t_3 = \frac{2qT_s}{\sqrt{3}\cos\phi_i}\sin\theta_O\sin\theta(60° - \theta_i)$$

$$t_4 = \frac{2qT_s}{\sqrt{3}\cos\phi_i}\sin\theta_O\sin\theta_i \tag{8.12}$$

where q is the voltage transfer ratio, ϕ_i is the input displacement angle chosen to achieve the desired input power factor (with $\phi_i = 0$, a. maximum value of $q = 0.866$ is obtained), θ_O and θ_i are the phase displacement angles of the output voltage and input current vectors, respectively, whose values are limited within the 0–60° range. The on-time of the zero vector is:

$$t_o = T_s - \sum_{i=1}^{4} t_i \tag{8.13}$$

The integral value of the reference vector is calculated over one sample time interval as the sum of the products of the two adjacent vectors and their on-time ratios. The process is repeated at every sample instant.

Control Implementation and Comparison of the Two Methods

Both methods need a digital signal processor (DSP)-based system for their implementation. In one scheme for the Venturini method, the programmable timers, as available, are used to time out the PWM gating signals. The processor calculates the six-switch duty cycles in each sampling interval, converts them to integer counts, and stores them in the memory for the next sampling period. In the SVM method, an EPROM is used to store the selected sets of active and zero vectors and the DSP calculates the on-times of the vectors. Then with an identical procedure as in the other method, the timers are loaded with the vector on-times to generate PWM waveforms through suitable output ports. The total computation time of the DSP for the SVM method has been found to be much less than that of the Venturini method. Comparison of the two schemes shows that while in the SVM method the switching losses are lower, the Venturini method shows better performance in terms of input current and output voltage harmonics.

8.2 TRADITIONAL MODELING FOR AC/AC (AC/DC/AC) CONVERTERS

Carefully investigating AC/AC (AC/DC/AC) converters in PWM inversion process, we can see that in each pulse-width $T = 1/f_\Delta$ the modulation ratio m_a is proportional to the control signal $v_C(t)$. If the frequency ratio m_f is large enough, the value of the control signal $v_C(t)$ in a sampling period T can be considered a constant value. The output voltage value is proportional to the input control signal. The corresponding output current value is an increasing or decreasing wave. The corresponding waveforms have been shown in Figure 8.3. In general condition the load is an R–L circuit with the time constant $\tau = L/R$, which is usually larger than the sampling interval T. Therefore, the output current is continuous and is generally accumulated interval by interval. The expression in per-unit system can be written as:

$$i_{O\text{-}k} = i_{O\text{-}(k-1)}(1 \pm e^{-t/T}) \tag{8.14}$$

where $i_{O\text{-}k}$ is the kth-step output current and $i_{O\text{-}(k-1)}$ is the previous step output current.

By per-unit system the voltage transfer gain is unity. The transfer function in the time domain is an exponential function, and it has the following form in the s-domain:

$$G(s) = \frac{1}{1 + sT} \tag{8.15}$$

In digital control system, all DC/AC PWM inverter is treated as an FOH has the transfer function in the z-domain:

$$G(z) = \frac{z}{z - 1/e} \tag{8.16}$$

It means the DC/AC PWM inverter is the first-order-element that possesses one zero at $z = 0$ and one pole at $z = 1/e$, which is located on the unit-cycle. The zero and pole in the z-plane are shown in Figure 8.19. Therefore, a rectifier is a critical stable element. In industrial applications, closed-loop control is required to increase the stability margin.

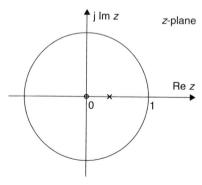

Figure 8.19   Zero and pole of the FOH.

8.3 SINGLE-PHASE AC/AC CONVERTER

The single-phase AC/AC converters are shown in Figure 8.1. The load is an R–L circuit. The open-loop control block diagram is shown in Figure 8.20. The sampling interval is $T = 1/f$, f is the input frequency. If $f = 50$ Hz, $T = 20$ ms. This control can be implementing by a digital computer, which offers a pulse a cycle in 20 ms. The actuator is usually an R–L load. The final output parameter is the current I_O shown in Figure 8.20.

The closed-loop control block diagram is shown in Figure 8.21. The sampling interval is $T = 1/f$. A current controller is always requested in a closed-loop control system. It can be a proportional-plus-integral (PI) controller in digital form. This control can be implementing by a digital computer, which offers a firing pulse a cycle in 20 ms. The actuator is usually an R–L load. The final output parameter is the current I_O shown in Figure 8.21.

8.4 THREE-PHASE AC/AC VOLTAGE CONTROLLERS

Three-phase AC/AC converters are shown in Figure 8.7. The load is an $R–L$ circuit. The open-loop control block diagram is shown in Figure 8.20. The sampling interval is $T = 1/3f$, f is the input frequency. If $f = 50$ Hz, $T = 6.67$ ms. This control can be implementing by a digital computer, which offers a pulse a cycle in 6.67 ms. The actuator is usually an R–L load. The final output parameter is the current I_O shown in Figure 8.20.

The closed-loop control block diagram is shown in Figure 8.21. The sampling interval is $T = 1/3f$. A current controller is always requested in a closed-loop control system. It can be a PI controller in digital form. This control can be implementing by a digital computer, which offers a firing pulse a cycle in 6.67 ms. The actuator is usually an $R–L$ load. The final output parameter is the current I_O shown in Figure 8.21.

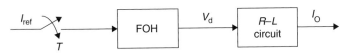

Figure 8.20   Open-loop control of the AC/AC PWM inverters.

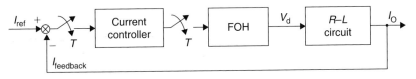

Figure 8.21   Closed-loop control of the AC/AC PWM inverters.

8.5 SISO CYCLOCONVERTERS

The SISO cycloconverter is shown in Figure 8.8. The load is an R–L circuit. The open-loop control block diagram is shown in Figure 8.20. The sampling interval is $T = 1/f$, f is the input frequency. If $f = 50$ Hz, $T = 20$ ms. This control can be implementing by a digital computer, which offers a pulse a cycle in 20 ms. The actuator is usually an R–L load. The final output parameter is the current I_O shown in Figure 8.20.

The closed-loop control block diagram is shown in Figure 8.21. The sampling interval is $T = 1/f$. A current controller is always requested in a closed-loop control system. It can be a PI controller in digital form. This control can be implementing by a digital computer, which offers a firing pulse a cycle in 20 ms. The actuator is usually an R–L load. The final output parameter is the current I_O shown in Figure 8.21.

8.6 TISO CYCLOCONVERTERS

TISO cycloconverter is shown in Figure 8.11. The load is an R–L circuit. The open-loop control block diagram is shown in Figure 8.20. The sampling interval is $T = 1/3f$, f is the input frequency. If $f = 50$ Hz, $T = 6.67$ ms. This control can be implementing by a digital computer, which offers a pulse a cycle in 6.67 ms. The actuator is usually an R–L load. The final output parameter is the current I_O shown in Figure 8.20.

The closed-loop control block diagram is shown in Figure 8.21. The sampling interval is $T = 1/3f$. A current controller is always requested in a closed-loop control system. It can be a PI controller in digital form. This control can be implementing by a digital computer, which offers a firing pulse a cycle in 6.67 ms. The actuator is usually an R–L load. The final output parameter is the current I_O shown in Figure 8.21.

8.7 TITO CYCLOCONVERTERS

TITO cycloconverter is shown in Figure 8.13. The load is an R–L circuit. The open-loop control block diagram is shown in Figure 8.20. The sampling interval is $T = 1/3f$, f is the input frequency. If $f = 50$ Hz, $T = 6.67$ ms. This control can be implementing by a digital computer, which offers a pulse a cycle in 6.67 ms. The actuator is usually an R–L load. The final output parameter is the current I_O shown in Figure 8.20.

The closed-loop control block diagram is shown in Figure 8.21. The sampling interval is $T = 1/3f$. A current controller is always requested in a closed-loop control system. It can be a PI controller in digital form. This control can be implementing by a digital computer, which offers a firing pulse a cycle in 6.67 ms. The actuator is usually an R–L load. The final output parameter is the current I_O shown in Figure 8.21.

8.8 AC/DC/AC PWM CONVERTERS

AC/DC/AC PWM converters are based on the AC/DC rectifiers and DC/AC inverters as an ASD shown in Figure 6.1. The load is an R–L circuit. The open-loop control block diagram is shown in Figure 8.20. The sampling interval is $T = 1/f_\Delta$, f_Δ is the triangle frequency. If the frequency $f = 400$ Hz, $T = 1/f = 2.5$ ms. This control can be implementing by a digital computer, which offers a pulse a cycle in 2.5 ms. The actuator is usually an *R–L* load. The final output parameter is the current I_O shown in Figure 8.20.

The closed-loop control block diagram is shown in Figure 8.21. The sampling interval is $T = 1/f_\Delta$. A current controller is always requested in a closed-loop control system. It can be a PI controller in digital form. This control can be implementing by a digital computer, which offers a firing pulse a cycle in 2.5 ms. The actuator is usually an *R–L* load. The final output parameter is the current I_O shown in Figure 8.21.

8.9 MATRIX CONVERTERS

Matrix converters are based on the AC/DC/AC converters shown in Figure 8.17. The load is an *R–L* circuit. The open-loop control block diagram is shown in Figure 8.20. The sampling interval is $T = 1/f_\Delta$, f_Δ is the triangle frequency. If $f = 400$ Hz, $T = 2.5$ ms. This control can be implementing by a digital computer, which offers a pulse a cycle in 2.5 ms. The actuator is usually an *R–L* load. The final output parameter is the current I_O shown in Figure 8.20.

The closed-loop control block diagram is shown in Figure 8.21. The sampling interval is $T = 1/f_\Delta$. A current controller is always requested in a closed-loop control system. It can be a PI controller in digital form. This control can be implementing by a digital computer, which offers a firing pulse a cycle in 2.5 ms. The actuator is usually an *R–L* load. The final output parameter is the current I_O shown in Figure 8.21.

FURTHER READING

1. Rombaut C., Seguier G. and Bausiere R., *Power Electronics Converters – AC/AC Converters*, McGraw-Hill, New York, 1987.
2. Lander C. W., *Power Electronics*, McGraw-Hill, London, 1993.
3. Dewan S. B. and Straughen A., *Power Semiconductor Circuits*, John Wiley, New York, 1975.
4. Hart D. W., *Introduction to Power Electronics*, Prentice-Hall, Englewood Cliffs, 1997.
5. Williams B. W., *Power Electric Devices, Drivers and Applications*, MacMillan, London, 1987.
6. Pelly B. R., *Thyristor Phase-Controlled Converters and Cycloconverters*, John Wiley, New York, USA, 1971.
7. McMurray W., *The Theory and Design of Cycloconverters*, MIT Press, Cambridge, 1972.
8. Syam P., Nandi P. K. and Chattopadhyay A. K., An improvement feedback technique to suppress sub-harmonics in a naturally commutated cycloconverter, *IEEE-Trans. IE*, Vol. 45, 1998, pp. 950–962.
9. Venturini M., A new sine-wave in sine-wave out converter technique eliminated reactor elements, *Proc. Powercon'80*, Vol. E3-1, 1980, pp. E3-1–E3-15.

10. Alesina A. and Venturini M., The generalized transformer: a new bidirectional waveform frequency converter with continuously adjustable input power factor, *Proc. IEEE-PESC'80*, 1980, pp. 242–252.

11. Alesina A. and Venturini M., Analysis and design of optimum amplitude nine-switch direct AC–AC converters, *IEEE-Trans. PEL*, Vol. 4, 1989, pp. 101–112.

12. Ziogas P. D., Khan S. I. and Rashid M., Some improved forced commutated cycloconverter structures, *IEEE-Trans. IA*, Vol. 21, 1985, pp. 1242–1253.

13. Ziogas P. D., Khan S. I. and Rashid M., Analysis and design of forced commutated cyclo-converter structures and improved transfer characteristics, *IEEE-Trans. IE*, Vol. 33, 1986, pp. 271–280.

14. Ishiguru A., Furuhashi T. and Okuma S., A novel control method of forced commutated cycloconverter using instantaneous values of input line voltages, *IEEE-Trans. IE*, Vol. 38, 1991, pp. 166–172.

15. Huber L., Borojevic D. and Burani N., Analysis, design and implementation of the space-vector modulator for commutated cycloconverters, *IEE-Proc. Part B*, Vol. 139, 1992, pp. 103–113.

16. Huber L. and Borojevic D., Space-vector modulated three-phase to three-phase matrix converter with input power factor correction, *IEEE-Trans. IA*, Vol. 31, 1995, pp. 1234–1246.

17. Zhang L., Watthanasarn C. and Shepherd W., Analysis and comparison of control strategies for AC–AC matrix converters, *IEE-Proc. EPA*, Vol. 145, 1998, pp. 284–294.

18. Das S. P. and Chattopadhyay A. K., Observer based stator flux oriented vector control of cycloconverter-fed synchronous motor drive, *IEEE-Trans. IA*, Vol. 33, 1997, pp. 943–955.

Chapter 9

Open-Loop Control for Digital Power Electronics

Open-loop control is the main control scheme for all digital control systems. Actually, a large number of the industrial applications of the converters are open-loop control systems. We have to carefully discuss these problems in this chapter.

9.1 INTRODUCTION

Digital control for the power electronics, especially for the switching circuits control, is the purpose of this book. We will discuss the fundamental problems in four main converters: AC/DC rectifiers, DC/AC inverters, DC/DC converters and AC/AC (and AC/DC/AC) converters. These problems are:

- Stability analysis
- Unity-step responses
- Impulse (interference) responses

These three main basic problems are the general characteristics of all control systems, including all digital control systems.

9.1.1 STABILITY ANALYSIS

Stability is one of the most important problems of the digital control systems. The fundamental stability criterion is zero-pole location adjustment. If a digital control system has all poles inside the unity-cycle in the z-plane, the system is stable. This

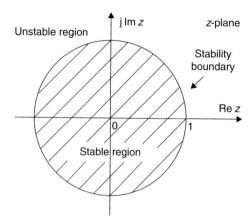

Figure 9.1   Stable/unstable region in the z-plane.

stability criterion is demonstrated in Figure 9.1. The stable region is inside the unity-cycle and the unstable region is outside the unity-cycle. If the pole is located on the cycle, it is the critical state of the stability.

Other stability criteria such as the Jury criterion are based on the theory. We will concentrate the fundamental method to judge our systems.

Stability of the digital control systems is also possibly adjusted in the s-domain. Considering the relation:

$$z = e^{Ts} \tag{9.1}$$

Solving for s in Equation (9.1), we obtain:

$$s = \frac{1}{T} \ln z \tag{9.2}$$

where T is the sampling interval. The fundamental stability criterion is zero-pole location adjustment. If a digital control system has all poles in the left-hand half plane (LHHP) on the s-plane, the system is stable. This stability criterion is demonstrated in Figure 9.2. The stable region is in LHHP and the unstable region is in the right-hand half plane (RHHP) on the s-plane. If the pole is located on the imaginary (vertical) axes, it is the critical state of the stability.

Converters Open-Loop Analysis

We discussed the four typical converters in previous chapters. Their mathematical models are the typical elements:

- A zero-order-hold (ZOH) for AC/DC rectifiers.
- A first-order-hold (FOH) for DC/AC inverters and AC/AC (including AC/DC/AC) converters.
- A second-order-hold (SOH) for DC/DC converters.

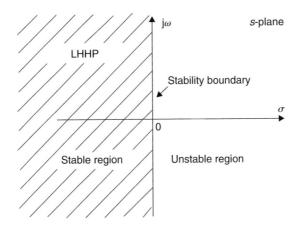

Figure 9.2 Stable/unstable region in the *s*-plane.

Usually, the existing individual converters are stable. All unstable converter topologies have been washed out by elimination process before published although some unstable converters can be used with closed-loop control.

By digital control theory, the ZOH, FOH and SOH are considered stable in open-loop control although the ZOH has a pole at $z = 1$ on the stability boundary.

Analysis of Converters with a First-Order Load

The existing converters are stable and applied in industrial applications. Converters are available to be used since they can provide power/energy to load. The load can be pure resistive load or be other forms such as inductive load, capacitive load and back electromagnetic force (EMF) load. In general, a first-order load is usually considered to be supplied by all types of converters. Therefore we have to discuss the stability change when the converters with a first-order load. Typical first-order load can be an R–L circuit or an R–C circuit. Its transfer function in per-unit system in the *s*-domain is:

$$G_1(s) = \frac{1}{1 + s\tau_1} \tag{9.3}$$

where τ_1 is the time constant of the first-order load. That is, $\tau_1 = L/R$ for an R–L circuit or $\tau_1 = RC$ for an R–C circuit.

Its transfer function in per-unit system in the *z*-domain is:

$$G_1(z) = \frac{z}{z - e^{-T/\tau_1}} \tag{9.4}$$

where T is the sampling interval. Definitely, $e^{-T/\tau_1} < 1$ since $T > 0$.

Analysis of Converters with a First-Order Load Plus an Integral Element

Furthermore, the industrial applications of all converters have always required the converters to provide the power/energy to a first-order load plus an integral element such as a DC motor drive system. In the case we have to consider the extra integral element added in the system.

Its transfer function in per-unit system in the s-domain is:

$$G_m(s) = \frac{1}{s\tau_m} \tag{9.5}$$

where τ_m is the integral time constant, i.e. $\tau_m = J = GD^2/375$ which is the rotor's inertia of a DC motor mechanical time constant. G is the rotor equivalent weight ($G = mg$), g is the gravitation acceleration ($g = 9.81$ m/s^2) and D is the rotor diameter.

Its transfer function in per-unit system in the z-domain is:

$$G_m(z) = \frac{z}{z - T/\tau_m} \tag{9.6}$$

where T is the sampling interval. Usually, the sampling interval T is smaller than the integral time constant τ_m, hence $T/\tau_m < 1$.

9.1.2 UNIT-STEP RESPONSES

Unit-step response is one of the most important problems of the digital control systems. An operation process that a digital control system operates from one steady state to another can be treated as the process with unit-step response. If a digital control system is stable unusually the corresponding unity-step response is stable and the transient process is completed in certain period. The control scheme is shown in Figure 9.3 with the output parameter $v_O(t)$ and input step signal $v_{in}(t)$. The unit-step response is presented in the s-domain by the Laplace transform:

$$V_O(s) = G(s)V_{in}(s) = G(s)\frac{1}{s} \tag{9.7}$$

where $G(s)$ is the converter transfer function and $V_{in}(s)$ is the Laplace transform function of the input unit-step function $V_{in}(s) = 1/s$.

Its response in the z-domain is:

$$V_O(z) = G(z)V_{in}(z) = G(z)\frac{z}{z-1} \tag{9.8}$$

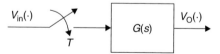

Figure 9.3   Open-loop control scheme.

where $G(z)$ is the converter transfer function and $V_{in}(z)$ is the Laplace transform function of the unit-step function $V_{in}(z) = z/(z-1)$.

Analysis of Converters with a First-Order Load

Unit-step response of the system consisting of converter with a first-order load is usually stable because the output from converter is stable. The control scheme is shown in Figure 9.4 with the output parameter $v_O(t)$ and input step signal $v_{in}(t)$. The unit-step response is presented in the s-domain by Laplace transform:

$$V_O(s) = G(s)G_1(s)V_{in}(s) = G(s)\frac{1}{1+s\tau_1}\frac{1}{s} \qquad (9.9)$$

where $G(s)$ is the converter transfer function, $G_1(s)$ is the first-order circuit transfer function and $V_{in}(s)$ is the Laplace transform function of the unit-step input signal $V_{in}(s) = 1/s$.

Its response in the z-domain is:

$$V_O(z) = G(z)G_1(z)V_{in}(z) = G(z)\frac{z}{z - \mathrm{e}^{-T/\tau_1}}\frac{z}{z-1} \qquad (9.10)$$

where $G(z)$ is the converter transfer function, $G_1(z)$ is the first-order circuit transfer function and $V_{in}(z)$ is the Laplace transform function of the unit-step function $V_{in}(z) = z/(z-1)$.

Analysis of Converters with a First-Order Load Plus an Integral Element

Unit-step response of the system consisting of converter with a first-order load plus an integral element is usually unstable because of the integral element. The control scheme is shown in Figure 9.5 with the output parameter $v_O(t)$ and input step signal $v_{in}(t)$. The unit-step response is presented in the s-domain by Laplace transform:

$$V_O(s) = G(s)G_1(s)G_m(s)V_{in}(s) = G(s)\frac{1}{1+s\tau_1}\frac{1}{s\tau_m}\frac{1}{s} \qquad (9.11)$$

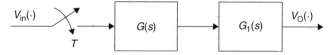

Figure 9.4 Open-loop control of converter with a first-order load.

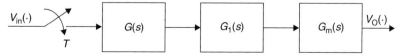

Figure 9.5 Open-loop control of converter with a first-order load plus an integral element.

where $G(s)$ is the converter transfer function, $G_1(s)$ is the first-order circuit transfer function, $G_m(s)$ is the integral element transfer function and $V_{in}(s)$ is the Laplace transform function of the unit-step function $V_{in}(s) = 1/s$.

Its response in the z-domain is:

$$V_O(z) = G(z)G_1(z)G_m(z)V_{in}(z) = G(z)\frac{z}{z - e^{-T/\tau_1}}\frac{z}{z - T/\tau_m}\frac{z}{z - 1} \tag{9.12}$$

where $G(z)$ is the converter transfer function, $G_1(z)$ is the first-order circuit transfer function, $G_m(z)$ is the integral element transfer function and $V_{in}(z)$ is the Laplace transform function of the unit-step function $V_{in}(z) = z/(z - 1)$.

9.1.3 IMPULSE RESPONSES

Interference response usually called the impulse response is one of the most important problems of the digital control systems. The interference signal randomly perturbs to the system as an impulse signal disturb system output parameter. Generally, the interference signal is added in the output point of the converter just likely the load suddenly vibrated. The system scheme is shown in Figure 9.6 with the output parameter $V_O(t)$ and interference signal $V_{int}(t) = U\delta(t)$. The impulse response is presented in the s-domain by Laplace transform:

$$V_O(s) = V_{int}(s) = U \tag{9.13}$$

where U is the interference signal. Its response in the z-domain is:

$$V_O(z) = V_{int}(z) = U \tag{9.14}$$

Analysis of Converters with a First-Order Load

Impulse response of the system consisting of converter with a first-order load is usually stable because the first-order circuit is stable. The control scheme is shown in Figure 9.7 with the output parameter is $V_O(t)$ and interference signal $V_{int}(t) = U\delta(t)$. The impulse response is presented in the s-domain by Laplace transform:

$$V_O(s) = G_1(s)V_{int}(s) = \frac{U}{1 + s\tau_1} \tag{9.15}$$

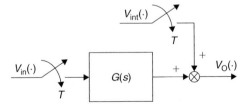

Figure 9.6 Open-loop control of converter (with interference signal).

where $G_1(s)$ is the first-order circuit transfer function, $V_{int}(s)$ is the Laplace transform function of the interference signal $V_{int}(s) = U$.

Its response in the z-domain is:

$$V_O(z) = G_1(z)V_{int}(z) = \frac{z}{z - e^{-T/\tau_1}}U \qquad (9.16)$$

where $G_1(z)$ is the first-order circuit transfer function and $V_{in}(z)$ is the Laplace transform function of the interference signal $V_{int}(z) = U$.

Analysis of Converters with a First-Order Load Plus an Integral Element

Impulse response of the system consisting of converter with a first-order load plus an integral element is usually unstable because of the integral element. The control scheme is shown in Figure 9.8 with the output parameter $v_O(t)$ and interference signal $v_{int}(t) = U\delta(t)$. The impulse response is presented in the s-domain by Laplace transform:

$$V_O(s) = G_1(s)G_m(s)V_{int}(s) = \frac{U}{1 + s\tau_1}\frac{1}{s\tau_m} \qquad (9.17)$$

where $G_1(s)$ is the first-order circuit transfer function, $G_m(s)$ is the integral element transfer function and $V_{int}(s)$ is the Laplace transform function of the interference signal $V_{int}(s) = U$.

Its response in the z-domain is:

$$V_O(z) = G_1(z)G_m(z)V_{int}(z) = \frac{z}{z - e^{-T/\tau_1}}\frac{z}{z - T/\tau_m}U \qquad (9.18)$$

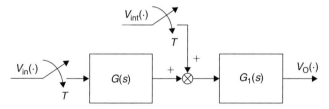

Figure 9.7 Open-loop control of converter with a first-order load (with interference signal).

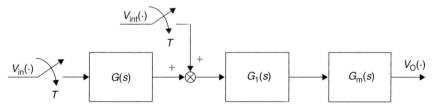

Figure 9.8 Open-loop control of converter with a first-order load plus an integral element (with interference signal).

where $G_1(z)$ is the first-order circuit transfer function and $G_m(z)$ is the integral element transfer function of the interference signal $V_{int}(z) = U$.

9.2 STABILITY ANALYSIS

The stability analysis of the open-loop systems with the four converters is discussed in detail as follows.

9.2.1 AC/DC RECTIFIERS

The mathematical model for power AC/DC Rectifiers is a ZOH. Its transfer function is assumed $u(t)$ in the time-domain, and that in the s-domain is:

$$G(s) = \frac{1}{s} \tag{9.19}$$

Its transfer function in the z-domain is:

$$G(z) = \frac{z}{z-1} \tag{9.20}$$

AC/DC Rectifiers Open-Loop Analysis

Naturally, all existing AC/DC rectifiers applied in industrial applications are stable although they have the pole located on the stability boundary. The block diagram is shown in Figure 9.3 with $G(\cdot)$ to be a ZOH. From Equation (9.20), we have got the location of the zero and pole of a ZOH, which are shown in Figure 9.9.

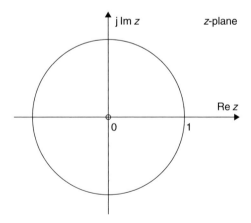

Figure 9.9   Open-loop control of power AC/DC rectifiers.

Analysis of AC/DC Rectifiers with a First-Order Load

The mathematical model for power AC/DC rectifiers is a ZOH. The open-loop control scheme for a ZOH with a first-order load is shown in Figure 9.4. The system transfer function in the *s*-domain is:

$$G(s)G_1(s) = \frac{1}{s}\frac{1}{1+s\tau_1} \tag{9.21}$$

Its transfer function in the *z*-domain is:

$$G(z)G_1(z) = \frac{z}{z-1}\frac{z}{z-e^{-T/\tau_1}} \tag{9.22}$$

From Equation (9.22), we have got two zeros at original point and two poles, one is inside the unity-cycle and another on the unity-cycle. Therefore, this open-loop control system is considered stable. The locations of the zeros and poles of a ZOH with a first-order circuit are shown in Figure 9.10.

Analysis of Rectifiers with a First-Order Load Plus an Integral Element

The mathematical model for power AC/DC rectifiers is a ZOH. The open-loop control scheme for a ZOH with a first-order circuit plus an integral element is shown in Figure 9.5. The system transfer function in the *s*-domain is:

$$G(s)G_1(s)G_m(s) = \frac{1}{s}\frac{1}{1+s\tau_1}\frac{1}{s\tau_m} = \frac{1}{s^2\tau_m(1+s\tau_1)} \tag{9.23}$$

Its transfer function in the *z*-domain is:

$$G(z)G_1(z)G_m(z) = \frac{z}{z-1}\frac{z}{z-T/\tau_m}\frac{z}{z-e^{-T/\tau_1}} \tag{9.24}$$

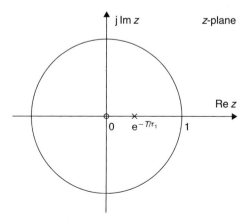

Figure 9.10 Open-loop control of power AC/DC rectifiers with a first-order load.

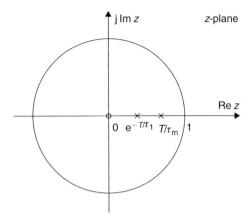

Figure 9.11 Open-loop control of power AC/DC rectifiers with a first-order load plus an integral element.

From Equation (9.24), we have got three zeros at original point and three poles. One pole is inside the unity-cycle and one on the unity-cycle, and another is uncertain. Therefore, this open-loop control system may be considered unstable. The locations of the zeros and poles of a ZOH with a first-order circuit are shown in Figure 9.11.

9.2.2 DC/AC INVERTERS AND AC/AC (AC/DC/AC) CONVERTERS

We discussed the mathematical model for power DC/AC inverters and AC/AC (AC/DC/AC) converters to be an FOH in Chapters 4, 6 and 8. Its transfer function in the s-domain is:

$$G(s) = \frac{1}{1 + sT} \qquad (9.25)$$

Its transfer function in the z-domain is:

$$G(z) = \frac{z}{z - 1/e} \qquad (9.26)$$

Open-Loop Stability Analysis for DC/AC Inverters and AC/AC (AC/DC/AC) Converters

Naturally, all existing DC/AC inverters and AC/AC (AC/DC/AC) converters applied in industrial applications are stable. The block diagram is shown in Figure 9.3 with $G(\cdot)$ to be an FOH. From Equation (9.26), we have got the location of the zero and pole of an FOH, which are shown in Figure 9.12.

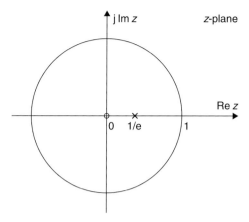

Figure 9.12 Open-loop control of DC/AC inverters.

Open-Loop Stability Analysis for DC/AC Inverters and AC/AC (AC/DC/AC) Converters with a First-Order Load

The open-loop control scheme for an FOH with a first-order load is shown in Figure 9.4. The system transfer function in the s-domain is:

$$G(s)G_1(s) = \frac{1}{1+sT}\frac{1}{1+s\tau_1} \tag{9.27}$$

Its transfer function in the z-domain is:

$$G(z)G_1(z) = \frac{z}{z-1/e}\frac{z}{z-e^{-T/\tau_1}} \tag{9.28}$$

From Equation (9.28), we have got two zeros at original point $z=0$, and two poles at $z=1/e$ and $z=e^{-T/\tau_1}$ inside the unity-cycle. Therefore, this open-loop control system is considered stable. The locations of the zeros and poles of an FOH with a first-order circuit are shown in Figure 9.13.

Open-Loop Stability Analysis for DC/AC Inverters and AC/AC (AC/DC/AC) Converters with a First-Order Load Plus an Integral Element

The open-loop control scheme for an FOH with a first-order circuit plus an integral element is shown in Figure 9.5. The system transfer function in the s-domain is:

$$G(s)G_1(s)G_m(s) = \frac{1}{1+sT}\frac{1}{1+s\tau_1}\frac{1}{s\tau_m} \tag{9.29}$$

Its transfer function in the z-domain is:

$$G(z)G_1(z)G_m(z) = \frac{z}{z-1/e}\frac{z}{z-T/\tau_m}\frac{z}{z-e^{-T/\tau_1}} \tag{9.30}$$

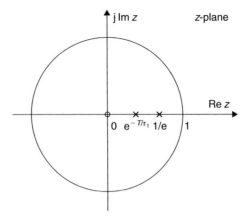

Figure 9.13   Open-loop control of DC/AC inverters with a first-order load.

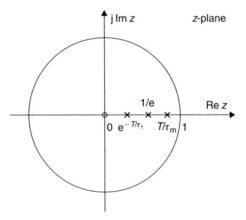

Figure 9.14   Open-loop control of DC/AC inverters with a first-order load plus an integral element.

From Equation (9.30), we have got three zeros at original point and three poles. One pole is inside the unity-cycle and others are uncertain. Therefore, this open-loop control system may be considered unstable. The locations of the zeros and poles of an FOH with a first-order circuit are shown in Figure 9.14.

9.2.3 DC/DC CONVERTERS

The mathematical model for power DC/DC converters is an SOH. Its transfer function in the s-domain is:

$$G(s) = \frac{1}{1 + s\tau + s^2 \tau \tau_d} \tag{9.31}$$

There are four conditions of the damping time constant τ_d related to the time constant τ:

- $\tau_d = 0$
- $\tau_d < 0.25\tau$
- $\tau_d = 0.25\tau$
- $\tau_d > 0.25\tau$

Its transfer functions in the s-domain and in the z-domain are:

$$G(s) = \frac{1}{1 + s\tau} \qquad \text{for } \tau_d = 0 \qquad (9.32)$$

$$G(z) = \frac{z}{z - e^{-T/\tau}} \qquad \text{for } \tau_d = 0 \qquad (9.33)$$

with one zero at $z = 0$ and one pole at $z = e^{-T/\tau}$ in the z-plane. Definitely, the pole is inside the unity-cycle, $e^{-T/\tau} < 1$ since $T > 0$.

$$G(s) = \frac{1/\tau\tau_d}{(s + \sigma_1)(s + \sigma_2)} \qquad \text{for } \tau_d < 0.25\tau \qquad (9.34)$$

$$G(z) = \frac{1/\tau\tau_d}{(\sigma_2 - \sigma_1)} \left(\frac{z}{z - e^{-\sigma_1 T}} - \frac{z}{z - e^{-\sigma_2 T}} \right) \qquad \text{for } \tau_d < 0.25\tau \qquad (9.35)$$

where $\sigma_{1,2} = 1/2\tau_d(1 \pm \sqrt{4 - \tau_d/\tau})$. There are one zero at $z = 0$ and two poles at $z_1 = e^{-\sigma_1 T}$ and $z_2 = e^{-\sigma_2 T}$. Both poles are inside the unit-cycle since $T > 0$.

$$G(s) = \frac{1/\tau\tau_d}{(s + \sigma)^2} \qquad \text{for } \tau_d = 0.25\tau \qquad (9.36)$$

$$G(z) = \frac{4Tze^{-(2/\tau)^T}}{(z - e^{-(2/\tau)^T})^2} \qquad \text{for } \tau_d = 0.25\tau \qquad (9.37)$$

where $\sigma = 1/2\tau_d = 2/\tau$. There are one zero at $z = 0$ and one double-folded pole at $z_1 = e^{-2T/\tau}$. The double-folded pole is inside the unit-cycle since $T > 0$.

$$G(s) = \frac{1/\tau\tau_d}{(s + \sigma)^2 + \omega^2} \qquad \text{for } \tau_d > 0.25\tau \qquad (9.38)$$

$$G(z) = \frac{2}{\sqrt{4\tau_d/\tau - 1}} \frac{ze^{-aT} \sin \omega T}{z^2 - 2ze^{-aT} \cos \omega T + e^{-2aT}} \qquad \text{for } \tau_d > 0.25\tau \qquad (9.39)$$

where $a = \sigma = 1/2\tau_d$ and $\omega = \sqrt{4\tau\tau_d - \tau^2}/2\tau\tau_d$. There are one zero at $z = 0$ and one couple of conjugated complex poles at $z_1 = e^{-aT}(\cos \omega T + j \sin \omega T)$ and $z_2 = e^{-aT}(\cos \omega T - j \sin \omega T)$. Both poles are inside the unit-cycle since $T > 0$.

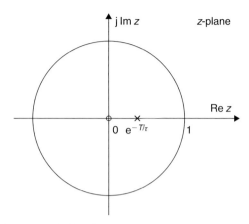

Figure 9.15 Open-loop control of DC/DC converters with $\tau_d = 0$.

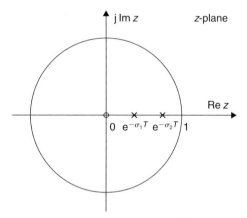

Figure 9.16 Open-loop control of DC/DC converters with $\tau_d < 0.25\tau$.

Converters Open-Loop Analysis

Naturally, all existing DC/DC converters applied in industrial applications are stable. The block diagram is shown in Figure 9.3 with $G(\cdot)$ to be an SOH. From Equation (9.33) for the condition of $\tau_d = 0$, we have got the location of the zero and pole of an SOH, which are shown in Figure 9.15.

From Equation (9.35) for the condition of $\tau_d < 0.25\tau$, we have got the location of the zero and pole of an SOH, which are shown in Figure 9.16.

From Equation (9.37) for the condition of $\tau_d = 0.25\tau$, we have got the location of the zero and pole of an SOH, which are shown in Figure 9.17.

From Equation (9.39) for the condition of $\tau_d > 0.25\tau$, we have got the location of the zero and pole of an SOH, which are shown in Figure 9.18.

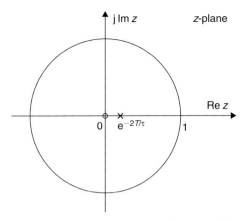

Figure 9.17 Open-loop control of DC/DC converters with $\tau_d = 0.25\tau$.

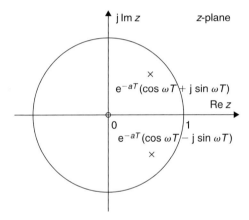

Figure 9.18 Open-loop control of DC/DC converters with $\tau_d > 0.25\tau$.

Analysis of Converters with a First-Order Load

The open-loop control scheme for an SOH with a first-order load is shown in Figure 9.4. The system transfer function in the s-domain is:

$$G(s)G_1(s) = \frac{1}{1 + s\tau + s^2\tau\tau_d}\frac{1}{1 + s\tau_1} \tag{9.40}$$

where τ_1 is the time constant of the first-order load. Its transfer functions in the s-domain and in the z-domain are:

$$G(s)G_1(s) = \frac{1}{1 + s\tau}\frac{1}{1 + s\tau_1} \qquad \text{for } \tau_d = 0 \tag{9.41}$$

$$G(z)G_1(z) = \frac{z}{z - e^{-T/\tau}}\frac{z}{z - e^{-T/\tau_1}} \qquad \text{for } \tau_d = 0 \tag{9.42}$$

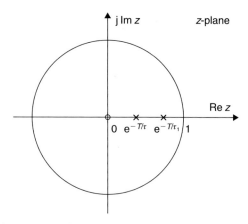

Figure 9.19   Open-loop control of DC/DC converters ($\tau_d = 0$) with a first-order load.

From Equation (9.42), we have got one double-folded zeros at the original point $z = 0$, and two poles at $z = e^{-T/\tau}$ and $z = e^{-T/\tau_1}$ inside the unity-cycle since $T > 0$. Therefore, this open-loop control system is considered stable. The locations of the zeros and poles of an SOH ($\tau_d = 0$) with a first-order circuit are shown in Figure 9.19.

The corresponding transfer functions ($\tau_d < 0.25\tau$) in the s-domain and in the z-domain are:

$$G(s)G_1(s) = \frac{1/\tau\tau_d}{(s + \sigma_1)(s + \sigma_2)} \frac{1}{1 + s\tau_1} \quad \text{for } \tau_d < 0.25\tau \tag{9.43}$$

$$G(z)G_1(z) = \frac{1/\tau\tau_d}{(\sigma_2 - \sigma_1)} \left(\frac{z}{z - e^{-\sigma_1 T}} - \frac{z}{z - e^{-\sigma_2 T}} \right) \frac{z}{z - e^{-T/\tau_1}} \quad \text{for } \tau_d < 0.25\tau \tag{9.44}$$

From Equation (9.44), we have got one double-folded zeros at the original point $z = 0$ and three poles at $z = e^{-\sigma_1 T}$, $z = e^{-\sigma_2 T}$ and $z = e^{-T/\tau_1}$ inside the unity-cycle since $T > 0$. Therefore, this open-loop control system is considered stable. The locations of the zeros and poles of an SOH ($\tau_d < 0.25\tau$) with a first-order circuit are shown in Figure 9.20.

The corresponding transfer functions ($\tau_d = 0.25\tau$) in the s-domain and in the z-domain are:

$$G(s)G_1(s) = \frac{4/\tau^2}{(s + \sigma)^2} \frac{1}{1 + s\tau_1} \quad \text{for } \tau_d = 0.25\tau \tag{9.45}$$

$$G(z)G_1(z) = \frac{4Tze^{-2T/\tau}}{(z - e^{-2T/\tau})^2} \frac{z}{z - e^{-T/\tau_1}} \quad \text{for } \tau_d = 0.25\tau \tag{9.46}$$

From Equation (9.46), we have got one double-folded zeros at the original point $z = 0$ and three poles at $z = e^{-2T/\tau}$ in double-folded and $z = e^{-T/\tau_1}$ inside the unity-cycle

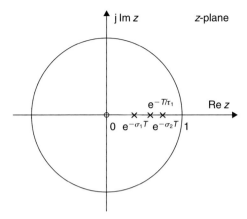

Figure 9.20 Open-loop control of DC/DC converters ($\tau_d < 0.25\tau$) with a first-order load.

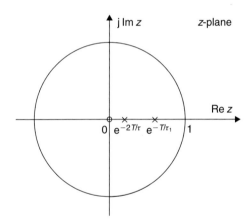

Figure 9.21 Open-loop control of DC/DC converters ($\tau_d = 0.25\tau$) with a first-order load.

since $T > 0$. Therefore, this open-loop control system is considered stable. The locations of the zeros and poles of an SOH ($\tau_d = 0.25\tau$) with a first-order circuit are shown in Figure 9.21.

The corresponding transfer functions ($\tau_d > 0.25\tau$) in the s-domain and in the z-domain are:

$$G(s)G_1(s) = \frac{1/\tau\tau_d}{(s+\sigma)^2 + \omega^2} \frac{1}{1 + s\tau_1} \quad \text{for } \tau_d > 0.25\tau \tag{9.47}$$

$$G(z)G_1(z) = \frac{2}{\sqrt{4\tau_d/\tau - 1}} \frac{ze^{-aT}\sin\omega T}{z^2 - 2ze^{-aT}\cos\omega T + e^{-2aT}} \frac{z}{z - e^{-T/\tau_1}} \quad \text{for } \tau_d > 0.25\tau \tag{9.48}$$

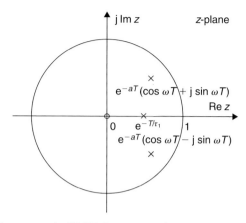

Figure 9.22   Open-loop control of DC/DC converters ($\tau_d > 0.25\tau$) with a first-order load.

where $a = \sigma = 1/2\tau_d$ and $\omega = \sqrt{4\tau\tau_d - \tau^2}/2\tau\tau_d$. From Equation (9.48), we have got one double-folded zeros at the original point $z = 0$ and three poles at $z = e^{-T/\tau_1}$ and one couple of conjugated complex poles $z_1 = e^{-aT}(\cos \omega T + j \sin \omega T)$ and $z_2 = e^{-aT}(\cos \omega T - j \sin \omega T)$ inside the unity-cycle since $T > 0$. Therefore, this open-loop control system is considered stable. The locations of the zeros and poles of an SOH ($\tau_d > 0.25\tau$) with a first-order circuit are shown in Figure 9.22.

Analysis of DC/DC Converters with a First-Order Load Plus an Integral Element

The open-loop control scheme for an SOH with a first-order circuit plus an integral element is shown in Figure 9.5. The system transfer function in the s-domain is:

$$G(s)G_1(s)G_m(s) = \frac{1}{1 + s\tau + s^2\tau\tau_d} \frac{1}{1 + s\tau_1} \frac{1}{s\tau_m} \qquad (9.49)$$

Its transfer functions in the s-domain and in the z-domain are:

$$G(s)G_1(s)G_m(s) = \frac{1}{1 + s\tau} \frac{1}{1 + s\tau_1} \frac{1}{s\tau_m} \qquad \text{for } \tau_d = 0 \qquad (9.50)$$

$$G(z)G_1(z)G_m(z) = \frac{z}{z - e^{-T/\tau}} \frac{z}{z - e^{-T/\tau_1}} \frac{z}{z - T/\tau_m} \qquad \text{for } \tau_d = 0 \qquad (9.51)$$

From Equation (9.51), we have got one triple-folded zeros at the original point $z = 0$ and three poles at $z = e^{-T/\tau}$, $z = e^{-T/\tau_1}$ and $z = T/\tau_m$ inside the unity-cycle since $\tau_m > T > 0$. Therefore, this open-loop control system is considered stable. The

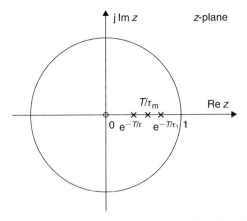

Figure 9.23 Open-loop control of DC/DC converters ($\tau_d = 0$) with a first-order load plus an integral element.

locations of the zeros and poles of an SOH ($\tau_d = 0$) with a first-order circuit are shown in Figure 9.23.

$$G(s)G_1(s)G_m(s) = \frac{1/\tau\tau_d}{(s+\sigma_1)(s+\sigma_2)}\frac{1}{1+s\tau_1}\frac{1}{s\tau_m} \quad \text{for } \tau_d < 0.25\tau \quad (9.52)$$

$$G(z)G_1(z)G_m(z) = \frac{1/\tau\tau_d}{(\sigma_2-\sigma_1)}\left(\frac{z}{z-e^{-\sigma_1 T}} - \frac{z}{z-e^{-\sigma_2 T}}\right)$$

$$\times \frac{z}{z-e^{-T/\tau_1}}\frac{z}{z-T/\tau_m} \quad \text{for } \tau_d < 0.25\tau \quad (9.53)$$

From Equation (9.53), we have got one triple-folded zeros at the original point $z = 0$ and four poles at $z = e^{-\sigma_1 T}$, $z = e^{-\sigma_2 T}$, and e^{-T/τ_1} and $z = T/\tau_m$ inside the unity-cycle since $\tau_m > T > 0$. Therefore, this open-loop control system is considered stable. The locations of the zeros and poles of an SOH ($\tau_d < 0.25\,\tau$) with a first-order circuit are shown in Figure 9.24:

$$G(s)G_1(s)G_m(s) = \frac{4/\tau^2}{(s+\sigma)^2}\frac{1}{1+s\tau_1}\frac{1}{s\tau_m} \quad \text{for } \tau_d = 0.25\tau \quad (9.54)$$

$$G(z)G_1(z)G_m(z) = \frac{4Tze^{-2T/\tau}}{(z-e^{-2T/\tau})^2}\frac{z}{z-e^{-T/\tau_1}}\frac{z}{z-T/\tau_m} \quad \text{for } \tau_d = 0.25\tau \quad (9.55)$$

From Equation (9.55), we have got one triple-folded zeros at the original point $z = 0$ and four poles at $z = e^{-2T/\tau}$ in double-folded, $z = e^{-T/\tau_1}$ and $z = T/\tau_m$ inside the unity-cycle since $\tau_m > T > 0$. Therefore, this open-loop control system is considered stable. The locations of the zeros and poles of an SOH ($\tau_d = 0.25\tau$) with a first-order

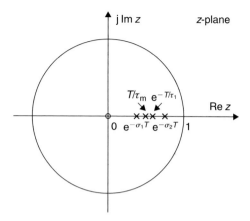

Figure 9.24 Open-loop control of DC/DC converters ($\tau_d < 0.25\tau$) with a first-order load plus an integral element.

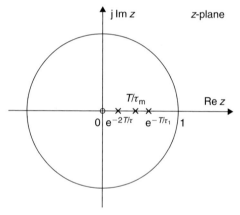

Figure 9.25 Open-loop control of DC/DC converters ($\tau_d = 0.25\tau$) with a first-order load plus an integral element.

circuit are shown in Figure 9.25:

$$G(s)G_1(s)G_m(s) = \frac{1/\tau\tau_d}{(s+\sigma)^2 + \omega^2} \frac{1}{1+s\tau_1} \frac{1}{s\tau_m} \quad \text{for } \tau_d > 0.25\tau \tag{9.56}$$

$$G(z)G_1(z)G_m(z) = \frac{2}{\sqrt{4\tau_d/\tau - 1}} \frac{ze^{-aT}\sin\omega T}{z^2 - 2ze^{-aT}\cos\omega T + e^{-2aT}} \frac{z}{z - e^{-T/\tau_1}}$$

$$\times \frac{z}{z - T/\tau_m} \quad \text{for } \tau_d > 0.25\tau \tag{9.57}$$

where $a = \sigma = 1/2\tau_d$ and $\omega = \sqrt{4\tau\tau_d - \tau^2}/2\tau\tau_d$.

From Equation (9.57), we have got one triple-folded zeros at the original point $z = 0$ and four poles at $z = e^{-T/\tau_1}$ and one couple of conjugated complex poles at

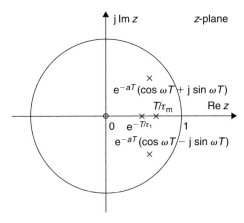

Figure 9.26 Open-loop control of DC/DC converters ($\tau_d > 0.25\tau$) with a first-order load plus an integral element.

$z_1 = e^{-aT}(\cos \omega T + j \sin \omega T)$ and $z_2 = e^{-aT}(\cos \omega T - j \sin \omega T)$, and $z = T/\tau_m$ inside the unity-cycle since $\tau_m > T > 0$. Therefore, this open-loop control system is considered stable. The locations of the zeros and poles of an SOH ($\tau_d > 0.25\tau$) with a first-order circuit are shown in Figure 9.26.

9.3 UNIT-STEP FUNCTION RESPONSES

The unit-step function response analysis of the open-loop systems with the four converters is discussed in detail as follows. The input signal is a unit-step function $V_{in}(s) = 1/s$ for Figures 9.3–9.5.

9.3.1 AC/DC RECTIFIERS

The mathematical model for power AC/DC rectifiers is a ZOH. Its transfer function in the *s*-domain is:

$$G(s) = \frac{1}{s} \tag{9.19}$$

Its transfer function in the *z*-domain is:

$$G(z) = \frac{z}{z - 1} \tag{9.20}$$

AC/DC Rectifiers Open-Loop Analysis

The block diagram is shown in Figure 9.3 with $G(s)$ to be a ZOH. The unit-step function response in the *s*-domain is:

$$V_O(s) = G(s)V_{in}(s) = \frac{1}{s^2} \tag{9.58}$$

The unit-step function response in the time domain is:

$$v_O(t) = t \tag{9.59}$$

It is a linear rising line, so that it is not stable. The unit-step function response in the z-domain:

$$V_O(z) = G(z)V_{in}(z) = \frac{Tz}{(z-1)^2} \tag{9.60}$$

This unit-step function response is not stable.

Analysis of AC/DC Rectifiers with a First-Order Load

The open-loop control scheme for a ZOH with a first-order load is shown in Figure 9.4. The unit-step function response in the s-domain is:

$$V_O(s) = G(s)G_1(s)V_{in}(s) = \frac{1}{s^2}\frac{1}{1+s\tau_1} \tag{9.61}$$

The unit-step function response in the time domain is:

$$v_O(t) = t - \tau_1(1 - e^{-t/\tau_1}) \tag{9.62}$$

It is nearly a linear rising line, so that it is not stable. The unit-step function response in the z-domain:

$$V_O(z) = G(z)G_1(z)V_{in}(z) = \frac{Tz}{(z-1)^2} - \frac{(1 - e^{-T/\tau_1})z/\tau_1}{(z-1)(z-e^{-T/\tau_1})} \tag{9.63}$$

This unit-step function response is not stable.

Analysis of Rectifiers with a First-Order Load Plus an Integral Element

The open-loop control scheme for a ZOH with a first-order circuit plus an integral element is shown in Figure 9.5. The unit-step function response in the s-domain is:

$$V_O(s) = G(s)G_1(s)G_m(s)V_{in}(s) = \frac{1}{s^2}\frac{1}{1+s\tau_1}\frac{1}{s\tau_m} \tag{9.64}$$

The unit-step function response in the time domain is:

$$v_O(t) = \frac{1}{2\tau_m}(t^2 - 2\tau_1 t + 2\tau_1^2 - 2\tau_1^2 e^{-t/\tau_1}) \tag{9.65}$$

It is nearly a linear rising line, so that it is not stable. The unit-step function response in the z-domain:

$$V_O(z) = G(z)G_1(z)G_m(z)V_{in}(z) = \frac{Tz}{\tau_m(z-1)^2}\left(\frac{T}{z-1} + \frac{T-2\tau_1}{2}\right) \tag{9.66}$$

This unit-step function response is not stable.

9.3.2 DC/AC INVERTERS AND AC/AC (AC/DC/AC) CONVERTERS

The mathematical model for power DC/AC inverters and AC/AC (AC/DC/AC) converters is an FOH. Its transfer function in the *s*-domain is:

$$G(s) = \frac{1}{1 + sT} \tag{9.25}$$

Its transfer function in the *z*-domain is:

$$G(z) = \frac{z}{z - 1/e} \tag{9.26}$$

Open-Loop Unit-Step Response Analysis

The block diagram is shown in Figure 9.3 with $G(\cdot)$ to be an FOH. The unit-step function response in the *s*-domain is:

$$V_O(s) = G(s)V_{in}(s) = \frac{1}{s(1 + Ts)} \tag{9.67}$$

The unit-step function response in the time domain is:

$$v_O(t) = 1 - e^{-t/T} \tag{9.68}$$

It is an exponential function (first-inertial element function), so that it is stable. The unit-step function response in the *z*-domain:

$$V_O(z) = G(z)V_{in}(z) = \frac{z(1 - 1/e)}{(z - 1)(z - 1/e)} = \frac{0.632z}{(z - 1)(z - 0.368)} \tag{9.69}$$

This unit-step function response is stable.

Analysis of an FOH with a First-Order Load

The open-loop control scheme for an FOH with a first-order load is shown in Figure 9.4. The unit-step function response in the *s*-domain is:

$$V_O(s) = G(s)G_1(s)V_{in}(s) = \frac{1}{s(1 + sT)} \frac{1}{1 + s\tau_1} \tag{9.70}$$

The unit-step function response in the time domain is:

$$v_O(t) = 1 + \frac{Te^{-t/T} - \tau_1 e^{-t/\tau_1}}{\tau_1 - T} \tag{9.71}$$

It is nearly an exponential function, so that it is stable. The unit-step function response in the z-domain:

$$V_O(z) = G(z)G_1(z)V_{in}(z) = \frac{z}{(z-1)} + \frac{z}{\tau_1 - T}\left(\frac{T}{z - 1/e} - \frac{\tau_1}{z - e^{-T/\tau_1}}\right) \quad (9.72)$$

This unit-step function response is stable.

Analysis of an FOH with a First-Order Load Plus an Integral Element

The open-loop control scheme for an FOH with a first-order circuit plus an integral element is shown in Figure 9.5. The unit-step function response in the s-domain is:

$$V_O(s) = G(s)G_1(s)G_m(s)V_{in}(s) = \frac{1}{s(1+sT)}\frac{1}{1+s\tau_1}\frac{1}{s\tau_m} \quad (9.73)$$

The unit-step function response in the time domain is:

$$v_O(t) = \frac{1}{\tau_m}\left[t - (\tau_1 + T) - \frac{(\tau_1 - T)e^{-t/T}}{\tau_1^2} + \frac{(\tau_1 - T)e^{-t/\tau_1}}{T^2}\right] \quad (9.74)$$

It is nearly a linear rising line, so that it is not stable. The unit-step function response in the z-domain:

$$V_O(z) = G(z)G_1(z)G_m(z)V_{in}(z)$$

$$= \frac{1}{\tau_m}\left[-\frac{(\tau_1 + T)z}{z - 1} + \frac{Tz}{(z-1)^2} - \frac{T^2 z}{(\tau_1 - T)(z - 1/e)} + \frac{\tau_1^2 z}{(\tau_1 - T)(z - e^{-T/\tau_1})}\right] \quad (9.75)$$

This unit-step function response is not stable.

9.3.3 DC/DC CONVERTERS

The mathematical model for power DC/DC converters is an SOH. Its transfer function in the s-domain is:

$$G(s) = \frac{1}{1 + s\tau + s^2\tau\tau_d} \quad (9.31)$$

There are four conditions of the damping time constant τ_d related to the time constant τ:

- $\tau_d = 0$
- $\tau_d < 0.25\tau$
- $\tau_d = 0.25\tau$
- $\tau_d > 0.25\tau$

We list its transfer functions in the s-domain and in the z-domain below:

$$G(s) = \frac{1}{1 + \tau s} \qquad \text{for } \tau_d = 0 \qquad (9.32)$$

$$G(z) = \frac{z}{z - e^{-T/\tau}} \qquad \text{for } \tau_d = 0 \qquad (9.33)$$

$$G(s) = \frac{1/\tau\tau_d}{(s + \sigma_1)(s + \sigma_2)} \qquad \text{for } \tau_d < 0.25\tau \qquad (9.34)$$

$$G(z) = \frac{1/\tau\tau_d}{(\sigma_2 - \sigma_1)} \left(\frac{z}{z - e^{-\sigma_1 T}} - \frac{z}{z - e^{-\sigma_2 T}} \right) \qquad \text{for } \tau_d < 0.25\tau \qquad (9.35)$$

$$G(s) = \frac{1/\tau\tau_d}{(s + \sigma)^2} \qquad \text{for } \tau_d = 0.25\tau \qquad (9.36)$$

$$G(z) = \frac{4Tze^{(-2/\tau)^T}}{(z - e^{(-2/\tau)^T})^2} \qquad \text{for } \tau_d = 0.25\tau \qquad (9.37)$$

$$G(s) = \frac{1/\tau\tau_d}{(s + \sigma)^2 + \omega^2} \qquad \text{for } \tau_d > 0.25\tau \qquad (9.38)$$

$$G(z) = \frac{2}{\sqrt{4\tau_d/\tau - 1}} \frac{ze^{-aT} \sin \omega T}{z^2 - 2ze^{-aT} \cos \omega T + e^{-2aT}} \qquad \text{for } \tau_d > 0.25\tau \qquad (9.39)$$

Converters Open-Loop Analysis

The block diagram is shown in Figure 9.3 with $G(\cdot)$ to be an SOH. Since there are four conditions, we analyze them one by one.

The Condition of $\tau_d = 0$

The unit-step function responses in the s-domain and in the z-domain for the condition of $\tau_d = 0$ are:

$$V_O(s) = G(s)V_{in}(s) = \frac{1}{s(1 + s\tau)} \qquad \text{for } \tau_d = 0 \qquad (9.76)$$

$$V_O(z) = G(z)V_{in}(z) = \frac{z(1 - e^{-T/\tau})}{(z - 1)(z - e^{-T/\tau})} \qquad \text{for } \tau_d = 0 \qquad (9.77)$$

The unit-step response in the time domain is:

$$v_O(t) = 1 - e^{-t/\tau} \qquad (9.78)$$

This is an exponential function, which is stable.

The Condition of $\tau_d < 0.25\tau$

The unit-step function response in the s-domain and in the z-domain for the condition of $\tau_d < 0.25\tau$ are:

$$V_O(s) = G(s)V_{in}(s) = \frac{1/\tau\tau_d}{s(s+\sigma_1)(s+\sigma_2)} \quad \text{for } \tau_d < 0.25\tau \tag{9.79}$$

$$V_O(z) = G(z)V_{in}(z)$$

$$= \frac{z}{(z-1)} + \frac{\sigma_2 z}{(\sigma_1-\sigma_2)(z-e^{-\sigma_1 T})} - \frac{\sigma_1 z}{(\sigma_1-\sigma_2)(z-e^{-\sigma_2 T})} \quad \text{for } \tau_d < 0.25\tau \tag{9.80}$$

The unit-step response in the time domain is:

$$v_O(t) = 1 + \frac{\sigma_2 e^{-\sigma_1 t}}{\sigma_1 - \sigma_2} - \frac{\sigma_1 e^{-\sigma_2 t}}{\sigma_1 - \sigma_2} \tag{9.81}$$

This is an exponential function, which is stable.

The Condition of $\tau_d = 0.25\tau$

The unit-step function response in the s-domain and in the z-domain for the condition of $\tau_d = 0.25\tau$ are:

$$V_O(s) = G(s)V_{in}(s) = \frac{4/\tau^2}{s(s+\tau/2)^2} \quad \text{for } \tau_d = 0.25\tau \tag{9.82}$$

$$V_O(z) = G(z)V_{in}(z) = \frac{z}{z-1} - \frac{z}{z-e^{-2T/\tau}} - \frac{(2T/\tau)e^{-2T/\tau}z}{(z-e^{-2T/\tau})^2} \quad \text{for } \tau_d = 0.25\tau \tag{9.83}$$

The unit-step response in the time domain is:

$$v_O(t) = 1 - \left(1 + \frac{2t}{\tau}\right)e^{-2t/\tau} \tag{9.84}$$

The Condition of $\tau_d > 0.25\tau$

The unit-step function response in the s-domain and in the z-domain for the condition of $\tau_d > 0.25\tau$ are:

$$V_O(s) = G(s)V_{in}(s) = \frac{1/\tau\tau_d}{s[(s+\sigma)^2 + \omega^2]} \quad \text{for } \tau_d > 0.25\tau \tag{9.85}$$

$$V_O(z) = G(z)V_{in}(z) = \frac{z}{z-1} - \frac{z^2 - ze^{-aT}\sec\phi\cos(\omega T+\phi)}{z^2 - 2ze^{-aT}\cos\omega T + e^{-2aT}} \quad \text{for } \tau_d > 0.25\tau \tag{9.86}$$

where $\phi = \tan^{-1}(-a/\omega)$. The unit-step response in the time domain is:

$$v_O(t) = 1 - e^{-at} \sec \phi \cos(\omega t - \phi) \qquad (9.87)$$

This is an exponential function, which is stable.

Analysis of DC/DC Converters with a First-Order Load

The open-loop control scheme for an SOH with a first-order load is shown in Figure 9.4.

The Condition of $\tau_d = 0$

The unit-step function responses in the *s*-domain and in the *z*-domain for the condition of $\tau_d = 0$ are:

$$V_O(s) = G(s)G_1(s)V_{in}(s) = \frac{1}{s(1+s\tau)(1+s\tau_1)} \quad \text{for } \tau_d = 0 \qquad (9.88)$$

$$V_O(z) = G(z)G_1(z)V_{in}(z) = \frac{z}{(z-1)} + \frac{\tau z}{(\tau_1 - \tau)(z - e^{-T/\tau})}$$

$$- \frac{\tau_1 z}{(\tau_1 - \tau)(z - e^{-T/\tau_1})} \quad \text{for } \tau_d = 0 \qquad (9.89)$$

The unit-step response in the time domain is:

$$v_O(t) = 1 - e^{-t/\tau}v_O(t) = 1 + \frac{\tau e^{-t/\tau}}{\tau_1 - \tau} - \frac{\tau_1 e^{-t/\tau_1}}{\tau_1 - \tau} \qquad (9.90)$$

This is an exponential function, which is stable.

The Condition of $\tau_d < 0.25\tau$

The unit-step function response in the *s*-domain and in the *z*-domain for the condition of $\tau_d < 0.25\tau$ are:

$$V_O(s) = G(s)G_1(s)V_{in}(s) = \frac{1/\tau\tau_d}{s(s+\sigma_1)(s+\sigma_2)(1+s\tau_1)} \quad \text{for } \tau_d < 0.25\tau \qquad (9.91)$$

$$V_O(z) = G(z)G_1(z)V_{in}(z) = \frac{z}{(z-1)} - \frac{K_1 z}{z - e^{-\sigma_1 T}} - \frac{K_2 z}{z - e^{-\sigma_2 T}} - \frac{K_3 z}{z - e^{-T/\tau_1}}$$

$$\text{for } \tau_d < 0.25\tau \qquad (9.92)$$

where $K_1 = \dfrac{\sigma_2/\tau_1}{\sigma_1^2 - \sigma_1\sigma_2 + \frac{\sigma_2 - \sigma_1}{\tau_1}}$

$K_2 = \dfrac{\sigma_1/\tau_1}{\sigma_2^2 - \sigma_1\sigma_2 + \frac{\sigma_1 - \sigma_2}{\tau_1}}$

$K_3 = \dfrac{\sigma_1\sigma_2}{\frac{1}{\tau_1^2} - \frac{\sigma_1}{\tau_1} - \frac{\sigma_2}{\tau_1} + \sigma_2\sigma_1}$

The unit-step response in the time domain is:

$$v_O(t) = 1 - K_1 e^{-\sigma_1 t} - K_2 e^{-\sigma_2 t} - K_3 e^{-t/\tau_1} \tag{9.93}$$

This is an exponential function, which is stable.

The Condition of $\tau_d = 0.25\tau$

The unit-step function response in the s-domain and in the z-domain for the condition of $\tau_d = 0.25\tau$ are:

$$V_O(s) = G(s)G_1(s)V_{in}(s) = \frac{4/\tau^2}{s(s + 2/\tau)^2(1 + s\tau_1)} \quad \text{for } \tau_d = 0.25\tau \tag{9.94}$$

$$V_O(z) = G(z)G_1(z)V_{in}(z) = \frac{z}{(z - 1)} + \frac{K_1 z}{z - e^{-2T/\tau}} + \frac{K_2 T z\, e^{-2T/\tau}}{(z - e^{-2T/\tau})^2} - \frac{K_3 z}{z - e^{-T/\tau_1}}$$

for $\tau_d = 0.25\tau$ \hfill (9.95)

where $K_1 = \dfrac{\tau(4\tau_1 - \tau)}{(2\tau_1 - \tau)^2}$

$K_2 = \dfrac{2}{2\tau_1 - \tau}$

$K_3 = \dfrac{4\tau_1/\tau}{2\tau_1 - \tau}$

The unit-step response in the time domain is:

$$v_O(t) = 1 + K_1 e^{-2t/\tau} + K_2 t e^{-2t/\tau} - K_3 e^{-t/\tau_1} \tag{9.96}$$

The Condition of $\tau_d > 0.25\tau$

The unit-step function response in the s-domain and in the z-domain for the condition of $\tau_d > 0.25\tau$ are:

$$V_O(s) = G(s)G_1(s)V_{in}(s) = \frac{1/\tau\tau_d}{s\left[(s+\sigma)^2 + \omega^2\right](1+s\tau_1)} \quad \text{for } \tau_d > 0.25\tau \quad (9.97)$$

$$V_O(z) = G(z)G_1(z)V_{in}(z) = \frac{z}{z-1} - \frac{K_1 z}{z - e^{-T/\tau_1}} + \frac{K_2 z^2 - K_3 z e^{-aT}\cos(\omega T - \phi)}{z^2 - 2z e^{-\sigma T}\cos\omega T + e^{-2\sigma T}}$$

for $\tau_d > 0.25\tau$ (9.98)

where $\quad \phi = \tan^{-1}\left[\dfrac{\sigma^2 - \omega^2 - \sigma/\tau_1}{(2\sigma - 1/\tau_1)\omega}\right]$

$$K_1 = \frac{\sigma^2 + \omega^2}{\sigma^2 + \omega^2 + 1/\tau_1^2 - 2\sigma/\tau_1}$$

$$K_2 = \frac{(2\sigma - 1/\tau_1)/\tau_1}{\sigma^2 + \omega^2 + 1/\tau_1^2 - 2\sigma/\tau_1}$$

$$K_3 = \frac{\sqrt{(\sigma^2 + (-2\sigma/\tau_1))^2 + (\omega/\tau_1)^2} \times 1/\omega\tau_1}{\sigma^2 + \omega^2 + 1/\tau_1^2 - 2\sigma/\tau_1}$$

The unit-step response in the time domain is:

$$v_O(t) = 1 - K_2 e^{-\sigma t} + K_3 \cos(\omega t + \phi) \quad (9.99)$$

This is an exponential function, which is stable.

Analysis of DC/DC Converters with a First-Order Load Plus an Integral Element

The open-loop control scheme for an SOH with a first-order circuit plus an integral element is shown in Figure 9.5.

The Condition of $\tau_d = 0$

The unit-step function responses in the s-domain and in the z-domain for the condition of $\tau_d = 0$ are:

$$V_O(s) = G(s)G_1(s)G_m(s)V_{in}(s) = \frac{1}{s^2\tau_m(1+s\tau)(1+s\tau_1)} \quad \text{for } \tau_d > 0.25\tau \quad (9.100)$$

$$V_O(z) = G(z)G_1(z)G_m(z)V_{in}(z) = -\frac{z}{z-1} + \frac{K_1 T z}{(z-1)^2} - \frac{K_2 z}{z - e^{-T/\tau}} + \frac{K_3 z}{z - e^{-T/\tau_1}}$$

for $\tau_d > 0.25\tau$ (9.101)

where $K_1 = \dfrac{\tau + \tau_1}{\tau_m}$

$K_2 = \dfrac{\tau^2}{(\tau_1^2 - \tau^2)\tau_m}$

$K_3 = \dfrac{\tau_1^2}{(\tau_1^2 - \tau^2)\tau_m}$

The unit-step response in the time domain is:

$$v_O(t) = 1 - e^{-t/\tau} v_O(t) = -1 + K_1 t - K_2 e^{-t/\tau} + K_3 e^{-t/\tau_1} \tag{9.102}$$

This is nearly a linear rising function, which is not stable.

The Condition of $\tau_d < 0.25\tau$

The unit-step function response in the s-domain and in the z-domain for the condition of $\tau_d < 0.25\tau$ are:

$$V_O(s) = G(s)G_1(s)G_m(s)V_{in}(s) = \dfrac{1/\tau\tau_d}{s^2 \tau_m (s + \sigma_1)(s + \sigma_2)(1 + s\tau_1)} \quad \text{for } \tau_d < 0.25\tau \tag{9.103}$$

$$V_O(z) = G(z)G_1(z)V_{in}(z) = -\dfrac{z}{z-1} + \dfrac{K_1 T z}{(z-1)^2} + \dfrac{K_2 z}{z - e^{-\sigma_1 T}} + \dfrac{K_3 z}{z - e^{-\sigma_2 T}}$$

$$+ \dfrac{K_4 z}{z - e^{-T/\tau_1}} \quad \text{for } \tau_d < 0.25\tau \tag{9.104}$$

where $K_1 = \dfrac{1/\tau_m}{\sigma_1 + \sigma_2 + \dfrac{1}{\tau_1}}$

$K_2 = \dfrac{1/\tau_m}{\left(\dfrac{\sigma_2 \tau_1}{\sigma_1^2} - \dfrac{\tau_1}{\sigma_1} - \dfrac{\sigma_2}{\sigma_1} + 1\right)\left(\dfrac{\tau_1}{\sigma_1} + 1 + \dfrac{\sigma_2}{\sigma_1}\right)}$

$K_3 = \dfrac{1/\tau_m}{\left(\dfrac{\sigma_1 \tau_1}{\sigma_2^2} - \dfrac{\tau_1}{\sigma_2} - \dfrac{\sigma_1}{\sigma_2} + 1\right)\left(\dfrac{\tau_1}{\sigma_2} + 1 + \dfrac{\sigma_1}{\sigma_2}\right)}$

$K_4 = \dfrac{1/\tau_m}{\left(\dfrac{\sigma_1 \sigma_2}{\tau_1^2} - \dfrac{\sigma_2}{\tau_1} - \dfrac{\sigma_1}{\tau_1} + 1\right)\left(\dfrac{\sigma_1}{\tau_1} + 1 + \dfrac{\sigma_2}{\tau_1}\right)}$

The unit-step response in the time domain is:

$$v_O(t) = -1 + K_1 t + K_2 e^{-\sigma_1 t} + K_3 e^{-\sigma_2 t} + K_4 e^{-t/\tau_1} \tag{9.105}$$

This is nearly a linear rising function, which is not stable.

The Condition of $\tau_d = 0.25\tau$

The unit-step function response in the s-domain and in the z-domain for the condition of $\tau_d = 0.25\tau$ are:

$$V_O(s) = G(s)G_1(s)G_m(s)V_{in}(s) = \frac{4/\tau^2}{s^2\tau_m(s + 2/\tau)^2(1 + s\tau_1)} \quad \text{for } \tau_d = 0.25\tau$$

$$(9.106)$$

$$V_O(z) = G(z)G_1(z)V_{in}(z)G_m(z) = \frac{z}{(z-1)} + \frac{K_1 Tz}{(z-1)^2} - \frac{K_2 z}{z - e^{-2T/\tau}} - \frac{K_3 Tz\, e^{-2T/\tau}}{(z - e^{-2T/\tau})^2}$$

$$+ \frac{K_4 z}{z - e^{-T/\tau_1}} \quad \text{for } \tau_d = 0.25\tau \qquad (9.107)$$

where $K_1 = \dfrac{1}{(\tau + \tau_1)\tau_m}$

$$K_2 = \frac{\tau^2 \tau_1}{2(\tau + \tau_1)(2\tau_1 - \tau)}$$

$$K_3 = \frac{\tau\tau_1^2}{(\tau + \tau_1)(2\tau_1 - \tau)\tau_m}$$

$$K_4 = \frac{4\tau_1^2/\tau}{(\tau + \tau_1)(2\tau_1 - \tau)^2\tau_m}$$

The unit-step response in the time domain is:

$$v_O(t) = 1 + K_1 t - K_2 e^{-2t/\tau} - K_3 t e^{-2t/\tau} + K_4 e^{-t/\tau_1} \qquad (9.108)$$

This is nearly a linear rising function, which is not stable.

The Condition of $\tau_d > 0.25\tau$

The unit-step function response in the s-domain and in the z-domain for the condition of $\tau_d = 0.25\tau$ are:

$$V_O(s) = G(s)G_1(s)G_m(s)V_{in}(s) = \frac{1/\tau\tau_d}{s^2\tau_m[(s + \sigma)^2 + \omega^2](1 + s\tau_1)} \quad \text{for } \tau_d > 0.25\tau$$

$$(9.109)$$

$$V_O(z) = G(z)G_1(z)G_m(z)V_{in}(z) = -\frac{z}{z-1} + \frac{K_1 Tz}{(z-1)^2} + \frac{K_2 z}{z - e^{-T/\tau_1}}$$

$$- \frac{K_3 z^2 - K_4 z e^{-aT}\cos(\omega T + \phi)}{z^2 - 2z e^{-\sigma T}\cos\omega T + e^{-2\sigma T}} \quad \text{for } \tau_d > 0.25\tau$$

$$(9.110)$$

where $\quad \phi = \tan^{-1}\left[\dfrac{\sigma^3 - 3\sigma\omega^2 - \sigma^2/\tau_1 + \omega^2/\tau_1}{(2\sigma^2 - \omega^2 - 2\sigma/\tau_1)\omega}\right]$

$$K_1 = \frac{\sigma^2 + \omega^2}{\tau_1 \tau_m}$$

$$K_2 = \frac{\left(\sigma^2 + \omega^2\right)^2 / \tau_m}{1/\tau_1^2 - 2\sigma/\tau_1 + \sigma^2 + \omega^2}$$

$$K_3 = \frac{(\sigma^2 + \omega^2)^2(2\sigma^2 - \omega^2 - 2\sigma/\tau_1) \times 1/\tau_1^2 \tau_m}{\sigma^6 - \frac{2\sigma^5}{\tau_1} + \frac{\sigma^4}{\tau_1^2} + 3\sigma^4\omega^2 - \frac{4\sigma^3\omega^2}{\tau_1} + \frac{2\sigma^2\omega^2}{\tau_1^2} + 3\sigma^2\omega^4 - \frac{2\sigma\omega^4}{\tau_1} + \frac{\omega^4}{\tau_1^2} + \omega^6}$$

$$K_4 = \frac{(\sigma^2 + \omega^2)^2\sqrt{(2\sigma^2 - \omega^2 - \frac{2\sigma}{\tau_1})^2\omega^2 + (\sigma^3 - 3\sigma\omega^2 - \frac{\sigma^2}{\tau_1} + \frac{\omega^2}{\tau_1})^2} \times \frac{1}{\omega\tau_1^2\tau_m}}{\sigma^6 - \frac{2\sigma^5}{\tau_1} + \frac{\sigma^4}{\tau_1^2} + 3\sigma^4\omega^2 - \frac{4\sigma^3\omega^2}{\tau_1} + \frac{2\sigma^2\omega^2}{\tau_1^2} + 3\sigma^2\omega^4 - \frac{2\sigma\omega^4}{\tau_1} + \frac{\omega^4}{\tau_1^2} + \omega^6}$$

The unit-step response in the time domain is:

$$v_O(t) = -1 + K_1 t + K_2 e^{-\sigma t} + K_3 e^{-t/\tau_1} + K_4 \cos(\omega t - \phi) \tag{9.111}$$

This is nearly a linear rising function, which is not stable.

9.4 IMPULSE RESPONSES

The impulse responses analysis of the open-loop systems with the four converters is discussed in this section. The interference signal is a unit-delta function $V_{int}(s) = U\delta(t)$ for Figures 9.6–9.8. To simplify the problem and concentrate the impulse response analysis, the input signal is assumed a constant $V_{in}(s) = 0$. Therefore, the converters cannot improve the impulse responses in the open-loop control since the interference signal is added in the output of the DC/DC converters.

9.4.1 IMPULSE RESPONSE OF THE CONVERTER OPEN-LOOP SYSTEMS

Refer to Figure 9.6 the impulse responses in the s-domain and in the z-domain are:

$$V_O(s) = V_{int}(s) = U \tag{9.112}$$

$$V_O(z) = V_{int}(z) = U \tag{9.113}$$

Its transfer function in the time domain is:

$$v_O(t) = U\delta(t) \tag{9.114}$$

9.4.2 IMPULSE RESPONSE OF THE CONVERTER WITH A FIRST-ORDER CIRCUIT

Refer to Figure 9.7 the impulse responses in the s-domain and in the z-domain are:

$$V_O(s) = V_{int}(s)G_1(s) = \frac{U}{1 + s\tau_1} \tag{9.115}$$

$$V_O(z) = V_{int}(z)G_1(z) = \frac{Uz}{z - e^{-T/\tau_1}} \tag{9.116}$$

Its transfer function in the time domain is:

$$v_O(t) = Ue^{-t/\tau_1} \tag{9.117}$$

9.4.3 IMPULSE RESPONSE OF THE CONVERTER WITH A FIRST-ORDER CIRCUIT PLUS AN INTEGRAL ELEMENT

Refer to Figure 9.8 the impulse responses in the s-domain and in the z-domain are:

$$V_O(s) = V_{int}(s)G_1(s)G_m(s) = \frac{U}{(1 + s\tau_1)s\tau_m} \tag{9.118}$$

$$V_O(z) = G_1(z)G_m(z)V_{int}(z) = \frac{Uz(1 - e^{-T/\tau_1})}{\tau_m(z - 1)(z - e^{-T/\tau_1})} \tag{9.119}$$

Its transfer function in the time domain is:

$$v_O(t) = \frac{U}{\tau_m}(1 - e^{-t/\tau_1}) \tag{9.120}$$

9.5 SUMMARY

From the analysis in this chapter, we understood the open-loop control systems have many drawbacks in the aspects: stability, unit-step response and impulse response. Closed-loop control can overcome these problems, and obtain good performances in all aspects.

We use various controllers in the closed-loop control systems. The controllers function not only enhancing the converters' characteristics, but also improving the load's technical features. Therefore, good system qualification can be achieved.

FURTHER READING

1. D'Azzo J. J. and Houpis C. H., *Linear Control System Analysis and Design: Conversional and Modern*, McGraw-Hill, New York, 1988.

2. Kuo B. C., *Discrete-Data Control Systems*, Prentice-Hall, Englewood Cliffs, New Jersey, 1970.
3. Mayback P., *Stochastic Models Estimation and Control*, xx edn. Academic Press, New York, 1982.
4. Jacquot R., *Modern Digital Control Systems*, Dekker, New York, 1981.
5. Boyce C., *Microprocessor and Microcomputers Basics*, Prentice-Hall, Englewood Cliffs, New Jersey, 1979.
6. Klingman E., *Microprocessor System Design*, Prentice-Hall, Englewood Cliffs, New Jersey, 1977.
7. Mano M., *Computer Systems Architecture*, Prentice-Hall, Englewood Cliffs, New Jersey, 1982.
8. Taub H., *Microprocessors*, McGraw-Hill, New York, 1982.
9. Tou J. C., *Modern Control Systems*, McGraw-Hill, New York, 1976.
10. Franklin G. F. and Powell D. C., *Digital Control of Dynamic Systems*, 2nd edn. Addison-Wesley, Reading, MA, 1988.
11. Garrett H., *Analog Systems for Microprocessor and Minicomputers*, Reston, Reston, VA, 1978.
12. VanDoren A., *Data Acquisition Systems*, Reston, Reston, VA, 1982.

Chapter 10

Closed-Loop Control for Digital Power Electronics

After the discussion of open-loop control in Chapter 9, we will carefully discuss the closed-loop control in this chapter.

Although all existing converters applied in industrial applications are stable, they have been working in variable state (stable but shifted) because of the interferences. Particularly, some loads, such as the motor inertia in motor drive systems, are naturally unstable elements. Closed-loop control is necessary to keep converters working in steady state to satisfy the industrial requirements.

10.1 INTRODUCTION

Closed-loop control is applied in most industrial applications to keep converters working in steady state to satisfy the industrial requirements. Traditionally, the proportional-plus-integral (PI) control and proportional-plus-integral-plus-differential (PID) control are very popular in closed-loop control systems.

10.1.1 PI CONTROLLER

A PI controller can be constructed by analog form using operational amplifier (OA) as shown in Figure 10.1. The input signal is $v_{in}(t)$, which inputs into the OA via a resistor R_0, and the OA output signal is $v_O(t)$. The feedback circuit of the OA is an R–C circuit. Its transfer function in time domain is:

$$\frac{v_O(t)}{v_{in}(t)} = \frac{R + \frac{1}{j\omega C}}{R_0} = \frac{R}{R_0}\left(1 + \frac{1}{j\omega RC}\right) = \frac{R}{R_0}\frac{1 + j\omega RC}{j\omega RC} \tag{10.1}$$

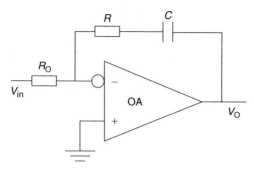

Figure 10.1   Analog PI controller using OA.

Its transfer function in the s-domain is:

$$G_{pi}(s) = \frac{V_O(s)}{V_{in}(s)} = \frac{R}{R_0}\frac{1+sRC}{sRC} = p\frac{1+s\tau}{s\tau} \tag{10.2}$$

where p is the proportional transfer gain, $p = R/R_0$, and τ is the integral time constant, $\tau_i = RC$.

The transfer function in the s-domain can be written in two items as:

$$G_{pi}(s) = p + \frac{p}{s\tau_i} = p + \frac{p_i}{s} \tag{10.3}$$

where $p_i = p/\tau_i$. The first item corresponds to the proportional operation and the second item corresponds to the integral operation with the integral gain, $p_i = p/\tau_i$.

The transfer function in the z-domain can be written in two items as well:

$$G_{pi}(z) = p + p_i\frac{z}{z-1} \tag{10.4}$$

Stability Analysis

From the transfer functions (10.3) and (10.4), we can recognize that the PI controller is an unstable element with the pole on the stability boundary in the s-plane, and on the unity-cycle in the z-plane. Usually, the pole on the stability boundary in the s-plane can be treated as in the right-hand-half-plane (RHHP) in the s-domain.

Unit-Step-Function Responses

The PI controller has the step-function response in the time domain and is shown in Figure 10.2. We can split the waveform in two parts: proportional part and integral part. The proportional part is a constant value which is equal to p at any time. The integral part is a linear line proportional to the time.

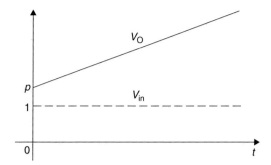

Figure 10.2 Input and output signals of PI controller in the time domain.

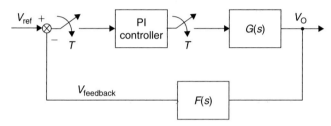

Figure 10.3 Closed-loop control system with a PI controller.

Closed-Loop Control

The PI controller has the main role in a closed-loop control system. The block diagram is shown in Figure 10.3. The regulated converter has its transfer function $G(\cdot)$ and the feedback element with transfer function $F(\cdot)$. The closed-loop transfer function of the whole system is:

$$G_C(s) = \frac{G_{pi}(s)G(s)}{1 + G_{pi}(s)G(s)F(s)} \tag{10.5}$$

This is very popular form for all closed-loop control systems. If the condition

$$G_{pi}(s)G(s)F(s) \gg 1 \tag{10.6}$$

is satisfied. The closed-loop transfer function is determined by the feedback network:

$$G_C(s) = \frac{G_{pi}(s)G(s)}{1 + G_{pi}(s)G(s)F(s)} \approx \frac{1}{F(s)} \tag{10.7}$$

10.1.2 PROPORTIONAL-PLUS-INTEGRAL-PLUS-DIFFERENTIAL CONTROLLER

A proportional-plus-integral-plus-differential (PID) controller can be constructed by analog form using OA as shown in Figure 10.4. The input signal is $v_{in}(t)$, which inputs

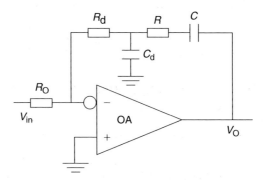

Figure 10.4   Analog PID controller using OA.

into the OA via a resistor R_0, and the OA output signal is $v_O(t)$. The feedback circuit of the OA is an R–C–R_d–C_d circuit. Its transfer function in time domain is:

$$G_{pid}(t) = \frac{v_O(t)}{v_{in}(t)} = \frac{R}{R_0}\frac{(1 + j\omega RC)(1 + j\omega R_d C_d) + j\omega R_d C}{j\omega RC} \qquad (10.8)$$

If the differential resistant R_d is small, i.e. $R_d \ll R$, we can have following expression:

$$G_{pid}(t) = \frac{v_O(t)}{v_{in}(t)} = \frac{R}{R_0}\frac{(1 + j\omega RC)(1 + j\omega R_d C_d)}{j\omega RC} \qquad (10.9)$$

Its transfer function in the s-domain is:

$$G_{pid}(s) = \frac{V_O(s)}{V_{in}(s)} = \frac{R}{R_0}\frac{(1 + sRC)(1 + sR_d C_d)}{sRC} = p\frac{(1 + s\tau_i)(1 + s\tau_d)}{s\tau_i} \qquad (10.10)$$

where p is the proportional transfer gain, $p = R/R_0$, τ_i is the integral time constant $\tau_i = RC$ and τ_d is the differential (off-set) time constant $\tau_d = R_d C_d$. Usually, the integral time constant $\tau_i = RC$ is greater than the differential (off-set) time constant $\tau_d = R_d C_d$, i.e. $\tau_i = \tau_d$. We can rewrite the transfer function in the s-domain in three items as:

$$G_{pid}(s) = p\frac{1 + s\tau_i + s^2\tau_i\tau_d}{s\tau_i} \qquad (10.11)$$

We can write it in three items as well:

$$G_{pid}(s) = p + \frac{p_i}{s} + p_d s \qquad (10.12)$$

where $p_d = p\tau_d$. The first item corresponds to the proportional operation, the second item corresponds to the integral operation with the integral gain p_i and the third item corresponds to the differential operation with the differential gain $p_d = p\tau_d$.

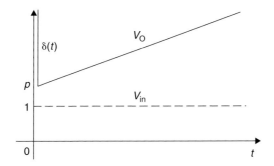

Figure 10.5 Input and output signals of PID controller in the time domain.

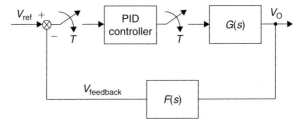

Figure 10.6 Closed-loop control system with a PID controller.

The transfer function in the z-domain can be written in two items as well:

$$G(z) = p + p_i \frac{z}{z-1} + p_d \frac{z-1}{z} \tag{10.13}$$

Stability Analysis

From the transfer functions (10.11)–(10.13) we can recognize that the PID controller is an unstable element with the pole is on the stability boundary in the s-domain and is on the unity-cycle in the z-domain.

Unit-Step-Function Responses

The PID controller has the step-function response in the time domain and is shown in Figure 10.5. We can split the waveform into three parts: proportional part, integral part and differential part. The proportional part is a constant value which is equal to p at any time. The integral part is a linear line proportional to the time. The differential part is a delta function in the time response.

Closed-Loop Control

The PID controller has the main role in a closed-loop control system. The block diagram is shown in Figure 10.6. The regulated converter has its transfer function $G(s)$ and the feedback element with transfer function $F(s)$. The closed-loop transfer function of the whole system is:

$$G_C(s) = \frac{G_{\text{pid}}(s)G(s)}{1 + G_{\text{pid}}(s)G(s)F(s)} \qquad (10.14)$$

This is very popular form for all closed-loop control systems. If the condition

$$G_{\text{pid}}(s)G(s)F(s) \gg 1$$

is satisfied, the closed-loop transfer function is determined by the feedback network:

$$G_C(s) = \frac{G_{\text{pid}}(s)G(s)}{1 + G_{\text{pid}}(s)G(s)F(s)} \approx \frac{1}{F(s)} \qquad (10.15)$$

Particularly the closed transfer function is much different from open-loop transfer function. Most closed-loop control system use PI controller rather than PID controller since PID controller has two forward items to cause unstable.

10.2 PI CONTROL FOR AC/DC RECTIFIERS

PI control is a typical method to improve system characteristics. Applying a PI controller to AC/DC rectifiers can obtain good characteristics. We now discuss the closed-loop control system of the AC/DC rectifiers with a PI controller in this section. The feedback network is assumed as a unity element (i.e. it is not very large). Therefore, the characteristics of the closed-loop system are still depending on the AC/DC rectifiers.

10.2.1 STABILITY ANALYSIS

Figure 10.7 shows the closed-loop control system block diagram of the converter $G(s)$ with a PI controller. The converter $G(s)$ is a zero-order hold (ZOH) simulating AC/DC rectifiers. The feedback network is assumed as a unity element. The system closed-loop transfer function in the s-domain is:

$$G_C(s) = \frac{G_{\text{pi}}(s)G(s)}{1 + G_{\text{pi}}(s)G(s)} = \frac{p\frac{1+s\tau_i}{s\tau_i}}{1 + p\frac{1+s\tau_i}{s\tau_i}} = \frac{1 + s\tau_i}{1 + p_C s\tau_i} \qquad (10.16)$$

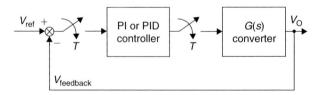

Figure 10.7   PI/PID controlled closed-loop control system of a converter.

where p_C is the closed-loop equivalent proportional gain, $p_C = (p+1)/p$. The closed-loop transfer function is stable. The pole $(-1/p_C\tau_i)$ is further away from the boundary with the comparison to the original PI controller's pole $(-1/\tau_i)$ since $p_C = (p+1)/p > 1$. There is a zero and is the closed-loop transfer function $(-1/\tau_i)$. Therefore, the closed-loop control system has higher stable margin to be stable, and quick response for the step response, since there is an off-set item. The stable state output voltage is still the same since the gain is unity in per-unit system.

The system closed-loop transfer function in the z-domain is:

$$G_C(z) = \mathbf{Z}[G_C(s)] = \mathbf{Z}\left[\frac{1 + s\tau_i}{1 + p_C s\tau_i}\right] = \frac{1}{p_C} + \left(1 - \frac{1}{p_C}\right)\frac{z}{z - e^{-T/p_C\tau_i}} \qquad (10.17)$$

The poles of this system are in the unity-cycle. Therefore, this system is stable. The location of the zero and pole is shown in Figure 10.8.

Analysis of Rectifiers with a First-Order Load

The closed-loop PI control system of the converter $G(s)$ with a first-order load is shown in Figure 10.9. The converter $G(s)$ is a ZOH simulating AC/DC rectifiers. The

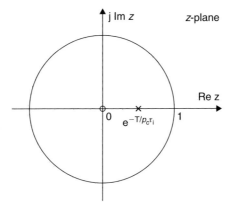

Figure 10.8 The locations of the zero and pole of the PI control closed-loop control system in the z-plane.

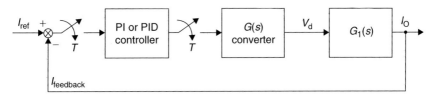

Figure 10.9 PI/PID controlled closed-loop control system of a converter with a first-order load.

closed-loop PI control transfer function of the AC/DC rectifiers with a first-order load in the s-domain is:

$$G_C(s) = \frac{G_{pi}(s)G(s)G_1(s)}{1 + G_{pi}(s)G(s)G_1(s)} = \frac{p\frac{1+s\tau_i}{s\tau_i}\frac{1}{1+s\tau_1}}{1 + p\frac{1+s\tau_i}{s\tau_i}\frac{1}{1+s\tau_1}} = \frac{p(1+s\tau_i)}{s\tau_i(1+s\tau_1) + p(1+s\tau_i)}$$

(10.18)

This is a second-order transfer function with two poles in the left-hand half-plane (LHHP), so that this system is stable. If we carefully select the integral time constant $\tau_i = \tau = L/R$, the items $(1 + s\tau_i)$ in the numerator and the items $(1 + s\tau_1)$ in the denominator can be eliminated each other. Therefore, the closed-loop transfer function can be rewritten as:

$$G_C(s) = \frac{p\frac{1}{s\tau_i}}{1 + p\frac{1}{s\tau_i}} = \frac{p}{s\tau_i + p} = \frac{1}{1 + s\frac{\tau_i}{p}} = \frac{1}{1 + s\tau_e}$$

(10.19)

where $\tau_e = \tau_i/p$ is the equivalent time constant. It means that the closed-loop transfer function is a transfer function of an equivalent first-order element with a smaller time constant $\tau_e = \tau_i/p$ (usually $p > 1$). The transfer function in the z-domain is:

$$G_C(z) = \mathbf{Z}[G_C(s)] = \mathbf{Z}\left[\frac{1}{1 + s\frac{\tau_i}{p}}\right] = \frac{z}{z - e^{-T/\tau_e}}$$

(10.20)

The system has a pole (e^{-T/τ_e}) located inside the unity-cycle further away from the unity-cycle with comparison to the original pole (e^{-T/τ_i}). The location of the zero and pole is shown in Figure 10.10.

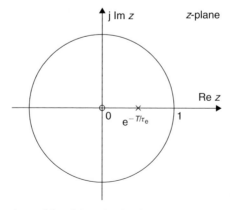

Figure 10.10 The locations of the zero and pole of the PI/PID controlled closed-loop control system of a converter with a first-order load in the z-plane.

Analysis of Rectifiers with a First-Order Load Plus an Integral Element

The closed-loop PI control system of the converter $G(s)$ with a first-order load plus an integral element is shown in Figure 10.11. The converter $G(s)$ is a ZOH simulating AC/DC rectifiers. The closed-loop PI control transfer function of the AC/DC rectifiers with a first-order load plus an integral element in the s-domain is:

$$G_C(s) = \frac{G_{pi}(s)G(s)G_1(s)G_m(s)}{1 + G_{pi}(s)G(s)G_1(s)G_m(s)} = \frac{p\frac{1+s\tau_i}{s\tau_i}\frac{1}{1+s\tau_1}\frac{1}{s\tau_m}}{1 + p\frac{1+s\tau_i}{s\tau_i}\frac{1}{1+s\tau_1}\frac{1}{s\tau_m}}$$

$$= \frac{p(1 + s\tau_i)}{s^2\tau_i\tau_m(1 + s\tau_1) + p(1 + s\tau_i)} \qquad (10.21)$$

This is a third-order transfer function with three poles, so that normally this system is unstable. If we carefully select the integral time constant $\tau_i = \tau_1 = L/R$, the items $(1 + s\tau_i)$ in the numerator and the items $(1 + s\tau_1)$ in the denominator can be eliminated each other. Therefore, the closed-loop transfer function can be rewritten as:

$$G_C(s) = \frac{p\frac{1}{s\tau_i}\frac{1}{s\tau_m}}{1 + p\frac{1}{s\tau_i}\frac{1}{s\tau_m}} = \frac{p}{s^2\tau_i\tau_m + p} = \frac{1}{1 + s^2\frac{\tau_i\tau_m}{p}} = \frac{1}{1 + s^2\tau_a^2} \qquad (10.22)$$

where τ_a is the auxiliary time constant, $\tau_a = \sqrt{\tau_i\tau_m/p}$. It means that the closed-loop transfer function is an equivalent second-order element with a pair of complex poles located on the stability boundary. Usually, this system is considered unstable.

The transfer function in the z-domain is:

$$G_C(z) = \mathbf{Z}[G_C(s)] = \mathbf{Z}\left[\frac{1}{1 + s\frac{\tau_i\tau_m}{p}}\right] = \frac{Tze^{-T/\tau_a}}{(z - e^{-T/\tau_a})^2} \qquad (10.23)$$

where τ_a is the auxiliary time constant, $\tau_a = \sqrt{\tau_i\tau_m/p}$. This system has double-folded poles at the location (e^{-T/τ_a}) inside the stability boundary, the unity-cycle and further away from the unity-cycle with comparison to the original pole (e^{-T/τ_i}). It means that the system in analog control is unstable, but it is stable in digital control. The location of the zero and pole is shown in Figure 10.12.

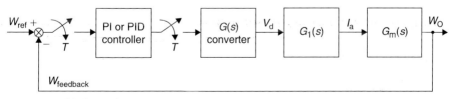

Figure 10.11 PI control closed-loop control system of a converter with a first-order load plus an integral element.

This system is a DC motor variable speed-control system. In order to keep it stable in analog control, usually set double closed-loop control in industrial applications. Considering the motor back electromagnetic force (EMF), we can redraw the system block diagram is shown in Figure 10.13.

The inner-loop control can yields the transfer function $G_{\text{inner}}(s)$ in a first-order circuit. The out-loop control can finally yields the whole system transfer function $G_C(s)$ as a first-order circuit. Therefore, this system becomes stable:

$$G_{\text{inner}}(s) = \frac{p\frac{1}{s\tau_i}}{1 + p\frac{1}{s\tau_i}} = \frac{p}{s\tau_i + p} = \frac{1}{1 + s\frac{\tau_i}{p}} \qquad (10.24)$$

The whole closed-loop transfer function is:

$$G_C(s) = \frac{G_{\text{pi}}(s)G_{\text{inner}}(s)G_i'(s)}{1 + G_{\text{pi}}(s)G_{\text{inner}}(s)G_i'(s)} = \frac{p\frac{1+s\tau_i}{s\tau_i}\frac{1}{1+s\frac{\tau_i}{p}}\frac{1}{1+s\tau_m}}{1 + p\frac{1+s\tau_i}{s\tau_i}\frac{1}{1+s\frac{\tau_i}{p}}\frac{1}{1+s\tau_m}} = \frac{p}{s\tau_m + p} = \frac{1}{1 + s\tau_m'}$$

$$(10.25)$$

where the equivalent time constant τ_m' is equal to τ_m/p. Therefore, the closed-loop transfer function is stable with only one pole at $-1/\tau_m'$ in the LHHP.

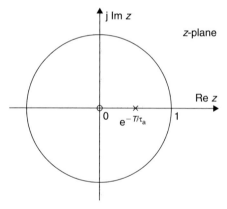

Figure 10.12 The locations of the zero and pole of the PI control closed-loop control system of a converter with a first-order load plus an integral element in the z-plane.

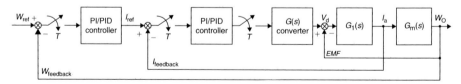

Figure 10.13 Double PI/PID controlled closed-loop control system of AC/DC rectifiers with a first-order load plus an integral element.

10.2.2 Unity-Step Responses

Refer to Figure 10.7 showing the closed-loop control system block diagram of the converter $G(s)$ with a PI controller. The converter $G(s)$ is a ZOH simulating AC/DC rectifiers. The feedback network is assumed as a unity element. The input signal is a unity-step function, $V_{in}(s) = 1/s$. Referring to (10.16), the output signal of the closed-loop system in the s-domain is:

$$V_O(s) = G_C(s)V_{in}(s) = \frac{1}{s}\frac{G_{pi}(s)G(s)}{1+G_{pi}(s)G(s)} = \frac{1}{s}\frac{1+s\tau_i}{1+p_C s\tau_i} \tag{10.26}$$

The unit-step response of the closed-loop transfer function is stable. The pole $(-1/p_C\tau_i)$ is further away from the boundary with the comparison to the original PI controller's pole $(-1/\tau_i)$ since $p_C = (p+1)/p > 1$. There is a zero and is the closed-loop transfer function $(-1/\tau_i)$. Therefore, the closed-loop control system has higher stable margin to be stable, and quick response for the step response, since there is an off-set item. The unit-step response in the time domain is:

$$v_O(t) = \frac{1}{p_C} + \left(1 - \frac{1}{p_C}\right)(1 - e^{-t/p_C\tau_i}) \tag{10.27}$$

The output signal of the closed-loop system is a constant plus an exponential function, so that it is stable.

The output signal of the closed-loop system in the z-domain is:

$$G_C(z) = \mathbf{Z}\left[\frac{1}{s}G_C(s)\right] = \mathbf{Z}\left[\frac{1}{s}\frac{1+s\tau_i}{1+p_C s\tau_i}\right] = \frac{z}{z-1}\left[\frac{1}{p_C} + \left(1 - \frac{1}{p_C}\right)\frac{1 - e^{-T/p_C\tau_i}}{z - e^{-T/p_C\tau_i}}\right] \tag{10.28}$$

The poles of this system are in the unity-cycle. Therefore, this system is stable.

Analysis of Rectifiers with a First-Order Load

The closed-loop PI control system of the converter $G(s)$ with a first-order load is shown in Figure 10.9. The converter $G(s)$ is a ZOH simulating AC/DC rectifiers. The input signal is a unit-step function $I_{in}(s) = 1/s$. We still select the integral time constant $\tau_i = \tau_1 = L/R$. Referring to (10.19), the unit-step response of the closed-loop PI control transfer function of the AC/DC rectifiers with a first-order load in the s-domain is:

$$I_O(s) = G_C(s)I_{in}(s) = \frac{1}{s}\frac{G_{pi}(s)G(s)G_1(s)}{1+G_{pi}(s)G(s)G_1(s)} = \frac{1}{s}\frac{1/(1+s\tau_e)}{1+1/(1+s\tau_e)} = \frac{1}{s}\frac{0.5}{1+0.5s\tau_e} \tag{10.29}$$

This is a second-order transfer function, and it is stable. Therefore, the unit-step response of the closed-loop control system in the time domain is:

$$i_O(t) = 0.5(1 - e^{-2t/\tau_e}) \tag{10.30}$$

It means that the unit-step response of the closed-loop control system is an exponential function, so that it is stable. The unit-step response will have quicker settling process since it has a smaller time constant.

The transfer function in the z-domain is:

$$I_O(z) = \mathbf{Z}\left[\frac{1}{s}\frac{0.5}{1+0.5s\tau_e}\right] = \frac{0.5z(1-e^{-2T/\tau_e})}{(z-1)(z-e^{-2T/\tau_e})} \tag{10.31}$$

The system has a pole (e^{-2T/τ_e}) located inside the unity-cycle further away from the unity-cycle with comparison to the original pole (e^{-2T/τ_i}).

Analysis of Rectifiers with a First-Order Load Plus an Integral Element

The closed-loop PI control system of the converter $G(s)$ with a first-order load plus an integral element is shown in Figure 10.11. The converter $G(s)$ is a ZOH simulating AC/DC rectifiers. The input signal is a unit-step function $\Omega_{in}(s) = 1/s$. We still select the integral time constant $\tau_i = \tau_1 = L/R$. Referring to (10.22), the unit-step response of the closed-loop PI control system of the AC/DC rectifiers with a first-order load plus an integral element in the s-domain is:

$$\Omega_O(s) = G_C(s)\Omega_{in}(s) = \frac{1}{s}\frac{p\frac{1+s\tau_i}{s\tau_i}\frac{1}{1+s\tau_1}\frac{1}{s\tau_m}}{1+p\frac{1+s\tau_i}{s\tau_i}\frac{1}{1+s\tau_1}\frac{1}{s\tau_m}} = \frac{1}{s}\frac{p}{s^2\tau_i\tau_m+p} = \frac{1}{s}\frac{1}{1+s^2\tau_a^2} \tag{10.32}$$

This is a third-order transfer function with three poles, so that normally this system is unstable. The unit-step response of the closed-loop control system in the time domain is:

$$\omega_O(t) = 1 - \cos\frac{t}{\tau_a} \tag{10.33}$$

where τ_a is the auxiliary time constant, $\tau_a = \sqrt{\tau_i\tau_m/p}$. It means that the unit-step response of the closed-loop control system is unstable with constant amplitude oscillation.

The transfer function in the z-domain is:

$$\Omega_O(z) = \mathbf{Z}\left[\frac{1}{s}\frac{1}{1+s^2\tau_a^2}\right] = \frac{z}{z-1}\frac{z^2-z\cos\frac{T}{\tau_a}}{z^2-2z\cos\frac{T}{\tau_a}+1} \tag{10.34}$$

where τ_a is the auxiliary time constant, $\tau_a = \sqrt{\tau_i\tau_m/p}$. The system response in the z-domain has three poles inside the unity-cycle. It means that the system in analog control is unstable, but it is stable in digital control.

This system is a DC motor variable speed-control system. In order to keep it stable in analog control, usually set double closed-loop control in industrial applications. The considering the motor back EMF, we can redraw the system block diagram is shown in Figure 10.13.

The inner-loop control can yields the transfer function $G_{\text{inner}}(s)$ in a first-order circuit. The out-loop control can finally yields the whole system transfer function $G_C(s)$ as a first-order circuit. Therefore, this system becomes stable. The inner-closed-loop transfer function is (10.24) and the whole closed-loop transfer function is (10.25). The unit-step response of the closed-loop control system in the s-domain is:

$$\Omega_O(s) = \frac{1}{s}\frac{1}{1+s\frac{\tau_m}{p}} = \frac{1}{s}\frac{1}{1+s\tau'_m} \tag{10.35}$$

The unit-step response of the double closed-loop control system in the time domain is:

$$\omega_O(t) = 1 - e^{-t/\tau'_m} \tag{10.36}$$

where the equivalent time constant τ'_m is equal to τ_m/p. Therefore, the unit-step response of the closed-loop control system is stable.

The transfer function in the z-domain is:

$$\Omega_O(z) = \mathbf{Z}\left[\frac{1}{s}\frac{1}{1+s\tau'_m}\right] = \frac{z}{z-1}\frac{1-e^{-T/\tau'_m}}{z-e^{-T/\tau'_m}} \tag{10.37}$$

The system response in the z-domain has two poles inside the unity-cycle. It means that the system is stable in digital control.

10.2.3 IMPULSE RESPONSES

Refer to Figure 10.14 showing the closed-loop control system block diagram of the converter $G(s)$ with a PI controller. The converter $G(s)$ is a ZOH simulating AC/DC rectifiers. The feedback network is assumed as a unity element. The interference signal $V_{\text{int}}(s) = U$ is a unit delta function. To simplify the analysis, the input signal is assumed $V_{\text{in}}(s) = 0$. Therefore, the equivalent block diagram is shown in Figure 10.15 for the

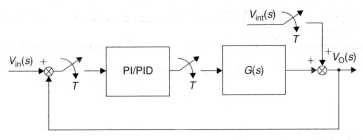

Figure 10.14 PI control closed-loop control system of a converter applying an impulse response.

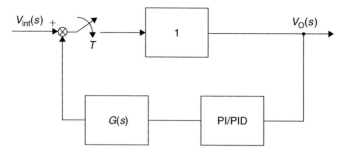

Figure 10.15 Equivalent block diagram of the PI/PID controlled closed-loop control system of a converter applying an impulse response.

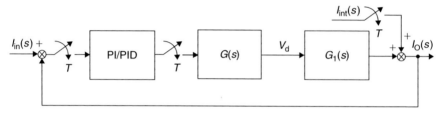

Figure 10.16 PI/PID controlled closed-loop control system of a converter with a first-order load applying an impulse response.

analysis of the impulse response. The output signal of the closed-loop system in the s-domain is:

$$V_O(s) = \frac{V_{int}(s)}{1 + G_{pi}(s)G(s)} = \frac{U}{1 + p\frac{1+s\tau_i}{s\tau_i}} = U\frac{s\tau_i}{p + (1+p)s\tau_i} \tag{10.38}$$

The impulse response in the time domain is:

$$v_O(t) = U\delta(t)\left[\frac{1}{1+p} - \frac{p}{1+p}(1 - e^{-t/(1+p)\tau_i})\right] \tag{10.39}$$

The impulse response is stable and that of the closed-loop system in the z-domain is:

$$V_O(z) = \mathbf{Z}\left[\frac{Us\tau_i}{1 + ps\tau_i}\right] = \frac{U}{p}\left(1 - \frac{z}{z - e^{-T/p\tau_i}}\right) \tag{10.40}$$

The pole of this system is in the unity-cycle. Therefore, this impulse response is stable.

Analysis of Rectifiers with a First-Order Load

Refer to Figure 10.16 showing the closed-loop control system block diagram of the PI controlled converter $G(s)$ with a first-order circuit. The converter $G(s)$ is a ZOH simulating AC/DC rectifiers. The feedback network is assumed as a unity element.

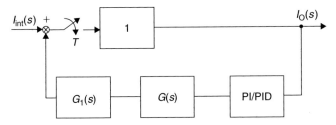

Figure 10.17 Equivalent block diagram of the PI control closed-loop control system of a converter with a first-order load applying an impulse response.

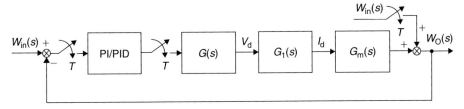

Figure 10.18 PI control closed-loop control system of a converter with a first-order load plus an integral element applying an impulse response.

The interference signal $I_{int}(s) = U$ as a unit delta function. To simplify the analysis, the input signal is assumed $I_{in}(s) = 0$. We still select the PI controller's integral time constant $\tau_i = \tau_1$. Therefore, the equivalent block diagram is shown in Figure 10.17 for the analysis of the impulse response. The output signal of the closed-loop system in the s-domain is:

$$I_O(s) = \frac{U}{1 + G_{pi}(s)G(s)G_1(s)} = \frac{U}{1 + p\frac{1}{s\tau_i}} = \frac{Us\tau_i}{p + s\tau_i} \tag{10.41}$$

The impulse response in the time domain is:

$$i_O(t) = U\delta(t)(1 - e^{-t/p\tau_i}) \tag{10.42}$$

The impulse response is stable and that of the closed-loop system in the z-domain is:

$$I_O(z) = \mathbf{Z}\left[\frac{Us\tau_i}{p + s\tau_i}\right] = U\left(1 - \frac{z}{z - e^{-pT/\tau_i}}\right) \tag{10.43}$$

The pole of this system is in the unity-cycle. Therefore, this impulse response is stable.

Analysis of Rectifiers with a First-Order Load Plus an Integral Element

Refer to Figure 10.18 showing the closed-loop control system block diagram of the PI controlled converter $G(s)$ with a first-order circuit plus an integral element. The converter $G(s)$ is a ZOH simulating AC/DC rectifiers. The feedback network is assumed as

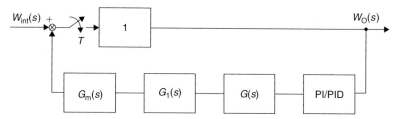

Figure 10.19 Equivalent block diagram of the PI control closed-loop control system of a converter with a first-order load plus an integral element applying an impulse response.

a unity element. The interference signal $\Omega_{int}(s) = U$ as a unit delta function. To simplify the analysis, the input signal is assumed $\Omega_{in}(s) = 0$. We still select the PI controller's integral time constant $\tau_i = \tau_1$. Therefore, the equivalent block diagram is shown in Figure 10.19 for the analysis of the impulse response. The output signal of the closed-loop system in the s-domain is:

$$\Omega_O(s) = \frac{U}{1 + G_{pi}(s)G(s)G_1(s)G_m(s)} = \frac{U}{1 + p\frac{1}{s^2 \tau_i \tau_m}} = \frac{Us^2 \tau_i \tau_m}{p + s^2 \tau_i \tau_m}$$

$$= U \left(1 - \frac{1}{1 + s^2 \tau_a^2} \right) \quad (10.44)$$

where $\tau_a^2 = \tau_i \tau_m / p$. The impulse response in the time domain is:

$$\omega_O(t) = U\delta(t) \left(1 + \sin \frac{t}{\tau_a} \right) \quad (10.45)$$

The impulse response is unstable. The impulse response of the closed-loop system in the z-domain is:

$$\Omega_O(z) = \mathbf{Z} \left[\frac{Us \tau_i \tau_m}{p + s \tau_i \tau_m} \right] = U \left(1 - \frac{z \sin T/\tau_a}{z^2 - 2z \cos T/\tau_a + 1} \right) \quad (10.46)$$

The poles of this system are in the unity-cycle. Therefore, this impulse response is stable.

10.3 PI CONTROL FOR DC/AC INVERTERS AND AC/AC (AC/DC/AC) CONVERTERS

PI control is a typical method to improve the system characteristics. Applying a PI controller to DC/AC inverters and AC/AC (AC/DC/AC) converters can obtain good characteristics. We now discuss the closed-loop control system of the DC/AC inverters and AC/AC (AC/DC/AC) converters with a PI controller in this section. The feedback

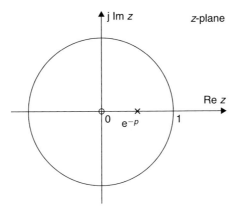

Figure 10.20 The locations of the zero and pole of the PI control closed-loop control system (FOH) in the z-plane.

network is assumed as a unity element (i.e. it is not very large). Therefore, the characteristics of the closed-loop system are still depending on the DC/AC inverters and AC/AC (AC/DC/AC) converters.

10.3.1 STABILITY ANALYSIS

Refer to Figure 10.7 showing the closed-loop control system block diagram of the converter $G(s)$ with a PI controller. The converter $G(s)$ is a first-order hold (FOH) with the sampling interval T to simulate the DC/AC inverters and AC/AC (AC/DC/AC) converters. The feedback network is assumed as a unity element. We still select the PI controller's integral time constant $\tau_i = T$. The system closed-loop transfer function in the s-domain is:

$$G_C(s) = \frac{G_{pi}(s)G(s)}{1 + G_{pi}(s)G(s)} = \frac{p\frac{1+s\tau_i}{s\tau_i}\frac{1}{1+sT}}{1 + p\frac{1+s\tau_i}{s\tau_i}\frac{1}{1+sT}} = \frac{1}{1 + s\frac{T}{p}} \tag{10.47}$$

The closed-loop transfer function is stable. The pole $(-p/T)$ is further away from the boundary with the comparison to the original PI controller's pole $(-1/\tau_i)$ since $p > 1$. Therefore, the closed-loop control system has higher stable margin to be stable, and quick response for the step response. The stable state output voltage is still the same since the gain is unity in per-unit system.

The system closed-loop transfer function in the z-domain is:

$$G_C(z) = \mathbf{Z}[G_C(s)] = \mathbf{Z}\left[\frac{1}{1 + s\frac{T}{p}}\right] = \frac{z}{z - e^{-p}} \tag{10.48}$$

The pole of this system is inside the unity-cycle. Therefore, this system is stable. The location of the zero and pole is shown in Figure 10.20.

Analysis of Rectifiers with a First-Order Load

The closed-loop PI control system of the converter $G(s)$ with a first-order load is shown in Figure 10.9. The converter $G(s)$ is an FOH simulating DC/AC inverters and AC/AC (AC/DC/AC) converters. The feedback network is assumed as a unity element. We still select the PI controller's integral time constant $\tau_i = \tau_1$. The system closed-loop transfer function in the s-domain is:

$$G_C(s) = \frac{G_{pi}(s)G(s)G_1(s)}{1 + G_{pi}(s)G(s)G_1(s)} = \frac{p \frac{1+s\tau_i}{s\tau_i} \frac{1}{1+s\tau_1} \frac{1}{1+sT}}{1 + p \frac{1}{s\tau_i} \frac{1}{1+s\tau_1} \frac{1}{1+sT}} = \frac{p}{s\tau_1(1+sT)+p} \tag{10.49}$$

This is a second-order transfer function with two poles in the LHHP, so that this system is stable. Since the time constant τ_1 of the first-order circuit is usually much greater than the sampling interval T (i.e. $\tau_1 \gg T$), we can ignore the item $s^2\tau_1 T$ in (10.49). Therefore, it can be rewritten as:

$$G_C(s) = \frac{G_{pi}(s)G(s)G_1(s)}{1 + G_{pi}(s)G(s)G_1(s)} = \frac{p}{p + s\tau_1} = \frac{1}{1 + s\tau_1/p} \tag{10.50}$$

It is likely a first-order element. It means that the closed-loop transfer function is a transfer function of an equivalent first-order element with a smaller time constant τ_1/p (usually $p > 1$). The transfer function in the z-domain is:

$$G_C(z) = \mathbf{Z}[G_C(s)] = \mathbf{Z}\left[\frac{1}{1 + s\frac{\tau_1}{p}}\right] = \frac{z}{z - e^{-pT/\tau_1}} \tag{10.51}$$

The system has a pole (e^{-pT/τ_1}) located inside the unity-cycle further away from the unity-cycle with comparison to the original pole (e^{-T/τ_1}). The location of the zero and pole is shown in Figure 10.21.

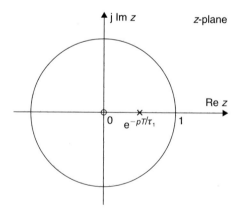

Figure 10.21   The locations of the zero and pole of the PI control closed-loop control system (FOH with a first-order circuit) in the z-plane.

Analysis of Rectifiers with a First-Order Load Plus an Integral Element

The closed-loop PI control system of the converter $G(s)$ with a first-order load plus an integral element is shown in Figure 10.11. The converter $G(s)$ is an FOH simulating DC/AC inverters and AC/AC (AC/DC/AC) converters. The feedback network is assumed as a unity element. We still select the PI controller's integral time constant $\tau_i = \tau_1$. The system closed-loop transfer function in the s-domain is:

$$G_C(s) = \frac{G_{pi}(s)G(s)G_1(s)G_m(s)}{1 + G_{pi}(s)G(s)G_1(s)G_m(s)} = \frac{p\frac{1+s\tau_i}{s\tau_i}\frac{1}{1+s\tau_1}\frac{1}{1+sT}\frac{1}{s\tau_m}}{1 + p\frac{1+s\tau_i}{s\tau_i}\frac{1}{1+s\tau_1}\frac{1}{1+sT}\frac{1}{s\tau_m}}$$

$$= \frac{p}{s^2\tau_1\tau_m(1+sT)+p} \qquad (10.52)$$

This is a third-order transfer function with three poles, so that normally this system is unstable. Since the time constant τ_1 of the first-order circuit is usually much greater than the sampling interval T (i.e. $\tau_1 \gg T$), we can ignore the item sT in (10.52). Therefore, it can be rewritten as:

$$G_C(s) = \frac{p}{s^2\tau_1\tau_m + p} = \frac{1}{1+s^2\frac{\tau_1\tau_m}{p}} = \frac{1}{1+s^2\tau_a^2} \qquad (10.53)$$

where τ_a is the auxiliary time constant, $\tau_a = \sqrt{\tau_1\tau_m/p}$. It means that the closed-loop transfer function is an equivalent second-order element with a pair of complex poles located on the stability boundary.

The transfer function in the z-domain is:

$$G_C(z) = \mathbf{Z}[G_C(s)] = \mathbf{Z}\left[\frac{1}{1+s\frac{\tau_1\tau_m}{p}}\right] = \frac{Tze^{-T/\tau_a}}{(z-e^{-T/\tau_a})^2} \qquad (10.54)$$

This system has double-folded poles at the location (e^{-T/τ_a}) inside the unity-cycle. It means that the system in analog control is unstable, but it is stable in digital control. The location of the zero and pole is shown in Figure 10.22.

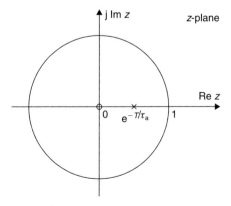

Figure 10.22 The locations of the zero and pole of the PI control closed-loop control system (FOH with a first-order circuit plus an integral element) in the z-plane.

10.3.2 UNIT-STEP RESPONSE FOR PI CONTROLLED DC/AC INVERTERS AND AC/AC (AC/DC/AC) CONVERTERS

Refer to Figure 10.7 showing the closed-loop control system block diagram of the converter $G(s)$ with a PI controller. The converter $G(s)$ is an FOH simulating DC/AC inverters and AC/AC (AC/DC/AC) converters. The feedback network is assumed as a unity element. The input signal is a unity-step function, $V_{in}(s) = 1/s$. Referring to (10.47), we still select $\tau_i = T$ and obtain the output signal of the closed-loop system in the s-domain:

$$V_O(s) = G_C(s)V_{in}(s) = \frac{1}{s}\frac{G_{pi}(s)G(s)}{1+G_{pi}(s)G(s)} = \frac{1}{s}\frac{1}{1+s\frac{T}{p}} \tag{10.55}$$

The unit-step response of the closed-loop transfer function is stable. The pole $(-p/T)$ is further away from the boundary with the comparison to the original PI controller's pole $(-1/\tau_i)$. The unit-step response in the time domain is:

$$v_O(t) = 1 - e^{-pt/T} \tag{10.56}$$

The output signal of the closed-loop system is an exponential function, so that it is stable.

The unit-step response of the closed-loop system in the z-domain is:

$$G_C(z) = \mathbf{Z}\left[\frac{1}{s}G_C(s)\right] = \mathbf{Z}\left[\frac{1}{s}\frac{1}{1+s\frac{T}{p}}\right] = \frac{z}{z-1}\frac{1-e^{-p}}{z-e^{-p}} \tag{10.57}$$

The pole of this system is in the unity-cycle. Therefore, this system is stable.

Analysis of an FOH with a First-Order Load

The closed-loop PI control system of the converter $G(s)$ with a first-order load is shown in Figure 10.9. The converter $G(s)$ is an FOH simulating DC/AC inverters and AC/AC (AC/DC/AC) converters. The input signal is a unit-step function $I_{in}(s) = 1/s$. We still select the integral time constant $\tau_i = \tau_1 = L/R$. Referring to (10.50) and considering $\tau_1 \gg T$, the unit-step response of the closed-loop PI control transfer function of the FOH with a first-order load in the s-domain is:

$$I_O(s) = G_C(s)I_{in}(s) = \frac{1}{s}\frac{G_{pi}(s)G(s)G_1(s)}{1+G_{pi}(s)G(s)G_1(s)} = \frac{1}{s}\frac{1}{1+s\tau_1/p} \tag{10.58}$$

This unit-step response is stable. Therefore, the unit-step response of the closed-loop control system in the time domain is:

$$i_O(t) = 1 - e^{-pt/\tau_1} \tag{10.59}$$

It means that the unit-step response of the closed-loop control system is an exponential function, so that it is stable. The unit-step response will have quicker settling process since it has a smaller time constant, τ_1/p.

The transfer function in the z-domain is:

$$I_O(z) = \mathbf{Z}\left[\frac{1}{s}\frac{1}{1+s\tau_1/p}\right] = \frac{z(1 - e^{-pT/\tau_1})}{(z-1)(z - e^{-pT/\tau_1})} \tag{10.60}$$

The system has a pole (e^{-pT/τ_1}) located inside the unity-cycle further away from the unity-cycle with comparison to the original pole (e^{-T/τ_1}).

Analysis of the FOH with a First-Order Load Plus an Integral Element

The closed-loop PI control system of the converter $G(s)$ with a first-order load plus an integral element is shown in Figure 10.11. The converter $G(s)$ is an FOH simulating DC/AC inverters and AC/AC (AC/DC/AC) converters. The input signal is a unit-step function $\Omega_{in}(s) = 1/s$. We still select the integral time constant $\tau_i = \tau_1 = L/R$. Referring to (10.53) and considering $\tau_1 \gg T$, the unit-step response of the closed-loop PI control system of the FOH with a first-order load plus an integral element in the s-domain is:

$$\Omega_O(s) = G_C(s)\Omega_{in}(s) = \frac{1}{s}\frac{1}{1 + s^2\tau_a^2} \tag{10.61}$$

This is a third-order transfer function with three poles, so that normally this unit-step response is unstable. The unit-step response of the closed-loop control system in the time domain is:

$$\omega_O(t) = 1 - \cos\frac{t}{\tau_a} \tag{10.62}$$

where τ_a is the auxiliary time constant, $\tau_a = \sqrt{\tau_1\tau_m/p}$. It means that the unit-step response of the closed-loop control system is unstable with constant amplitude oscillation.

The transfer function in the z-domain is:

$$\Omega_O(z) = \mathbf{Z}\left[\frac{1}{s}\frac{1}{1 + s^2\tau_a^2}\right] = \frac{z}{z-1}\frac{z^2 - z\cos\frac{T}{\tau_a}}{z^2 - 2z\cos\frac{T}{\tau_a} + 1} \tag{10.63}$$

The system response in the z-domain has three poles inside the unity-cycle. It means that the system in analog control is unstable, but it is stable in digital control.

10.3.3 IMPULSE RESPONSE FOR PI CONTROLLED DC/AC INVERTERS AND AC/AC (AC/DC/AC) CONVERTERS

Refer to Figure 10.14 showing the closed-loop control system block diagram of the converter $G(s)$ with a PI controller. The converter $G(s)$ is an FOH simulating DC/AC inverters and AC/AC (AC/DC/AC) converters. The feedback network is assumed as a unity element. The interference signal $V_{int}(s) = U$ is a unit delta function. To simplify the analysis, the input signal is assumed $V_{in}(s) = 0$. Therefore, the equivalent block diagram is shown in Figure 10.15 for the analysis of the impulse response. We still select the integral time constant $\tau_i = T$. The output signal of the closed-loop system in the s-domain is:

$$V_O(s) = \frac{V_{int}(s)}{1 + G_{pi}(s)G(s)} = \frac{U}{1 + p\frac{1+s\tau_i}{s\tau_i}\frac{1}{1+sT}} = U\frac{sT}{p+sT} = U\left(1 - \frac{1}{1+sT/p}\right)$$

(10.64)

The impulse response in the time domain is:

$$v_O(t) = U\delta(t)(1 - e^{-pt/T})$$

(10.65)

The impulse response is stable and that of the closed-loop system in the z-domain is:

$$V_O(z) = \mathbf{Z}\left[U\left(1 - \frac{1}{1+sT/p}\right)\right] = U\left(1 - \frac{z}{z - e^{-p}}\right)$$

(10.66)

The pole of this system is in the unity-cycle. Therefore, this impulse response is stable.

Analysis of the FOH with a First-Order Load

Refer to Figure 10.16 showing the closed-loop control system block diagram of the PI controlled converter $G(s)$ with a first-order circuit. The converter $G(s)$ is an FOH simulating DC/AC inverters and AC/AC (AC/DC/AC) converters. The feedback network is assumed as a unity element. The interference signal $I_{int}(s) = U$ as a unit delta function. To simplify the analysis, the input signal is assumed $I_{in}(s) = 0$. We still select the PI controller's integral time constant $\tau_i = \tau_1$ with the condition $\tau_1 \gg T$. Therefore, the equivalent block diagram is shown in Figure 10.17 for the analysis of the impulse response. Using the closed-loop transfer function (10.50), we obtain the output signal of the closed-loop system in the s-domain:

$$I_O(s) = \frac{U}{1 + G_{pi}(s)G(s)G_1(s)} = \frac{U}{1 + p\frac{1}{s\tau_1}\frac{1}{1+sT}} \approx \frac{Us\tau_1}{p+s\tau_1} = U\left(1 - \frac{1}{1+s\tau_1/p}\right)$$

(10.67)

The impulse response in the time domain is:

$$i_O(t) = U\delta(t)(1 - e^{-pt/\tau_1})$$

(10.68)

The impulse response is stable and that of the closed-loop system in the z-domain is:

$$I_O(z) = \mathbf{Z}\left[U\left(1 - \frac{1}{1+s\tau_1/p}\right)\right] = U\left(1 - \frac{z}{z - e^{-pT/\tau_1}}\right)$$

(10.69)

The pole of this system is in the unity-cycle. Therefore, this impulse response is stable.

Analysis of the FOH with a First-Order Load Plus an Integral Element

Refer to Figure 10.18 showing the closed-loop control system block diagram of the PI controlled converter $G(s)$ with a first-order circuit plus an integral element. The converter $G(s)$ is an FOH simulating DC/AC inverters and AC/AC (AC/DC/AC) converters. The feedback network is assumed as a unity element. The interference signal $\Omega_{int}(s) = U$ as a unit delta function. To simplify the analysis, the input signal is assumed $\Omega_{in}(s) = 0$. We still select the PI controller's integral time constant $\tau_i = \tau_1$ with the condition $\tau_1 \gg T$. Therefore, the equivalent block diagram is shown in Figure 10.19 for the analysis of the impulse response. Using the closed-loop transfer function (10.53), we obtain the output signal of the closed-loop system in the s-domain:

$$\Omega_O(s) = \frac{U}{1 + G_{pi}(s)G(s)G_1(s)G_m(s)} = \frac{U}{1 + p\frac{1}{s^2\tau_1\tau_m}\frac{1}{1+sT}} \approx \frac{Us^2\tau_1\tau_m}{p + s^2\tau_1\tau_m}$$

$$= U\left(1 - \frac{1}{1 + s^2\tau_a^2}\right) \qquad (10.70)$$

where $\tau_a^2 = \tau_1\tau_m/p$. The impulse response in the time domain is:

$$\omega_O(t) = U\delta(t)\left(1 + \sin\frac{t}{\tau_a}\right) \qquad (10.71)$$

The impulse response is unstable. The impulse response of the closed-loop system in the z-domain is:

$$\Omega_O(z) = \mathbf{Z}\left[\frac{Us\tau_i\tau_m}{p + s\tau_i\tau_m}\right] = U\left(1 - \frac{z\sin T/\tau_a}{z^2 - 2z\cos T/\tau_a + 1}\right) \qquad (10.72)$$

The poles of this system are in the unity-cycle. Therefore, this impulse response is stable.

10.4 PID CONTROL FOR DC/DC CONVERTERS

PID control is a typical method to improve system characteristics. Applying a PID controller to DC/DC converters can obtain good characteristics. We now discuss the closed-loop control system of the DC/DC converters with a PID controller in this section. The feedback network is assumed as a unity element (i.e. it is not very large). Therefore, the characteristics of the closed-loop system are still depending on the DC/DC converters.

10.4.1 Stability Analysis of PID Controlled DC/DC Converters

Refer to Figure 10.7 showing the closed-loop control system block diagram of the converter $G(s)$ with a PID controller. The converter $G(s)$ is a second-order hold (SOH) with the sampling interval T to simulate the DC/DC converters. The feedback network is assumed as a unity element. We still select the PID controller's integral time constant $\tau_i = \tau$ and differential time constant τ_d to be equal to the damping time constant τ_d of the converter $G(s)$. The system closed-loop transfer function in the s-domain is:

$$G_C(s) = \frac{G_{\text{pid}}(s)G(s)}{1 + G_{\text{pid}}(s)G(s)} = \frac{p\frac{1+s\tau_i+s^2\tau_i\tau_d}{s\tau_i}\frac{1}{1+s\tau+s^2\tau\tau_d}}{1 + p\frac{1+s\tau_i+s^2\tau_i\tau_d}{s\tau_i}\frac{1}{1+s\tau+s^2\tau\tau_d}} = \frac{1}{1+s\frac{\tau}{p}} \tag{10.73}$$

The closed-loop transfer function is stable. The pole $(-p/\tau)$ is further away from the boundary with the comparison to the original PID controller's pole $(-1/\tau)$ since $p > 1$. Therefore, the closed-loop control system has higher stable margin, and quick response for the step response. The stable state output voltage is still the same since the gain is unity in per-unit system.

The system closed-loop transfer function in the z-domain is:

$$G_C(z) = \mathbf{Z}[G_C(s)] = \mathbf{Z}\left[\frac{1}{1+s\frac{\tau}{p}}\right] = \frac{z}{z - e^{-pT/\tau}} \tag{10.74}$$

The pole of this system is inside the unity-cycle. Therefore, this system is stable. The location of the zero and pole is shown in Figure 10.23.

Analysis of DC/DC Converters with a First-Order Load

The closed-loop PID control system of the converter $G(s)$ with a first-order load is shown in Figure 10.9. The converter $G(s)$ is an SOH simulating DC/DC converters. The feedback network is assumed as a unity element. We still select the PID controller's integral time constant τ_i to be equal to the time constant τ, and the differential time

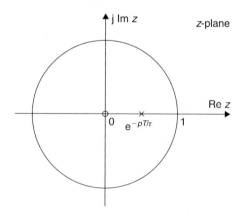

Figure 10.23 The locations of the zero and pole of the PI controlled closed-loop control system (SOH) in the z-plane.

constant τ_d to be equal to the damping time constant τ_d of the converter $G(s)$. The system closed-loop transfer function in the s-domain is:

$$G_C(s) = \frac{G_{pid}(s)G(s)G_1(s)}{1 + G_{pid}(s)G(s)G_1(s)} = \frac{p\frac{1+s\tau_i+s^2\tau_i\tau_d}{s\tau_i}\frac{1}{1+s\tau+s^2\tau\tau_d}\frac{1}{1+s\tau_1}}{1 + p\frac{1+s\tau_i+s^2\tau_i\tau_d}{s\tau_i}\frac{1}{1+s\tau+s^2\tau\tau_d}\frac{1}{1+s\tau_1}}$$

$$= \frac{p}{s\tau(1+s\tau_1)+p} \tag{10.75}$$

This is a second-order transfer function with two poles in the LHHP, so that this system is stable. Since the time constant τ of the converter $G(s)$ is usually greater than the time constant τ_1 of the first-order circuit (i.e. $\tau > \tau_1$), we can ignore the item $s\tau_1$ in (10.75). Therefore, it can be rewritten as:

$$G_C(s) = \frac{G_{pid}(s)G(s)G_1(s)}{1 + G_{pid}(s)G(s)G_1(s)} \approx \frac{1}{1+s\tau/p} \tag{10.76}$$

It is likely a first-order element. It means that the closed-loop transfer function is a transfer function of an equivalent first-order element with a smaller time constant τ/p (usually $p > 1$). The transfer function in the z-domain is:

$$G_C(z) = \mathbf{Z}[G_C(s)] = \mathbf{Z}\left[\frac{1}{1+s\frac{\tau}{p}}\right] = \frac{z}{z - e^{-pT/\tau}} \tag{10.77}$$

This system has a pole ($e^{-pT/\tau}$) located inside the unity-cycle further away from the unity-cycle. The location of the zero and pole is shown in Figure 10.24.

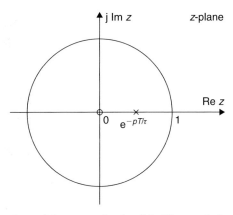

Figure 10.24 The locations of the zero and pole of the PI control closed-loop control system (SOH with a first-order circuit) in the z-plane.

Analysis of DC/DC Converters with a First-Order Load Plus an Integral Element

The closed-loop PID control system of the converter $G(s)$ with a first-order load plus an integral element is shown in Figure 10.11. The converter $G(s)$ is an SOH simulating DC/DC converters. The feedback network is assumed as a unity element. We still select the PI controller's integral time constant $\tau_i = \tau$ and the differential time constant τ_d to be equal to the time constant τ of converter $G(s)$. The system closed-loop transfer function in the s-domain is:

$$G_C(s) = \frac{G_{pi}(s)G(s)G_1(s)G_m(s)}{1 + G_{pi}(s)G(s)G_1(s)G_m(s)} = \frac{p\frac{1}{s\tau}\frac{1}{1+s\tau_1}\frac{1}{s\tau_m}}{1 + p\frac{1}{s\tau}\frac{1}{1+s\tau_1}\frac{1}{s\tau_m}} = \frac{p}{s^2\tau\tau_m(1 + s\tau_1) + p}$$

(10.78)

This is a third-order transfer function with three poles, so that normally this system is unstable. Since the time constant τ of the converter $G(s)$ is usually greater than the time constant τ_1 of the first-order circuit (i.e. $\tau > \tau_1$), we can ignore the item $s\tau_1$ in (10.78). Therefore, it can be rewritten as:

$$G_C(s) = \frac{p}{s^2\tau\tau_m + p} = \frac{1}{1 + s^2\frac{\tau\tau_m}{p}} = \frac{1}{1 + s^2\tau_a^2}$$

(10.79)

where τ_a is the auxiliary time constant, $\tau_a = \sqrt{\tau\tau_m/p}$. It means that the closed-loop transfer function is an equivalent second-order element with a pair of complex poles located on the stability boundary.

The transfer function in the z-domain is:

$$G_C(z) = Z[G_C(s)] = Z\left[\frac{1}{1 + s\frac{\tau\tau_m}{p}}\right] = \frac{Tze^{-T/\tau_a}}{(z - e^{-T/\tau_a})^2}$$

(10.80)

This system has double-folded poles at the location (e^{-T/τ_a}) inside the unity-cycle. It means that the system in analog control is unstable, but it is stable in digital control. The location of the zero and pole is shown in Figure 10.25.

10.4.2 UNIT-STEP RESPONSE FOR PID CONTROLLED DC/DC CONVERTERS

Refer to Figure 10.7 showing the closed-loop control system block diagram of the converter $G(s)$ with a PID controller. The converter $G(s)$ is an SOH simulating DC/DC converters. The feedback network is assumed as a unity element. The input signal is a unity-step function, $V_{in}(s) = 1/s$. Referring to (10.73), we still select the PID controller's integral time constant $\tau_i = \tau$ and differential time constant τ_d to be equal to the damping time constant τ_d of the converter $G(s)$. We then obtain the unit-step response of the

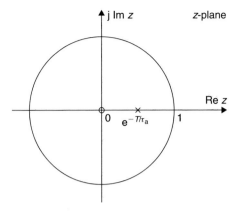

Figure 10.25 The locations of the zero and pole of the PI control closed-loop control system (SOH with a first-order circuit plus an integral element) in the z-plane.

closed-loop system in the s-domain:

$$V_O(s) = G_C(s)V_{in}(s) = \frac{1}{s}\frac{G_{pid}(s)G(s)}{1 + G_{pid}(s)G(s)} = \frac{1}{s}\frac{1}{1 + s\frac{\tau}{p}} \tag{10.81}$$

Since,

$$G_{pid}(s) = p\frac{1 + s\tau_i + s^2\tau_i\tau_d'}{s\tau_i}$$

$$G(s) = \frac{1}{1 + s\tau + s^2\tau\tau_d}$$

where τ_i is the integral time constant and τ_d' is the differential time constant of the PID controller, τ is the time constant and τ_d is the damping time constant of the DC/DC converter. We select $\tau_i = \tau$ and $\tau_d' = \tau_d$, and then obtain

$$G_{pid}(s)G(s) = p\frac{1 + s\tau_i + s^2\tau_i\tau_d'}{s\tau_i}\frac{1}{1 + s\tau + s^2\tau\tau_d} = \frac{p}{s\tau}$$

The unit-step response of the closed-loop transfer function is stable. The pole $(-p/\tau)$ is further away from the boundary with the comparison to the original PI controller's pole $(-1/\tau)$. The unit-step response in the time domain is:

$$v_O(t) = 1 - e^{-pt/\tau} \tag{10.82}$$

The output signal of the closed-loop system is an exponential function, so that it is stable.

The unit-step response of the closed-loop system in the z-domain is:

$$G_C(z) = \mathbf{Z}\left[\frac{1}{s}G_C(s)\right] = \mathbf{Z}\left[\frac{1}{s}\frac{1}{1+s\frac{\tau}{p}}\right] = \frac{z}{z-1}\frac{1-e^{-pT/\tau}}{z-e^{-pT/\tau}} \qquad (10.83)$$

The pole of this system is in the unity-cycle. Therefore, this system is stable.

Analysis of an SOH with a First-Order Load

The closed-loop PID control system of the converter $G(s)$ with a first-order load is shown in Figure 10.9. The converter $G(s)$ is an SOH simulating DC/DC converters. The input signal is a unit-step function $I_{in}(s) = 1/s$. We still select the PID controller's integral time constant $\tau_i = \tau$ and differential time constant τ_d to be equal to the damping time constant τ_d of the converter $G(s)$. Referring to (10.76) and considering $\tau > \tau_1$, the unit-step response of the closed-loop PID control transfer function of the SOH with a first-order load in the s-domain is:

$$I_O(s) = G_C(s)I_{in}(s) = \frac{1}{s}\frac{G_{pid}(s)G(s)G_1(s)}{1 + G_{pid}(s)G(s)G_1(s)} = \frac{1}{s}\frac{1}{1 + s\tau/p} \qquad (10.84)$$

This unit-step response is stable. Therefore, the unit-step response of the closed-loop control system in the time domain is:

$$i_O(t) = 1 - e^{-pt/\tau} \qquad (10.85)$$

It means that the unit-step response of the closed-loop control system is an exponential function, so that it is stable. The unit-step response will have quicker settling process since it has a smaller time constant, τ/p.

The transfer function in the z-domain is:

$$I_O(z) = \mathbf{Z}\left[\frac{1}{s}\frac{1}{1 + s\tau/p}\right] = \frac{z(1 - e^{-pT/\tau})}{(z-1)(z - e^{-pT/\tau})} \qquad (10.86)$$

The system has a pole $(e^{-pT/\tau})$ located inside the unity-cycle further away from the unity-cycle with comparison to the original pole $(e^{-T/\tau})$.

Analysis of the SOH with a First-Order Load Plus an Integral Element

The closed-loop PI control system of the converter $G(s)$ with a first-order load plus an integral element is shown in Figure 10.11. The converter $G(s)$ is an SOH simulating DC/DC converters. The input signal is a unit-step function $\Omega_{in}(s) = 1/s$. We still select the PID controller's integral time constant τ_i to be equal to the time constant τ, and the differential time constant τ_d to be equal to the dampling time constant τ_d of the converter $G(s)$. Referring to (10.79) and considering $\tau > \tau_1$, the unit-step response of the closed-loop PID control system of the SOH with a first-order load plus an integral element in the s-domain is:

$$\Omega_O(s) = G_C(s)\Omega_{in}(s) = \frac{1}{s}\frac{1}{1+s^2\tau_a^2} \tag{10.87}$$

This is a third-order transfer function with three poles, so that normally this unit-step response is unstable. The unit-step response of the closed-loop control system in the time domain is:

$$\omega_O(t) = 1 - \cos\frac{t}{\tau_a} \tag{10.88}$$

where τ_a is the auxiliary time constant, $\tau_a = \sqrt{\tau\tau_m/p}$. It means that the unit-step response of the closed-loop control system is unstable with constant amplitude oscillation.

The transfer function in the z-domain is:

$$\Omega_O(z) = \mathbf{Z}\left[\frac{1}{s}\frac{1}{1+s^2\tau_a^2}\right] = \frac{z}{z-1}\frac{z^2 - z\cos\frac{T}{\tau_a}}{z^2 - 2z\cos\frac{T}{\tau_a} + 1} \tag{10.89}$$

The system response in the z-domain has three poles inside the unity-cycle. It means that the system in analog control is unstable, but it is stable in digital control.

10.4.3 IMPULSE RESPONSE FOR PID CONTROLLED DC/DC CONVERTERS

Refer to Figure 10.14 showing the closed-loop control system block diagram of the converter $G(s)$ with a PID controller. The converter $G(s)$ is an SOH simulating DC/DC converters. The feedback network is assumed as a unity element. The interference signal $V_{int}(s) = U$ is a unit delta function. To simplify the analysis, the input signal is assumed $V_{in}(s) = 0$. Therefore, the equivalent block diagram is shown in Figure 10.15 for the analysis of the impulse response. We still select the PID controller's integral time constant $\tau_i = \tau$ and differential time constant τ_d to be equal to the damping time constant τ_d of the converter $G(s)$. The impulse response output signal of the closed-loop system in the s-domain is:

$$V_O(s) = \frac{V_{int}(s)}{1 + G_{pid}(s)G(s)} = \frac{U}{1 + p\frac{1+s\tau_i+s^2\tau_i\tau_d}{s\tau_i}\frac{1}{1+s\tau+s^2\tau\tau_d}} = U\frac{s\tau}{p+s\tau}$$

$$= U\left(1 - \frac{1}{1+s\tau/p}\right) \tag{10.90}$$

The impulse response in the time domain is:

$$v_O(t) = U\delta(t)(1 - e^{-pt/\tau}) \tag{10.91}$$

The impulse response is stable and that of the closed-loop system in the z-domain is:

$$V_O(z) = \mathbf{Z}\left[U\left(1 - \frac{1}{1 + s\tau/p}\right)\right] = U\left(1 - \frac{z}{z - e^{-pT/\tau}}\right) \qquad (10.92)$$

The pole of this system is in the unity-cycle. Therefore, this impulse response is stable.

Analysis of the SOH with a First-Order Load

Refer to Figure 10.16 showing the closed-loop control system block diagram of the PID controlled converter $G(s)$ with a first-order circuit. The converter $G(s)$ is an SOH simulating DC/DC converters. The feedback network is assumed as a unity element. The interference signal $I_{int}(s) = U$ as a unit delta function. To simplify the analysis, the input signal is assumed $I_{in}(s) = 0$. We still select the PID controller's integral time constant $\tau_i = \tau$ and differential time constant τ_d to be equal to the damping time constant τ_d of the converter $G(s)$ with the condition $\tau > \tau_1$. Therefore, the equivalent block diagram is shown in Figure 10.17 for the analysis of the impulse response. Using the closed-loop transfer function (10.76), we obtain the output signal of the closed-loop system in the s-domain:

$$I_O(s) = \frac{U}{1 + G_{pi}(s)G(s)G_1(s)} = \frac{U}{1 + p\frac{1}{s\tau}\frac{1}{1+s\tau_1}} \approx \frac{U s\tau}{p + s\tau} = U\left(1 - \frac{1}{1 + s\tau/p}\right)$$
$$(10.93)$$

The impulse response in the time domain is:

$$i_O(t) = U\delta(t)(1 - e^{-pt/\tau}) \qquad (10.94)$$

The impulse response is stable and that of the closed-loop system in the z-domain is:

$$I_O(z) = \mathbf{Z}\left[U\left(1 - \frac{1}{1 + s\tau/p}\right)\right] = U\left(1 - \frac{z}{z - e^{-pT/\tau}}\right) \qquad (10.95)$$

The pole of this system is in the unity-cycle. Therefore, this impulse response is stable.

Analysis of the SOH with a First-Order Load Plus an Integral Element

Refer to Figure 10.18 showing the closed-loop control system block diagram of the PI controlled converter $G(s)$ with a first-order circuit plus an integral element. The converter $G(s)$ is an SOH simulating DC/DC converters. The feedback network is assumed as a unity element. The interference signal $\Omega_{int}(s) = U$ as a unit delta function. To simplify the analysis, the input signal is assumed $\Omega_{in}(s) = 0$. We still select the PI controller's integral time constant $\tau_i = \tau$ and the differential time constant τ_d to be equal to the time constant τ of converter $G(s)$ with the condition $\tau > \tau_1$. Therefore, the equivalent block diagram is shown in Figure 10.19 for the analysis of the impulse response. Using the closed-loop transfer function (10.78), we obtain the output signal of the closed-loop system in the s-domain is:

$$\Omega_O(s) = \frac{U}{1 + G_{pi}(s)G(s)G_1(s)G_m(s)} = \frac{U}{1 + p\frac{1}{s^2\tau\tau_m}\frac{1}{1+s\tau_1}} \approx \frac{Us^2\tau\tau_m}{p + s^2\tau\tau_m}$$

$$= U\left(1 - \frac{1}{1 + s^2\tau_a^2}\right) \qquad (10.96)$$

where $\tau_a^2 = \tau\tau_m/p$. The impulse response in the time domain is:

$$\omega_O(t) = U\delta(t)\left(1 + \sin\frac{t}{\tau_a}\right) \qquad (10.97)$$

The impulse response is unstable. The impulse response of the closed-loop system in the z-domain is:

$$\Omega_O(z) = \mathbf{Z}\left[\frac{Us\tau\tau_m}{p + s\tau\tau_m}\right] = U\left(1 - \frac{z\sin T/\tau_a}{z^2 - 2z\cos T/\tau_a + 1}\right) \qquad (10.98)$$

The poles of this system are in the unity-cycle. Therefore, this impulse response is stable.

FURTHER READING

1. D'Azzo J. J. and Houpis C. H., *Feedback Control System Analysis and Synthesis*, 2nd edn., McGraw-Hill, New York, 1966.
2. Wylie Jr. C. R., *Advanced Engineering Mathematics*, McGraw-Hill, New York, 1951.
3. Rabiner L. R. and Gold B., *Theory and Applications of Digital Signal Processing*, Prentice-Hall, Englewood Cliffs, New Jersey, 1975.
4. Fletcher R. and Powell M. J. D., A rapid convergent decent method for minimization, *Comput J*, Vol. 6, No. 2, 1963, pp. 163–168.
5. Kranc G. M., Input–output analysis of multirate feedback systems, *IRE Trans Autocont*, November 1957, pp. 21–28.
6. Coffey T. C. and Williams I. J., Stability analysis of multiloop, multirate sampled systems, *AIAA J*, Vol. 4, No. 12, December 1966, pp. 2178–2190.
7. Boykin W. H. and Frazier B. D., Analysis of multiloop multirate sampled-data system, *AIAA J*, Vol. 13, No. 4, April 1975, pp. 453–456.
8. Sklansky J. and Ragazzine J. R., Analysis of error in sample data feedback systems, *AIEE Trans*, Vol. 74, Part II, 1955, pp. 65–71.
9. Stanford D.P., Stability for multirate sample data systems, *SIAM J Optim Cont*, Vol. 17, No. 3, 1979, pp. 390–399.
10. Bode H., *Network Analysis and Feedback Amplifier Design*, Van Nostrand, New York, 1945.
11. Mayr O., *The Origins of Automatic Feedback Control*, MIT Press, Cambridge, Massachusetts, 1970.
12. Truxal J. G., *Automatic Feedback Control System Synthesis*, McGraw-Hill, New York, 1955.

Chapter 11

Energy Factor Application in AC and DC Motor Drives

AC and DC motor variable-speed drive systems are the main projects of the "power electronics". Energy storage in AC and DC motors has recently been investigated and discussed.

AC motors can be supplied by fixed frequency (e.g. $f = 50$ or $60\,\text{Hz}$) power supply, but its speed cannot be changed. An AC motor variable-speed drive system performs in adjustable speed, so that its power supply frequency is changeable. AC motors working in variable-speed drive system are supplied by choppers, DC/AC pulse-width modulated (PWM) inverters and AC/AC (AC/DC/AC) converters. Therefore, this system is working in discrete state. We can call these AC motor power supplies as discrete AC power supply sources.

DC motors can be supplied by fixed DC voltage power supply, but its speed cannot be changed (under the condition of fixed field flux). A DC motor variable-speed drive system performs in adjustable speed, so that its power supply voltage is changeable. DC motors working in variable-speed drive system are usually supplied by choppers, AC/DC rectifiers and DC/DC converters. Therefore, this system is working in discrete state. We can call these DC motor power supplies as discrete DC power supply sources.

11.1 INTRODUCTION

AC and DC motors are an important equipment to convert the electrical energy to mechanical energy. Variable-speed drive system is the main project of the "power electronics". Energy storage in motors has recently been investigated and discussed. We can apply the energy factor (EF) to demonstrate the motor drive system's characteristics.

Although there are many AC motors, we choose the three-phase induction motor as the example in the next text. A three-phase induction motor is power supplied by a three-phase AC power supply source. The energy is supplied to the stator, and then transferred to the rotor. The main parts of the stored energy in an induction motor are kinetic energy since the stator inductance is comparably small.

Traditionally, the system characteristics of AC motor drive systems have been analyzed by cybernetics and analog control theory since the operation of AC motor drive systems was treated as a continuous process. If an induction motor working in a variable-speed AC motor drive system is supplied by discrete AC power supply sources, the stored energy in the induction motor is accumulated in period by period with the energy quantization.

As convenience we choose a permanent magnet (PM) DC motor as the example in the next text. A PM brushed DC is power supplied by a DC power supply source. The energy is supplied to the armature circuit via brushes and commutators, and then transferred to the rotor. The main parts of the stored energy in a DC motor are kinetic energy since the armature inductance is comparably small.

Traditionally, the system characteristics of DC motor drive systems have been analyzed by cybernetics and analog control theory since the operation of DC motor drive systems was treated as a continuous process. If a PM DC motor working in a variable-speed DC motor drive system is supplied by discrete DC power supply sources, the stored energy in the PM DC motor is accumulated in period by period with the energy quantization.

Although digital control theory has been applied in all motor drive systems for a long time, unfortunately the converters were not treated as correct models. We can think that all these systems are digital control systems and we have to apply the digital control theory to these systems. We will discuss these topics in this chapter. Most discussions are novel approaches of the development in digital power electronics.

11.2 ENERGY STORAGE IN MOTORS

Energy storage in motors was well discussed long time ago by the electromechanical theory. If a motor is supplied by a converter or other switching circuit the energy transfer from the source to the motor is by the quantization manner. The energy is pumped to the motor by energy quantum in each sampling interval, although the sampling interval T is small.

11.2.1 ENERGY STORAGE IN AC MOTOR

Energy storage in AC motor has two parts: mechanical stored energy (*MSE*) and electrical stored energy (*ESE*). Total stored energy (*SE*) is defined as:

$$SE = MSE + ESE \qquad (11.1)$$

Mechanical Energy Storage

Usually the mechanical stored energy of an AC motor includes few parts:

- The mechanical stored energy in the rotor.
- The mechanical stored energy in the joint and gear-box.
- The mechanical stored energy in the further equipment.

The first stored energy will respond to the motor inertia, J, and motor running speed, ω. The second and third parts of the stored energy will respond to the equivalent inertia, J', and motor running speed, ω'. Usually, we combine all inertia of whole mechanical system together and symbolize as J_e. We can consider all the mechanical stored energy as kinetic energy which is measured by:

$$MSE = \frac{1}{2} J_e \omega^2 \qquad (11.2)$$

where J_e is the equivalent inertia including motor rotor, jointer and gear-box, and the further mechanical equipment. It is measured in $\text{kg}\,\text{m}^2$, ω is the motor running speed and measured in rad/s.

To simplify the investigation we may ignore the other energy losses/storage in the motor such as hit, frictional and windage energy losses, which will affect the transient process.

Electrical Energy Storage

The electrical stored energy in an AC motor is considered in the stator circuit. Back electromagnetic force (EMF) is corresponding to the motor running speed. Assume the stator inductance is L and the stator current is I_s, the electrical stored energy is measured by:

$$ESE = \frac{1}{2} L I_s^2 \qquad (11.3)$$

where L_s is the stator inductance including the cable's inductance in H, I_s is the AC motor stator current in A.

11.2.2 ENERGY STORAGE IN DC MOTOR

Energy storage in a DC motor has two parts: mechanical stored energy (MSE) and electrical stored energy (ESE). Total stored energy is defined as:

$$SE = MSE + ESE \qquad (11.4)$$

Mechanical Energy Storage

Usually the mechanical stored energy of a DC motor includes few parts:

- The mechanical stored energy in the rotor.

- The mechanical stored energy in the joint and gear-box.
- The mechanical stored energy in the further equipment.

The first stored energy will respond to the DC motor inertia, J, and motor running speed, ω. The second and third parts of the stored energy will respond to the equivalent inertia, J', and motor running speed, ω'. Usually, we combine all inertia of whole mechanical system together and symbolize as J_e. We can consider all the mechanical stored energy is kinetic energy measured by:

$$MSE = \frac{1}{2}J_e\omega^2 \tag{11.5}$$

where J_e is the equivalent inertia including motor rotor, jointer and gear-box, and the further mechanical equipment. It is measured in kg m^2, ω is the DC motor running speed and measured in rad/s.

To simplify the investigation we may ignore the other energy losses/storage in the motor such as hit, frictional and windage energy losses, which will affect the transient process.

Electrical Energy Storage

The electrical stored energy in a DC motor is considered in the armature circuit and the back EMF. Assume the armature inductance is L_a and the armature current is I_a, and the field inductance is L_f and the field current is I_f, then the electrical stored energy is measured by:

$$ESE = \frac{1}{2}(L_a I_a^2 + L_f I_f^2) + EI_a \tag{11.6}$$

where E is the DC motor back EMF, L_a is the armature inductance including the cable's inductance in H, I_a is the motor armature current in A. L_f is the field inductance in H, I_f is the motor field current in A. A PM DC motor has no field winding, so the equation can be simplified as:

$$ESE = \frac{1}{2}L_a I_a^2 + EI_a \tag{11.7}$$

11.3 A DC/AC VOLTAGE SOURCE

We introduce an application of DC/AC PWM inverter in this section. It is called "zero-phase odd-harmonic repetitive controller for a single-phase PWM inverter". The proposed repetitive controller combines an odd-harmonic periodic generator with a non-casual phase lead compensation filter. It occupies less data memory than a conventional repetitive controller does. Moreover, it offers faster convergence of the tracking error, and yields very low-total harmonic distortion (THD) and low-tracking error. Analysis and design of the proposed system are presented. Experimental results with the proposed repetitive controller are presented to validate the approach. The drawback of the proposed controller is discussed and experimentally demonstrated.

High-performance constant-voltage-constant-frequency (CVCF) PWM inverters should accurately regulate the output AC voltage/current to the reference sinusoidal input with low THD and fast dynamic response. Nonlinear loads such as the rectifier loads that cause periodic distortion are major sources of THD. According to the internal model principle [1], repetitive control (RC) [2–5] provides an effective solution for tracking periodic reference signals, or eliminating periodic disturbances. In Ref. [6–10], RC found its promising usage in CVCF PWM converters for waveform compensation. In a conventional RC system, any reference signal with a fundamental period N can be exactly tracked by including a periodic signal generator $1/(z^N - 1)$ in the closed-loop system. Such a periodic signal generator needs at least N memory cells. A usual repetitive controller introduces infinite gain at both even and odd-harmonic frequencies. However, AC references and disturbances mainly contain odd-harmonic frequencies in the CVCF PWM inverters. In order to improve the convergence performance of RC systems, a new odd-harmonic periodic signal generator is proposed to update control output every $N/2$ sampling intervals [11]. On the other hand, although RC mathematically assures that the tracking errors asymptotically converge to zero when repetition goes to infinity, poor tracking accuracy and transient may be caused by poor design of the filter of repetitive controller in practical applications [14].

A discrete-time zero-phase odd-harmonic repetitive control scheme is proposed for a CVCF PWM inverter to achieve high tracking accuracy (low-THD and low-tracking error) and good dynamic response (fast monotonic convergence) in the presence of linear and nonlinear loads, and parameter uncertainties. The analysis and design of the proposed controller are discussed. Experimental results are presented to validate the proposed approach. Moreover, the drawback of the proposed controller is discussed.

11.3.1 Zero-Phase Odd-Harmonic Repetitive Control

Zero-phase odd-harmonic repetitive control is a novel approach of digital control methods. We assume that all harmonics are in odd orders.

Odd-Harmonic Periodic Signal Generator

In the discrete time domain, a conventional periodic signal generator can be written as:

$$G_r(z) = \frac{z^{-N}}{1 - z^{-N}} = \frac{1}{z^N - 1} \tag{11.8}$$

where $N = T_s/T$, T_s and T being the signal period and the sampling time, respectively. The generator in Equation (11.8) can eliminate the harmonics that are below the Nyquist frequency, $\omega \ (= \pi/T)$, by introducing infinite gain at both even and odd-harmonic frequencies [3]. For systems such as the CVCF PWM inverters, the references and disturbances mainly contain odd-harmonic frequencies. When a conventional repetitive controller is used, it updates the control output every N sampling intervals with at least N memory cells. In the following, we investigate a new odd-harmonic periodic signal generator, which occupies $N/2$ data memory cells and updates control output every $N/2$ sampling intervals.

In general, a discrete-time signal $x(n)$ with period $N \times T$ can be written as:

$$x(n + N) = x(n), \quad \forall n \in Z \tag{11.9}$$

Its Fourier series is as follows:

$$\begin{cases} x(n) = \displaystyle\sum_{k=0}^{N-1} c_k e^{j2\pi kn/(NT)} \\ c_k = \dfrac{1}{N} \displaystyle\sum_{n=0}^{N-1} x(n) e^{-j2\pi kn/(NT)} \end{cases} \tag{11.10}$$

In Equation (11.10), if the coefficients c_k are zero for even index ($k \bmod 2 = 0$) (including $k = 0$), then the signal is an odd-harmonic periodic signal. Moreover, if the period N is even, the discrete-time signal, $x(n)$, is an odd-harmonic signal if and only if $x(n + (N/2)) = x(n)$.

A discrete time odd-harmonic periodic signal generator has the following transfer function:

$$G_r(z) = -\frac{1}{z^{N/2} + 1} \tag{11.11}$$

From Equation (11.11), the generator has its poles at:

$$z = e^{j(2k+1)\frac{2\pi}{NT}} \quad k = 0, 1, \dots, \frac{N}{2} - 1 \tag{11.12}$$

Equation (11.12) means that Equation (11.11) has infinite gain at frequency $\omega_k = (2k + 1)2\pi/(NT)$ of all odd harmonics. Furthermore, if the odd-harmonic signal generator as in Equation (11.11) is incorporated in a system, it will achieve perfect asymptotic tracking or disturbance rejection for this class of periodic signals. As compared to traditional periodic signal generator as in Equation (11.8), all even harmonic poles $\omega_k = 4k\pi/(NT)$ are removed.

Odd-Harmonic Repetitive Control

Figure 11.1 shows the proposed discrete-time odd-harmonic repetitive control system under consideration, where $R(z)$ is the reference input, $Y(z)$ is the output, $E(z) = R(z) - Y(z)$ is the tracking error, $D(z)$ is the disturbance, $G_c(z)$ is the conventional controller, $G_s(z)$ is the plant, $G_r(z)$ is a feedforward plug-in odd-harmonic repetitive controller, k_r is the *repetitive control gain*, $U_r(z)$ is the output of the repetitive controller,

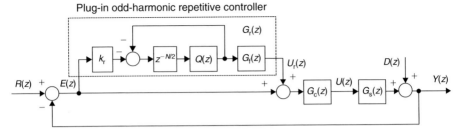

Figure 11.1   Plug-in discrete-time odd-harmonic repetitive control system.

$G_f(z)$ is a filter to obtain a stable overall closed-loop system, and $Q(z)$ is a low-pass filter to enhance the robustness of the overall system.

As shown in Figure 11.1, the proposed plug-in odd-harmonic repetitive control law can be expressed as:

$$U_r(z) = -Q(z)(z^{-N/2}U_r(z) + k_r z^{-N/2}G_f(z)E(z))   \tag{11.13}$$

From Figure 11.1, the transfer functions from $R(z)$ and $D(z)$ to $Y(z)$ in the overall closed-loop control system can be derived as:

$$\frac{Y(z)}{R(z)} = \frac{(1+G_r(z))G_c(z)G_s(z)}{1+(1+G_r(z))G_c(z)G_s(z)} = \frac{(1+z^{-N/2}Q(z)(1-k_rG_f(z)))H(z)}{1+z^{-N/2}Q(z)(1-k_rG_f(z)H(z))}   \tag{11.14}$$

$$\frac{Y(z)}{D(z)} = \frac{1+z^{-N/2}}{1+G_c(z)G_s(z)}\frac{1}{1+z^{-N/2}Q(z)(1-k_rG_f(z)H(z))}   \tag{11.15}$$

where

$$H(z) = \frac{G_c(z)G_s(z)}{1+G_c(z)G_s(z)}$$

From Equations (11.14) and (11.15), the overall closed-loop system is stable if the following conditions hold:

1. the roots of $1+G_c(z)G_s(z)=0$ are located inside the unit circle;
2. $\|Q(z)(1-k_rG_f(z)H(z))\| < 1 \quad \forall z = e^{j\omega}, \quad 0 < \omega < \dfrac{\pi}{T}$ (11.16)

It should be pointed out that the above stability conditions for an odd-harmonic RC system are the same as those for a conventional RC one [8]. It means that, if $Q(z)$, $G_f(z)$ and $H(z)$ are identical in both an odd-harmonic RC system and a conventional RC system, the stability range of repetitive control gain k_r are identical too. An odd-harmonic RC updates its output every $N/2$ sampling intervals; while a conventional RC renews its output every N sampling intervals. Therefore, if $Q(z)$, $G_f(z)$, $H(z)$ and k_r are identical for both two RC systems, the tracking error convergence rate of an odd-harmonic RC is about two times as fast as that of a conventional one.

The error transfer function of the overall system is:

$$G_e(z) = \frac{E(z)}{R(z) - D(z)} = \frac{1 + z^{-N/2}}{1 + G_c(z)G_s(z)} \frac{1}{1 + z^{-N/2}Q(z)(1 - k_r G_f(z)H(z))} \quad (11.17)$$

Thus, if the overall closed-loop system is asymptotically stable and the angular frequency, ω, of the reference input $R(t)$ and disturbance $D(t)$ approaches to $\omega_m = (2m + 1)2\pi/(NT)$, $m = 0, 1, 2, \ldots, (N/2) - 1$, then $z^{-N/2} \to 1$, thus:

$$\lim_{\omega \to \omega_m} \|e(j\omega)\| = 0 \quad (11.18)$$

According to Equation (11.18), if the frequencies of odd-harmonic references and/or disturbances are less than half of the sampling frequency (Nyquist frequency), steady-state zero-tracking error can be ensured by using the odd-harmonic repetitive controller $G_r(z)$. Theoretically, for CVCF PWM inverters, the odd-harmonic repetitive controller Equation (11.13) is a zero-tracking error control law, if there are no even-harmonics disturbances.

Phase Cancellation Compensation

From Equations (11.13) and (11.16), we observe that the performance of the repetitive controller $G_r(z)$ is determined by the design of $G_f(z)$, k_r and $Q(z)$. Within its stability range determined by Equation (11.16), larger k_r leads to smaller damping ratio with faster transients. In many cases, it is very difficult for all frequencies up to Nyquist frequency to satisfy the inequality in Equation (11.16), if $Q(z) = 1$. To enhance the robustness, a low-pass FIR filter $Q(z)$ [5] with $\|Q(z)\| \leq 1$ and zero phase shift, e.g. $Q(z) = \alpha_1 z + \alpha_0 + \alpha_1 z^{-1}$ with $2\alpha_1 + \alpha_0 = 1$ and $\alpha_0 > 0$, $\alpha_1 \geq 0$, can be introduced to cut out the frequencies which are not able to satisfy Equation (11.16), and relax the stability range for k_r. However, some high-frequency periodic disturbances cannot be eliminated exactly due to $Q(z)$. Hence $Q(z)$ will bring a trade-off between tracking accuracy and the system robustness. A well-designed $G_f(z)$ could make all frequencies up to Nyquist frequency to satisfy Equation (11.16) with $Q(z) = 1$, then yields high tracking accuracy with good transients.

Suppose $H(z)$ has frequency characteristics $H(j\omega) = N_h(\omega)\exp(j\theta_h(\omega))$ with $N_h(\omega)$ and $\theta_h(\omega)$ being its magnitude and phase; and $G_f(z)$ has frequency characteristics $G_f(j\omega) = N_f(\omega)\exp(j\theta_f(\omega))$ with $N_f(\omega)$ and $\theta_f(\omega)$ being its magnitude and phase. Using these characteristics, Equation (11.16) leads to [12–14]:

$$k_r N_h(\omega)N_f(\omega) < 2\cos(\theta_h + \theta_f) \quad (11.19)$$

Since k_r, $N_h(\omega)$ and $N_f(\omega)$ are positive, Equation (11.19) necessarily yields:

$$-90° < \theta_h(\omega) + \theta_f(\omega) < 90° \quad (11.20)$$

If $\theta_h(\omega) = -\theta_f(\omega)$ (i.e. "*zero phase*"), for all frequencies up to Nyquist frequency, Equation (11.16) will be satisfied with $k_r \in (0, 2/(N_h(\omega)N_f(\omega)))$. Since a CVCF PWM inverter is a minimal phase system with phase lag $\theta_h(\omega) > 0$, a phase lead compensation filter $G_f(z)$ with $\theta_f(\omega) < 0$ is needed. To obtain exact phase lead compensation, the poles-zeros cancellation [15] is the most direct approach, i.e. $G_f(z) = 1/H(z)$. Equation (11.16) will lead to:

$$0 < k_r < 2 \tag{11.21}$$

In practice, since $H(z)$ is only approximately known, our design effort is to cancel the phase lag as close as possible. Since the periodic signal generator in the repetitive controller brings a delay $z^{-N/2}$ or z^{-N}, the above non-casual phase lead filter $G_f(z)$ is realizable. Due to uncertainties and disturbances, $Q(z)$ will still be needed in the repetitive controllers in practice.

11.3.2 ZERO-PHASE ODD-HARMONIC REPETITIVE CONTROLLED PWM INVERTER

In order to apply this theory, we offer a particular zero-phase odd-harmonic repetitive controlled PWM inverter in this section.

Modeling of the System

Figure 11.2 shows the setup of the odd-harmonic repetitive controlled inverter. The dynamics of the inverter can be described as follows:

$$\begin{pmatrix} \dot{v}_c \\ \ddot{v}_c \end{pmatrix} = \begin{pmatrix} 0 & 1 \\ -\frac{1}{L_n C_n} & -\frac{1}{C_n R_n} \end{pmatrix} \begin{pmatrix} v_c \\ \dot{v}_c \end{pmatrix} + \begin{pmatrix} 0 \\ \frac{1}{L_n C_n} \end{pmatrix} v_{in}$$

$$v_{in} = \begin{cases} -v_{dc}, & \text{if } S_2 \ \& \ S_3 \text{ are on; } S_1 \ \& \ S_4 \text{ are off} \\ +v_{dc}, & \text{if } S_1 \ \& \ S_4 \text{ are on; } S_2 \ \& \ S_3 \text{ are off} \end{cases} \tag{11.22}$$

where v_c is the output voltage; i_O is the output current; v_{dc} is the DC bus voltage; the control input v_{in} is a PWM pulse; L_n, C_n, and R_n are the nominal component values of the inductor, the capacitor and the load, respectively.

For a linear system $\dot{x} = Ax + Bu$, its sampled-data equation can be expressed as:

$$x(k+1) = e^{AT}x(k) + \int e^{A(T-\tau)}Bu(\tau)d\tau$$

Therefore, the sampled-data form for Equation (11.22) can be approximately expressed as:

$$\begin{pmatrix} v_c(k+1) \\ \dot{v}_c(k+1) \end{pmatrix} = \begin{pmatrix} \varphi_{11} & \varphi_{12} \\ \varphi_{21} & \varphi_{22} \end{pmatrix} \begin{pmatrix} v_c(k) \\ \dot{v}_c(k) \end{pmatrix} + \begin{pmatrix} g_1 \\ g_2 \end{pmatrix} u(k) \tag{11.23}$$

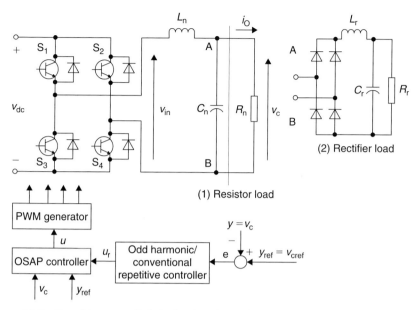

Figure 11.2 Repetitive controlled single-phase PWM inverter.

where the coefficients

$$\varphi_{11} = 1 - \frac{T^2}{2L_n C_n}$$

$$\varphi_{21} = -\frac{T}{L_n C_n} + \frac{T^2}{2L_n C_n^2 R_n}$$

$$\varphi_{12} = T - \frac{T^2}{2C_n R_n}$$

$$\varphi_{22} = 1 - \frac{T}{C_n R_n} - \frac{T^2}{2L_n C_n} + \frac{T^2}{2C_n^2 R_n^2}$$

$$g_1 = \frac{T}{2L_n C_n}$$

$$g_2 = \frac{1}{L_n C_n} - \frac{T}{2L_n C_n^2 R_n}$$

and the average active input

$$u(k) = v_{in}(k) \approx \left(\frac{2\Delta T(k)}{T} - 1 \right) v_{dcn}$$

with nominal DC bus voltage v_{dcn}. The two-level PWM switching waveform for v_{in} is shown in Figure 11.3.

The output equation can be expressed as:

$$y(k) = v_{\text{c}}(k) \qquad (11.24)$$

Zero-Phase Odd-Harmonic Control

Based on nominal values, the autoregressive-moving-average (ARMA) equation for the Equations (11.23) and (11.24) can be expressed as follows:

$$y(k+1) = -p_1 y(k) - p_2 y(k-1) + m_1 u(k) + m_2 u(k-1) \qquad (11.25)$$

where $p_1 = -(\varphi_{11} + \varphi_{22})$, $p_2 = \varphi_{11}\varphi_{22} - \varphi_{21}\varphi_{12}$, $m_1 = g_1$, $m_2 = g_2\varphi_{12} - g_1\varphi_{22}$.

In practice, the inverter parameters are $L = L_n + \Delta L$, $C = C_n + \Delta C$, $R = R_n + \Delta R$ and $v_{\text{dc}} = v_{\text{dcn}} + \Delta v$. Therefore, the ARMA equation for the actual plant should be:

$$y(k+1) = -a_1 y(k) - a_2 y(k-1) + b_1 u(k) + b_2 u(k-1) \qquad (11.26)$$

where $a_1 = p_1 + \Delta p_1$, $a_2 = p_2 + \Delta p_2$, $b_1 = m_1 + \Delta m_1$ and $b_2 = m_2 + \Delta m_2$.

If a one-sampling-ahead-preview (OSAP) control law [6]

$$u(k) = \frac{1}{m_1}[y_{\text{ref}}(k) - m_2 u(k-1) + p_1 y(k) + p_2 y(k-1)] \qquad (11.27)$$

is applied to the plant, then $H(z)$ without repetitive controller becomes [9]:

$$H(z) = \frac{b_1 + b_2 z^{-1}}{(z + a_1 + a_2 z^{-1})(m_1 + m_2 z^{-1}) - (p_1 + p_2 z^{-1})(b_1 + b_2 z^{-1})} \qquad (11.28)$$

Obviously, if $v_{\text{dc}} = v_{\text{dcn}}$, $L = L_n$, $C = C_n$ and $R = R_n$, a deadbeat response $H(z) = z^{-1}$ is achieved.

In practice, there are uncertainties in model parameters, such as $\Delta v = v_{\text{dc}} - v_{\text{dcn}}$, $\Delta L = L - L_n$, $\Delta C = C - C_n$, load disturbance $\Delta R = R - R_n$, and even un-modeled

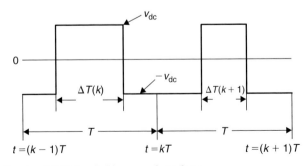

Figure 11.3   Two-level PWM switching waveform for v_{in}.

uncertainties. Hence, zero-tracking error cannot be achieved by OSAP controller. To overcome the uncertainties and disturbances, an odd-harmonic repetitive controller $G_r(z) = -(k_r z^{-N/2} Q(z))/(1 + z^{-N/2} Q(z) G_f(z))$ is plugged into the prior OSAP controlled inverter, where $N = f/f_s$ (even); f_s is the frequency of y_{ref}; $f = 1/T$ is the sampling frequency; $G_f(z) = 1/H(z)$ is the phase lead compensation filter; $Q(z) = \alpha_1 z + \alpha_0 + \alpha_1 z^{-1}$ with $2\alpha_1 + \alpha_0 = 1$ and $\alpha_1 \geq 0, \alpha_0 > 0$. If $\alpha_1 = 0, \alpha_0 = 1$, then $Q(z) = 1$. Of course, since it is impossible to get accurate $H(z)$ in practice, "zero phase" can be only roughly obtained. Furthermore, in case of compensating the un-modeled pure delays in the plant, a pure lead z^m can be added into the compensation filter as:

$$G_f(z) = z^m/H(z)$$

11.3.3 EXPERIMENTAL VERIFICATION

To validate the theoretical study, we have setup an experimental system in our laboratory. DSPACE DS1102 and Matlab/Simulink have been used in fast prototyping the experimental platform and collecting experimental data. For the odd-harmonic and/or conventional repetitive controlled PWM inverter shown in Figure 11.2, the sinverter parameters in our laboratory are setup as follows: $L_n = 20$ mH; $L = 30$ mH; $C_n = 45 \mu F$; $C = 50 \mu F$; $R_n = 15 \Omega$; $v_{dc} = 70$ V; $v_{dcn} = 80$ V; $y_{ref} = v_{Cref}$ is $(f_s = 50$ Hz$)$ 50 V (peak) sinusoidal voltage; $f = 1/T = 10$ kHz; $N = f/f_s = 200$; resistive load $R = 22 \Omega$; uncontrolled rectifier $L_r = 1$ mH; $C_r = 500 \mu F$ and $R_r = 22 \Omega$.

As shown in Figure 11.4, based on above parameters, all three poles of $H(z)$ are located inside the unit circle when load $R > 6 \Omega$. Hence, for the resistance load $R \in (6, \infty) \Omega$, the OSAP controlled inverter $H(z)$ is stable. To compensate the phase lag of $H(z)$ and un-modeled delay in the range of $(6, \infty) \Omega$, the filter $G_f(z)$ is chosen as

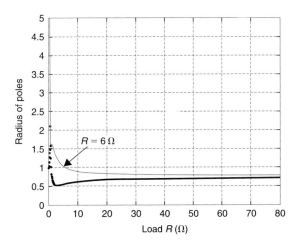

Figure 11.4 Radius of poles of $H(z)$.

$G_f(z) = z^3/(H(z)|_{R=30\,\Omega})$. In our experiments, repetitive control gain $k_r = 0.8 \in (0, 2)$; $Q(z) = (z + 2 + z^{-1})/4$.

Steady-State Response

Figures 11.5–11.7 show the steady-state responses of the output voltage $v_C(t)$, load current $i_O(t)$ of the OSAP controlled inverter with no load, resistive load and rectifier load, respectively. The results indicate that, OSAP offers low THD (up to 2.06%) output voltage under linear load (resistor and no load), but yields worse THD (7.91%) output voltage under nonlinear rectifier load. Note that, in the diagrams of output voltage

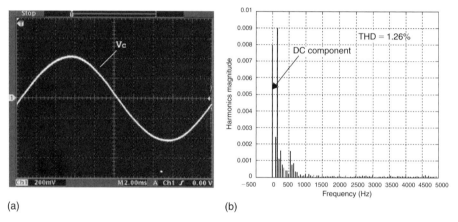

(a)   (b)

Figure 11.5 OSAP controlled steady-state response with under no load. (a) Voltage v_C (12 V/div) and (b) output voltage harmonics spectrum.

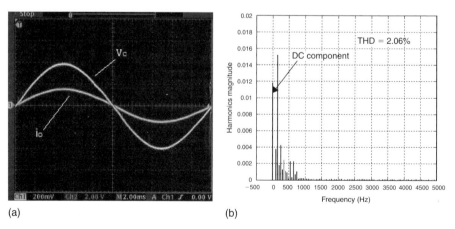

(a)   (b)

Figure 11.6 OSAP controlled steady-state response with under resistor load. (a) Voltage v_C (12 V/div) and i_O (2.85 A/div) and (b) output voltage harmonics spectrum.

harmonics spectrum, the fundamental frequency components (50 Hz, with normalized amplitude) are removed.

Figures 11.8–11.10 show the steady-state responses of the output voltage $v_c(t)$, load current $i_O(t)$ of the conventional RC controlled inverter with no load, resistive load and rectifier load, respectively. The results indicate that conventional RC control offers very low THD (<1%) output voltage under both linear load (resistor and no load) and nonlinear rectifier load.

Figures 11.11–11.13 show the steady-state responses of the output voltage $v_c(t)$, load current $i_O(t)$ of the odd-harmonic RC controlled inverter with no load, resistive load and rectifier load, respectively. The results indicate that odd-harmonic RC control also offers very low THD (up to 1.23%) output voltage under both linear load (resistor

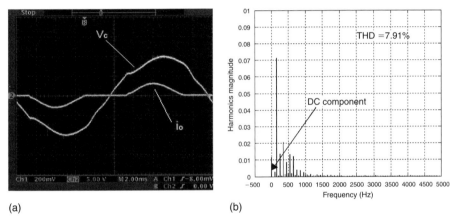

(a)   (b)

Figure 11.7 OSAP controlled steady-state response with under rectifier load. (a) Voltage v_C (12 V/div) and i_O (7.12 A/div) and (b) output voltage harmonics spectrum.

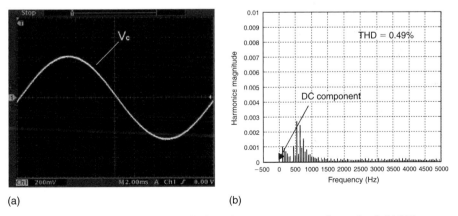

(a)   (b)

Figure 11.8 Conventional RC controlled steady-state response under no load. (a) Voltage v_C (12 V/div) and (b) output voltage harmonics spectrum.

and no load) and nonlinear rectifier load. Therefore, in terms of eliminating THD, odd-harmonic RC as well as conventional RC are efficient control schemes.

Transient Response

Figure 11.14 shows the transient responses of the tracking error $e = v_{cref}(t) - v_c(t)$ with conventional RC and odd-harmonic RC being, respectively, plugged into OSAP controlled inverter with different load at $t \approx 0$ s. Figure 11.14(a) shows that conventional RC controller successfully force the OSAP controlled tracking error (peak) from 4 V (under no load), 6 V (under resistor load), 10 V (under rectifier load) to about 0.5 V (no

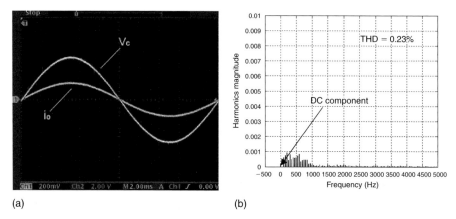

(a)   (b)

Figure 11.9 Conventional RC controlled steady-state response under resistor load. (a) Voltage v_C (12 V/div) and i_O (2.85 A/div) and (b) output voltage harmonics spectrum.

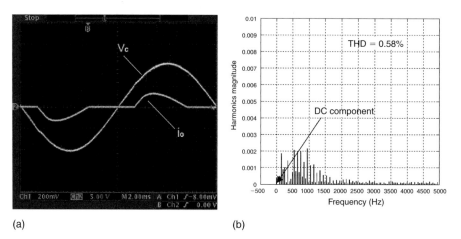

(a)   (b)

Figure 11.10 Conventional RC controlled steady-state response under rectifier load. (a) Voltage v_C (12 V/div) and i_O (7.12 A/div) and (b) output voltage harmonics spectrum.

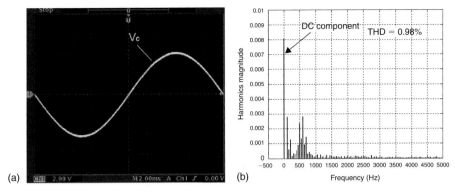

Figure 11.11   Odd-harmonic RC controlled steady-state response under no load. (a) Voltage v_C (12 V/div) and (b) output voltage harmonics spectrum.

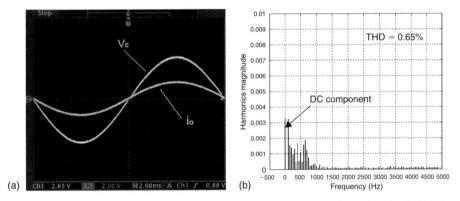

Figure 11.12   Odd-harmonic RC controlled steady-state response under resistor load. (a) Voltage v_C (12 V/div) and i_O (2.85 A/div) and (b) output voltage harmonics spectrum.

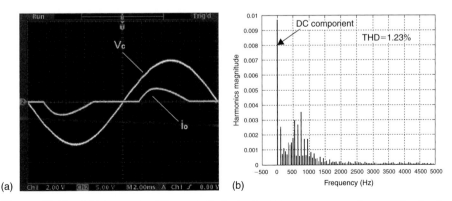

Figure 11.13   Odd-harmonic RC controlled steady-state response under rectifier load. (a) Voltage v_C (12 V/div) and i_O (7.12 A/div) and (b) output voltage harmonics spectrum.

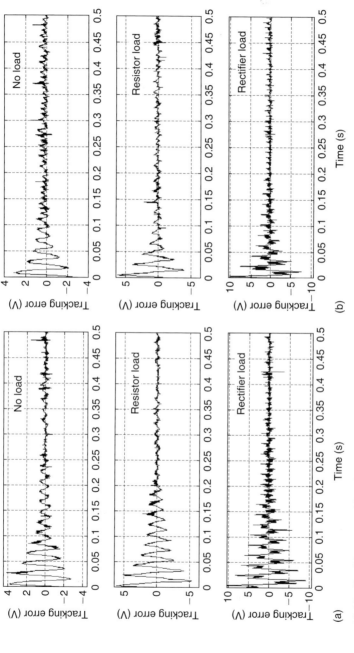

Figure 11.14 Tracking error histories with RC controller being plugged into OSAP control loop at time $t \approx 0\,\mathrm{s}$ under different loads. (a) Conventional RC and (b) odd-harmonic RC.

load), 0.5 V (resistor load) and 1 V (rectifier load) within about 0.2–0.25 s, respectively; whereas, Figure 11.14(b) shows that odd-harmonic RC controller successfully force the tracking error (peak) from 4 V (under no load), 6 V (under resistor load), 10 V (under rectifier load) to about 0.5 V (no load), 0.5 V (resistor load) and 1 V (rectifier load) within about 0.1–0.13 s, respectively. It is clear from the diagram that, under identical conditions, the tracking error convergence rate of an odd-harmonic RC is about two times as fast as that of a conventional one.

It should be pointed out that, as shown in Figure 11.14(b), there are minor DC voltage bias residues (even-harmonic components) in the odd-harmonic controlled steady-state tracking errors. Figure 11.15 shows the zoomed in steady-state tracking errors under different loads. In Figure 11.15(a), there are no obvious DC voltage bias residues in conventional RC controlled steady-state tracking errors; whereas, as shown in Figure 11.15(b), for odd-harmonic RC controller, there are about 0.2–0.25 V DC voltage bias residues in the tracking error under no load and resistor load, and about 0.5 V bias residue under rectifier load.

Sudden Step Load Change

Figures 11.16 and 11.17 show the odd-harmonic repetitive controller operates with sudden step load changes. It is clear from the diagrams that, odd-harmonic RC controlled output voltages do not vary too much (3–4 V), and recover from the sudden step load changes (between no load and resistor load, and between no load and rectifier load) within about 4–5 cycles (i.e. 80–100 ms).

11.3.4 SUMMARY

In summary, the results shown in Figures 11.14–11.17 indicate that an odd-harmonic RC offers significantly faster convergence rate than a conventional RC under different loads (linear load – resistor, and nonlinear load – rectifier) and parameter uncertainties. And our designed odd-harmonic RC controlled inverter provides very low THD output voltage, and is robust to sudden step load changes, too. However, odd-harmonic RC is not immune from even-harmonic disturbances. DC voltage bias residues in the odd-harmonic controlled converter may have a bad impact upon magnetic components, such as transformers and inductors. In case of high-power applications, we should pay much attention to the impact from the phenomena of DC voltage bias residues in odd-harmonic RC controlled converters.

In this section, a zero-phase odd-harmonic repetitive control scheme is proposed for the CVCF PWM inverters. The data memory occupied by the odd-harmonic periodic signal generator is only half of the conventional repetitive controller. It simplifies the implementation and offers faster convergence of the tracking error with shorter control update period. The well-design phase lead compensation filter $G_f(z)$ helps repetitive controller achieve high tracking accuracy by well phase lead compensation. Experimental results show that the proposed approach effectively eliminated odd-harmonic distortion under different loads (linear and nonlinear loads), as compared with a

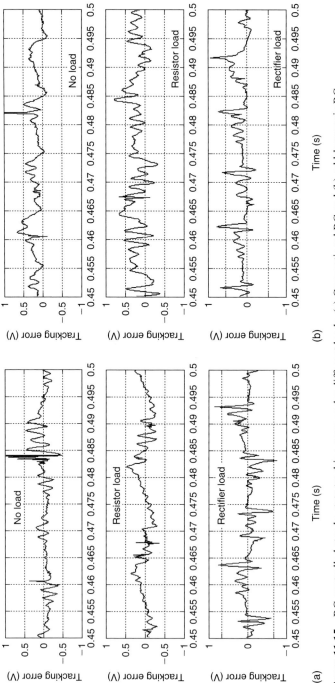

Figure 11.15 RC controlled steady-state tracking error under different loads. (a) Conventional RC and (b) odd-harmonic RC.

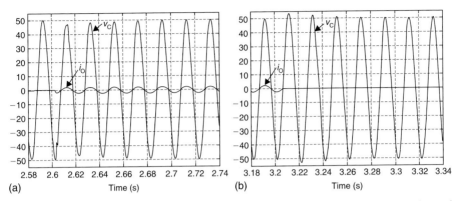

Figure 11.16 Odd-harmonic RC controlled response under resistor load change. (a) Load change $R = \infty \to 22\,\Omega$ and (b) load change $R = 22 \to \infty\,\Omega$.

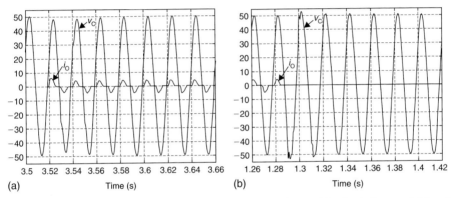

Figure 11.17 Odd-harmonic RC controlled response under rectifier load change. (a) Load change $R = \infty\,\Omega \to$ rectifier load and (b) load change rectifier load $\to R = \infty\,\Omega$.

conventional controller. And the proposed odd-harmonic RC is robust to sudden step load changes too. It is the drawback of an odd-harmonic RC system that even-harmonic residues may occur in the tracking error. The phenomena of even-harmonic residues is pointed out and experimentally demonstrated.

11.4 AN AC/DC CURRENT SOURCE

We introduce an application of AC/DC silicon controlled rectifier (SCR) rectifying current source in this section. It is called "digitally controlled AC/DC SCR rectifying current source". The proposed system is shown in Figure 11.18. The system combines a proportional integral (PI)-controller, an erasable programmable read-only memory (EPROM) look-up table stored the arccosine operation to offer the SCR firing angle, an SCR firing angle generator to yield firing pulse, a three-phase full-bridge SCR to

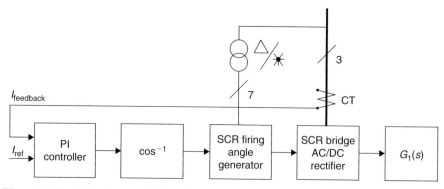

Figure 11.18   Digitally controlled AC/DC SCR rectifying current source.

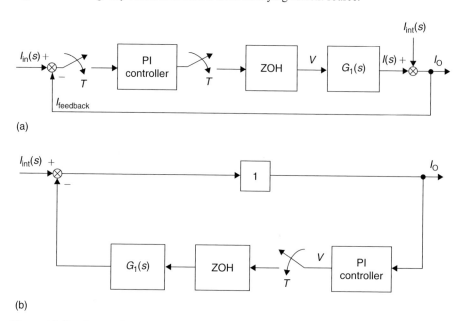

Figure 11.19   System block diagram of an AC/DC SCR rectifying current source. (a) System block diagram for stability and unit-step response. (b) Equivalent block diagram for disturbance analysis.

be simulated by a zero-order-hold (ZOH) and an L–R load ($R = 54\,\Omega$, $L = 200\,\text{mH}$ for the experiments).

The block diagram of the SCR current source with PI-controller is shown in Figure 11.19. The input signal is $I_{\text{in}}(s)$, the output current is $I_O(s)$ and the interference signal is $I_{\text{int}}(s)$ in Figure 11.19(a). Usually, when we analyze the disturbance response, the input signal is assumed no change. The equivalent block diagram for disturbance analysis is shown in Figure 11.19(b). The PI-controller is effective to keep the current source to have satisfied system stability, unit-step response and interference impulse response.

11.4.1 SYSTEM ARRANGEMENT

In order to simplify the analysis, we assume the system to be analyzed in per unit system with unity feedback. Therefore, there are three important elements in this system:

- PI-controller,
- ZOH,
- the first-order load.

We will describe them in detail one by one.

PI-Controller

The PI-controller is designed in digital form, its transfer function in the s-domain is:

$$G_{pi}(s) = K_p + \frac{K_i}{s} \tag{11.29}$$

The transfer function of the PI-controller in the z-domain is:

$$G_{pi}(z) = K_p + \frac{z}{z-1} K_i \tag{11.30}$$

where K_p is the proportional gain, and K_i is the integral gain. The PI-control algorithm has been implemented by a digital signal processor (DSP). The input of the PI-controller is the error between the input current reference and the current feedback signals. The output signal of the PI-controller is formed by a difference equation. For an output Y_k and an error E_k in kth step, we have the step operation:

$$Y_k = K_p E_k + \sum_{j=0}^{k} K_i E_j \tag{11.31}$$

or

$$Y_k = K_p \delta E_k + K_i E_k + Y_{k-1} \tag{11.32}$$

where Y_{k-1} is the output signal of the PI-controller in the $(k-1)$th step.

ZOH to Simulate the SCR

The AC/DC SCR is a three-phase thyristor bridge with six devices. The output DC voltage of the AC/DC SCR is determined by the firing angle α.

$$V_d = V_{dO} \cos \alpha \tag{11.33}$$

where V_{dO} is the maximum DC output voltage corresponding to the firing angle $\alpha = 0$. Since the thyristor is out of control once it starts conducting in a period until the current through it reduces to zero, the AC/DC thyristor bridge rectifier is inherently a sample-and-hold element in the control system. The system may, therefore, be implemented by a latch so the thyristor bridge rectifier is considered a ZOH in the algorithm. We use the ZOH to simulate the AC/DC SCR in the per-unit system. Its transfer function in the s-domain is:

$$G(s) = \frac{1 - e^{-Ts}}{s} \tag{11.34}$$

where T is the sampling interval, $T = 1/6f$. In a power supply network with the frequency of 50 Hz, the frequency is $f = 50$ Hz. So that $T = 1/300\,\mathrm{s} = 3.33\,\mathrm{ms}$.

The ZOH's transfer function in the z-domain is:

$$G(z) = \frac{z}{z - 1} \tag{11.35}$$

It means that the output value of the ZOH will keep a constant in a sampling interval T.

The First-Order Load

There is an R–L circuit with the time constant $\tau_1 = L/R$. Considering the given values: $R = 54\,\Omega$ and $L = 200\,\mathrm{mH}$, the time constant of the first-order load $\tau_1 = L/R = 3.7\,\mathrm{ms}$. Its transfer function in the s-domain is:

$$G_1(s) = \frac{1}{1 + s\tau_1} \tag{11.36}$$

Its transfer function in the z-domain is:

$$G_1(z) = \frac{z}{z - e^{-T/\tau_1}} \tag{11.37}$$

Considering the setting values:

$$G_1(z) = \frac{z}{z - e^{-3.33/3.7}} = \frac{z}{z - 0.4} \tag{11.38}$$

Disturbance Signal

In most situations the interference signal is a unit-delta function. It means that the signal disappears in a short time. If the interference is kept such as the load resistance R changed. The disturbance signal $I_{int}(s)$ is assumed as a unit-step function. Its transfer function in the s-domain is:

$$I_{int}(s) = \frac{1}{s} \tag{11.39}$$

The corresponding transfer function in the z-domain is:

$$I_{int}(z) = \frac{z}{z - 1} \tag{11.40}$$

11.4.2 SYSTEM STABILITY ANALYSIS

Referring to Figure 11.19(a), we have got the system closed-loop transfer function as:

$$G_C(s) = \frac{G_{pi}(s)G(s)G_1(s)}{1 + G_{pi}(s)G(s)G_1(s)} = \frac{\left(K_p + \frac{K_i}{s}\right)\frac{1-e^{-Ts}}{s}\frac{1}{1+s\tau_1}}{1 + \left(K_p + \frac{K_i}{s}\right)\frac{1-e^{-Ts}}{s}\frac{1}{1+s\tau_1}} \tag{11.41}$$

This is stable system if we carefully select the proportional gain K_p and the integral gain K_i to keep all poles are in the LHHP in the s-plane. We will concentrate our analysis in digital control. Therefore, the closed-loop transfer function in the z-domain is:

$$G_C(z) = \frac{G_{pi}(z)G(z)G_1(z)}{1 + G_{pi}(z)G(z)G_1(z)} = \frac{\left(K_p + \frac{z}{z-1}K_i\right)\frac{z}{z-1}\frac{z}{z-e^{-T/\tau_1}}}{1 + \left(K_p + \frac{z}{z-1}K_i\right)\frac{z}{z-1}\frac{z}{z-e^{-T/\tau_1}}} \tag{11.42}$$

This is a stable system if we carefully select the proportional gain K_p and the integral gain K_i to keep all poles inside the unit-cycle in the z-plane. The good digital control system can implement optimization according to the restriction [16–18]:

$$G_C(z) = \frac{1}{z} \tag{11.43}$$

Therefore, the open-loop transfer function in the z-domain is:

$$G_O(z) = G_{pi}(z)G(z)G_1(z) = \frac{1}{z-1} \tag{11.44}$$

We know the transfer functions of a ZOH and a first-order load is:

$$G(z)G_1(z) = \frac{1-a}{z-a} \tag{11.45}$$

where $a = e^{-T/\tau_1} = 0.4$. After optimization we got the proportional gain, K_p, and integral gain, K_i, to be:

$$K_p = \frac{a}{1-a} = 0.66 \quad \text{and} \quad K_i = 1 \tag{11.46}$$

11.4.3 UNIT-STEP RESPONSE ANALYSIS

Referring to Figure 11.19(a), we obtain the unit-step response with the input signal to be a unit-step function:

$$I_O(z) = G_C(z)I_{in}(z) = \frac{1}{z}\frac{z}{z-1} = \frac{1}{z-1} = \sum_{k=1}^{\infty} z^{-k} \tag{11.47}$$

The unit-step response is shown in Figure 11.20(a).

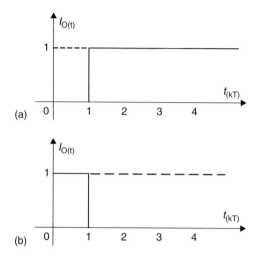

Figure 11.20 System responses. (a) Input unit-step response and (b) disturbance response.

11.4.4 IMPULSE RESPONSE ANALYSIS

Referring to Figure 11.19(b), we obtain the impulse response with the disturbance signal to be a unit-step function:

$$I_{O-d}(z) = \frac{I_{int}(z)}{1 + G_O(z)} = (1 - G_C(z))I_{int}(z) = \frac{z-1}{z}\frac{z}{z-1} = 1 \qquad (11.48)$$

The unit-step response is shown in Figure 11.20(b).

11.5 AC MOTOR DRIVES

AC motor drive is the one main method to transfer AC electrical energy to mechanical energy.

11.5.1 AC MOTOR SUPPLIED BY A CHOPPER

Figure 11.21 shows a single-phase AC motor supplied by a chopper. The power supply is a DC voltage source with DC link voltage V_d and the copping frequency is f, and the sampling period is $T = 1/f$. The stored energy is calculated by Equation (11.1). The pumping energy (*PE*) is:

$$PE = V_d I_d T \qquad (11.49)$$

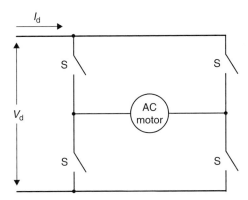

Figure 11.21   A single-phase AC motor supplied by a chopper.

where V_d is the DC link source voltage, I_d is the DC source current, T is the chopping period ($T = 1/f$, where f is the switching frequency). We can define the EF as:

$$EF = \frac{SE}{PE} = \frac{MSE + ESE}{PE} \tag{11.50}$$

We can estimate the transient process from motor stop to a certain speed. The settling time is about:

$$t_{\text{settling}} = EF \times T = \frac{MSE + ESE}{V_d I_d} \tag{11.51}$$

The settling time from one running speed to another speed is approximately estimated as:

$$\Delta t_{\text{settling}} = \Delta EF \times T = \frac{\Delta MSE + \Delta ESE}{V_d I_d} \tag{11.52}$$

where $\Delta t_{\text{settling}}$ is the transient settling time from one running speed to another, ΔEF is the EF variation between the two running states, ΔMSE is the mechanical stored energy variation between the two running states and ΔESE is the electrical stored energy variation between the two running states. This calculation is similar to the small signal analysis and calculation in Chapter 2.

11.5.2 AC MOTOR SUPPLIED BY A DC/AC INVERTER OR AC/AC CONVERTER

Figure 11.22 shows a three-phase AC motor supplied by a DC/AC inverter or AC/AC converter. This power supply system is likely a chopper supplying an AC motor. Therefore, the settling time is still estimated by Equations (11.51) and (11.52).

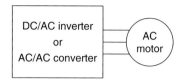

Figure 11.22 A three-phase AC motor supplied by a DC/AC inverter or AC/AC converter.

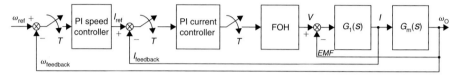

Figure 11.23 The system block diagram of the AC motor supplied by a FOH.

11.5.3 VARIABLE-SPEED AC MOTOR DRIVE SYSTEM SUPPLIED BY A FOH

Usually, AC motor variable-speed drive system has closed-loop control. The block diagram of the system is shown in Figure 11.23, which is a three-phase AC motor supplied by a DC/AC inverter or AC/AC converter. The system consists of a PI speed-controller $G_{pi1}(s)$, a PI current-controller $G_{pi2}(s)$, an first-order-hold (FOH) simulating a DC/AC inverter or AC/AC converter, AC motor stator circuit (a first-order circuit with the time constant $\tau_1 = L/R$) and an integral element with the integral time constant τ_m. The PI speed-controller $G_{pi1}(s)$ has the proportional gain p_1 and the integral time constant τ_{i1}, and the PI current-controller $G_{pi2}(s)$ has the proportional gain p_2 and the integral time constant τ_{i2}. The inner current closed-loop control transfer function is:

$$G_{C-i}(s) = \frac{G_{pi2}(s)G(s)G_1(s)}{1 + G_{pi2}(s)G(s)G_1(s)} = \frac{p_2 \frac{1+s\tau_{i2}}{s\tau_{i2}} \frac{1}{1+sT} \frac{1}{1+s\tau_1}}{1 + p_2 \frac{1+s\tau_{i2}}{s\tau_{i2}} \frac{1}{1+sT} \frac{1}{1+s\tau_1}} \tag{11.53}$$

We select the PI-control integral time constant τ_{i2} to be equal to the time constant τ_1 of the first-order circuit. Considering that the sampling interval T is very small, we obtain the closed-loop transfer function in the s-domain of the inner current loop as:

$$G_{C-i}(s) = \frac{p_2 \frac{1}{s\tau_1} \frac{1}{1+sT}}{1 + p_2 \frac{1}{s\tau_1} \frac{1}{1+sT}} \approx \frac{1}{1 + s\frac{\tau_1}{p_2}} \tag{11.54}$$

We can see that the inner current closed-loop transfer function is likely a first-order circuit with small time constant τ_1/p_2 (usually $p_2 > 1$). The corresponding transfer function in the z-domain is:

$$G_{C-i}(z) = \frac{z}{z - e^{-p_2 T/\tau_1}} \tag{11.55}$$

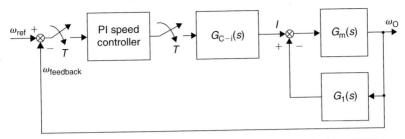

Figure 11.24 The outer system block diagram of the AC motor supplied by a FOH.

The simplified system block diagram is shown in Figure 11.24. The transfer function of the outer closed-loop transfer function in the s-domain is:

$$G_{C-s}(s) = \frac{G_{pi}(s)G_{C-i}(s)\frac{G_m(s)}{1+G_1(s)G_m(s)}}{1+G_{pi}(s)G_{C-i}(s)\frac{G_m(s)}{1+G_1(s)G_m(s)}} = \frac{p_1\frac{1+s\tau_{i1}}{s\tau_{i1}}\frac{1}{1+s\tau_1/p_2}\frac{1/s\tau_m}{1+\frac{1/s\tau_m}{1+s\tau_1}}}{1+p_1\frac{1+s\tau_{i1}}{s\tau_{i1}}\frac{1}{1+s\tau_1/p_2}\frac{1/s\tau_m}{1+\frac{1/s\tau_m}{1+s\tau_1}}}$$

$$(11.56)$$

Usually, the integral element has very large time constant, τ_m, which is much greater than the time constant of the first-order circuit, i.e. $\tau_m \gg \tau_1$. We select the PI speed-control integral time constant, τ_{i1}, to be equal to the time constant, τ_m, of the integral element. We obtain the closed-loop transfer function in the s-domain of the outer speed loop as:

$$G_{C-s}(s) \approx \frac{p_1\frac{1+s\tau_{i1}}{s\tau_{i1}}\frac{1}{1+s\tau_1/p_2}\frac{1}{1+s\tau_m}}{1+p_1\frac{1+s\tau_{i1}}{s\tau_{i1}}\frac{1}{1+s\tau_1/p_2}\frac{1}{1+s\tau_m}} = \frac{p_1\frac{1}{s\tau_m}\frac{1}{1+s\tau_1/p_2}}{1+p_1\frac{1}{s\tau_m}\frac{1}{1+s\tau_1/p_2}} \approx \frac{1}{1+s\tau_m/p_1}$$

$$(11.57)$$

We can see that the outer speed closed-loop transfer function is likely a first-order circuit with much small time constant τ_m/p_1 (usually $p_1 \gg 1$). It means that the AC motor has smaller time constant and quick response under the closed-loop control. The corresponding transfer function in the z-domain is:

$$G_{C-s}(z) = \frac{z}{z - e^{-p_1 T/\tau_m}}$$

$$(11.58)$$

This closed-loop control system is stable since the pole is inside the unity cycle in the z-plane. If the PWM DC/AC inverter with carrier frequency is $f_C = 4\,\text{kHz}$, the sampling interval is $T = 1/f_C = 0.25\,\text{ms}$. The AC induction motor has a mechanical integral time constant τ_m is 300 ms. The proportional gain p_1 of the PI speed-controller can be $p_1 = 20$. Therefore, the closed-loop transfer function is:

$$G_{C-s}(z) = \frac{z}{z - e^{-p_1 T/\tau_m}} = \frac{z}{z - e^{-20 \times 0.25/300}} = \frac{z}{z - 0.98}$$

$$(11.59)$$

The unit-step response is:

$$\Omega_O(z) = G_{C-s}(z)\frac{z}{z-1} = \frac{z}{z-0.98}\frac{z}{z-1} \tag{11.60}$$

This response in the time domain is:

$$\omega_O(t) = 50(e^{-t} - e^{-1.02t}) \tag{11.61}$$

We can see that the digital control system has very quick response. The interference impulse response in the *s*-domain is:

$$\Omega_O(s) = G_{C-s}(s)\frac{1}{s} = \frac{1/s}{1+s\tau_m/p_1} \tag{11.62}$$

The corresponding interference impulse response in the *z*-domain is:

$$\Omega_O(z) = G_{C-s}(z)\frac{z}{z-1} = \frac{z}{z-1}\frac{1-e^{-p_1T/\tau_m}}{z-e^{-p_1T/\tau_m}} \tag{11.63}$$

The corresponding interference impulse response in the time domain is:

$$\omega_O(t) = 1 - e^{-p_1 t/\tau_m} \tag{11.64}$$

It is clearly to illustrate that the AC motor speed response can be very quick if the proportional gain of the PI speed-controller p_1 is large. Normally, in industrial applications the proportional gain of the PI speed-controller p_1 is selected in the range of 10–30. In the mean time the integral time constant τ_{i1} of the PI speed-controller is selected to be equal to the motor integral time constant τ_m, i.e. $\tau_{i1} = \tau_m$. The optimization control can be completed.

11.6 DC MOTOR DRIVES

DC motor drive is the one main method to transfer DC electrical energy to mechanical energy.

11.6.1 DC MOTOR SUPPLIED BY A CHOPPER

Figure 11.25 shows a DC motor supplied by a chopper (assuming the field is a permanent magnet). The power supply is a DC voltage source with DC link voltage V_d and the copping frequency is f, and the sampling period is $T = 1/f$. The stored energy is a DC calculated by Equation (11.4). The pumping energy (*PE*) is:

$$PE = V_d I_d T \tag{11.65}$$

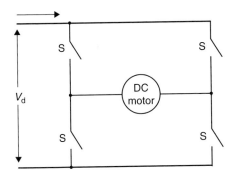

Figure 11.25   A PM DC motor supplied by a chopper.

where V_d is the DC link source voltage, I_d is the DC source current, T is the chopping period ($T = 1/f$, where f is the switching frequency). We can define the energy factor (*EF*) as:

$$EF = \frac{SE}{PE} = \frac{MSE + ESE}{PE} \tag{11.66}$$

We can estimate the transient process from motor stop to a certain speed. The settling time is about:

$$t_{settling} = EF \times T = \frac{MSE + ESE}{V_d I_d} \tag{11.67}$$

The settling time from one running speed to another speed is approximately estimated as:

$$\Delta t_{settling} = \Delta EF \times T = \frac{\Delta MSE + \Delta ESE}{V_d I_d} \tag{11.68}$$

where $\Delta t_{settling}$ is the transient settling time from one running speed to another, ΔEF is the energy factor variation between the two running states, ΔMSE is the energy mechanical stored energy variation between the two running states and ΔESE is the electrical stored energy variation between the two running states. This calculation is similar to the small signal analysis and calculation in Chapter 2.

11.6.2 DC MOTOR SUPPLIED BY AN AC/DC RECTIFIER

Figure 11.26 shows a PM DC motor supplied by an AC/DC rectifier. This power supply system is likely a chopper supplying an AC motor. Therefore, the settling time is still estimated by Equations (11.67) and (11.68).

Usually, DC motor variable-speed drive system with an AC/DC rectifier has closed-loop control. This problem has been well discussed in literature. Therefore, we will discuss the DC motor variable-speed drive system with a DC/DC converter in Section 11.6.3.

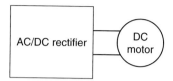

Figure 11.26   A PM DC motor supplied by an AC/DC rectifier.

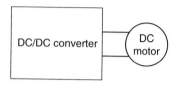

Figure 11.27   A PM DC motor supplied by a power DC/DC converter.

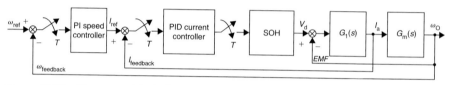

Figure 11.28   The system block diagram of a PM DC motor supplied by a SOH.

11.6.3 VARIABLE-SPEED DC PM MOTOR DRIVE SYSTEM SUPPLIED BY A SOH

Figure 11.27 shows a PM DC motor supplied by a DC/DC converter. This power supply system is likely a chopper supplying a DC motor. Therefore, the settling time is still estimated by Equations (11.67) and (11.68).

Usually, DC motor variable-speed drive system has closed-loop control. The block diagram of the system is shown in Figure 11.28, which is a PM DC motor supplied by a DC/DC converter. The system consists of a PI speed-controller $G_{pi1}(s)$, a PID current-controller $G_{pid2}(s)$, an second-order-hold (SOH) simulating a DC/DC converter, DC motor armature circuit (a first-order circuit with the time constant $\tau_1 = L_a/R_a$) and an integral element with the integral time constant τ_m. The PI speed-controller $G_{pi1}(s)$ has the proportional gain p_1 and the integral time constant τ_{i1}, and the PID current-controller $G_{pid2}(s)$ has the proportional gain p_2, the integral time constant τ_{i2} and the differential time constant τ_{d2}. The inner current closed-loop control transfer function is:

$$G_{C-i}(s) = \frac{G_{pid2}(s)G(s)G_1(s)}{1 + G_{pid2}(s)G(s)G_1(s)} = \frac{p_2 \dfrac{1 + s\tau_{i2} + s^2\tau_{i2}\tau_{d2}}{s\tau_{i2}} \dfrac{1}{1 + s\tau + s^2\tau\tau_d} \dfrac{1}{1 + s\tau_1}}{1 + p_2 \dfrac{1 + s\tau_{i2} + s^2\tau_{i2}\tau_{d2}}{s\tau_{i2}} \dfrac{1}{1 + s\tau + s^2\tau\tau_d} \dfrac{1}{1 + s\tau_1}}$$

$$(11.69)$$

We select the PID current-control integral time constant τ_{i2} and the differential time constant τ_{d2} to be equal to the time constant τ and damping time constant τ_d of the DC/DC converter, respectively. We obtain the closed-loop transfer function in the s-domain of the inner current loop as:

$$G_{C-i}(s) = \frac{p_2 \frac{1}{s\tau} \frac{1}{1+s\tau_1}}{1 + p_2 \frac{1}{s\tau} \frac{1}{1+s\tau_1}} = \frac{1}{1 + s\frac{\tau}{p_2} + s^2 \frac{\tau}{p_2} \tau_1} \tag{11.70}$$

We can see that the inner current closed-loop transfer function is likely a second-order circuit with small time constant τ/p_2 (usually we can select $p_2 < 1$) and an equivalent damping time constant τ_1. Usually the equivalent damping time constant τ_1 is smaller than the time constant τ/p_2. So that this inner closed-loop transfer function can be rewritten as:

$$G_{C-i}(s) = \frac{1}{1 + s\frac{\tau}{p_2}} \tag{11.71}$$

The corresponding transfer function in the z-domain is:

$$G_{C-i}(z) = \frac{z}{z - e^{-p_2 T/\tau}} \tag{11.72}$$

The simplified system block diagram is shown in Figure 11.29. The transfer function of the outer closed-loop transfer function in the s-domain is:

$$G_{C-s}(s) = \frac{G_{pi}(s)G_{C-i}(s)\frac{G_m(s)}{1+G_1(s)G_m(s)}}{1 + G_{pi}(s)G_{C-i}(s)\frac{G_m(s)}{1+G_1(s)G_m(s)}} = \frac{p_1 \frac{1+s\tau_{i1}}{s\tau_{i1}} \frac{1}{1+s\tau/p_2} \frac{1/s\tau_m}{1+\frac{1/s\tau_m}{1+s\tau_1}}}{1 + p_1 \frac{1+s\tau_{i1}}{s\tau_{i1}} \frac{1}{1+s\tau/p_2} \frac{1/s\tau_m}{1+\frac{1/s\tau_m}{1+s\tau_1}}} \tag{11.73}$$

Usually, the integral element has very large time constant τ_m, which is much greater than the time constant of the first-order circuit, i.e. $\tau_m \gg \tau_1$. We select the PI speed-control integral time constant τ_{i1} to be equal to the time constant τ_m of the integral element. We obtain the closed-loop transfer function in the s-domain of the outer speed loop as:

$$G_{C-s}(s) \approx \frac{p_1 \frac{1+s\tau_{i1}}{s\tau_{i1}} \frac{1}{1+s\tau/p_2} \frac{1}{1+s\tau_m}}{1 + p_1 \frac{1+s\tau_{i1}}{s\tau_{i1}} \frac{1}{1+s\tau/p_2} \frac{1}{1+s\tau_m}} = \frac{p_1 \frac{1}{s\tau_m} \frac{1}{1+s\tau/p_2}}{1 + p_1 \frac{1}{s\tau_m} \frac{1}{1+s\tau/p_2}} \approx \frac{1}{1 + \frac{s\tau_m}{p_1}} \tag{11.74}$$

We can see that the outer speed closed-loop transfer function is likely a second-order circuit with much small time constant τ_m/p_1 (usually $p_1 \gg 1$). It means that the PM DC motor has smaller time constant and quick response under the closed-loop control. The corresponding transfer function in the z-domain is:

$$G_{C-s}(z) = \frac{z}{z - e^{-p_1 T/\tau_m}} \tag{11.75}$$

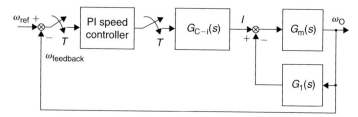

Figure 11.29   The outer system block diagram of the PM DC motor supplied by a SOH.

This closed-loop control system is stable since the pole is inside the unity cycle in the z-plane. If the power DC/DC converter with the switching frequency is $f = 20\,\text{kHz}$, the sampling interval is $T = 1/f = 50\,\mu\text{s}$. The PM DC motor has a mechanical integral time constant τ_m is $300\,\text{ms}$. The proportional gain p_1 of the PI speed-controller can be $p_1 = 20$. Therefore, the closed-loop transfer function is:

$$G_{C-s}(z) = \frac{z}{z - e^{-p_1 T/\tau_m}} = \frac{z}{z - e^{-20 \times 0.05/300}} = \frac{z}{z - 0.996} \tag{11.76}$$

The unit-step response is:

$$\Omega_O(z) = G_{C-s}(z)\frac{z}{z - 1} = \frac{z}{z - 0.996}\frac{z}{z - 1} \tag{11.77}$$

This response in the time domain is:

$$\omega_O(t) = 250(e^{-t} - e^{-1.004t}) \tag{11.78}$$

We can see that the digital control system has very quick response.

The interference impulse response in the s-domain is:

$$\Omega_O(s) = G_{C-s}(s)\frac{1}{s} = \frac{\frac{1}{s}}{1 + \frac{s\tau_m}{p_1}} \tag{11.79}$$

The corresponding interference impulse response in the z-domain is:

$$\Omega_O(z) = G_{C-s}(z)\frac{z}{z - 1} = -\frac{z}{z - 1}\frac{1 - e^{-p_1 T/\tau_m}}{z - e^{-p_1 T/\tau_m}} \tag{11.80}$$

The corresponding interference impulse response in the time domain is:

$$\omega_O(t) = 1 - e^{-p_1 t/\tau_m} \tag{11.81}$$

It is clearly to illustrate that the AC motor speed response can be very quick if the proportional gain of the PI speed-controller p_1 is large. Normally, in industrial applications the proportional gain of the PI speed-controller p_1 is selected in the range of 10–30. In the mean time the integral time constant τ_{i1} of the PI speed-controller is selected to be equal to the motor integral time constant τ_m, i.e. $\tau_{i1} = \tau_m$. The optimization control can be completed.

REFERENCES

1. Francis B. A. and Wonham W. M., The internal model principle of control theory, *Automatica*, Vol. 12, 1976, pp. 457–465.
2. Hara S., Yamamoto Y., Omata T. and Nakano M., Repetitive control system: a new type servo system for periodical exogenous signals, *IEEE Trans Auto Contr*, Vol. 33, 1988, pp. 659–667.
3. Tomizuka M., Tsao T. and Chew K., Analysis and synthesis of discrete-time repetitive controllers, *Trans ASME: J Dyn Syst, Measur Contr*, Vol. 110, 1988, pp. 271–280.
4. Cosner C., Anwar G. and Tomizuka M., Plug in repetitive control for industrial robotic manipulators, *Proc IEEE Int Conf Robot Automat*, 1990, pp. 1970–1975.
5. Broberg H. L. and Molyet R. G., Reduction of repetitive errors in tracking of periodic signals: theory and application of repetitive control, *Proc 1st IEEE Conf Contr Appl*, Dayton, OH, September 1992, pp. 1116–1121.
6. Haneyoshi T., Kawamura A. and Hoft R. G., Waveform compensation of PWM inverter with cyclic fluctuating loads, *IEEE Power Electr Special Conf*, 1987, pp. 745–751.
7. Tzou Y. Y., Ou R. S., Jung S. L. and Chang M. Y., High-performance programmable AC power sourepetitivee with low harmonic distortion using DSP-based repetitive control technique, *IEEE Trans PEL*, Vol. 12, 1997, pp. 715–725.
8. Zhou K., Wang D. and Low K. S., Periodic errors elimination in CVCF PWM DC/AC converter systems: a repetitive control approach, *IEE Proc Contr Theory Appl*, Vol. 147, No. 6, 2000, pp. 694–700.
9. Zhou K. and Wang D., Digital repetitive learning controlled three-phase PWM rectifier, *IEEE Trans PEL*, Vol. 18, No. 1, 2003, pp. 309–316.
10. Griñó R., Costa-Castelló R. and Fossas E., Digital control of a single-phase shunt active filter, *Proc 34th IEEE Power Electr Special Conf*, Acapulco, June 15–19, 2003.
11. Costa-Castelló R., Grinó R. and Fossas E., Odd-harmonic digital repetitive control of a single-phase current active filter, *IEEE Trans PEL*, Vol. 19, No. 4, 2004, pp. 1060–1068.
12. Ye Y. and Wang D., Better robot tracking accuracy with phase lead compensated ILC, *Proc 2003 IEEE Int Conf Robot Automat*, September 2003, pp. 4380–4485.
13. Oh S. J. and Longman R. W., Methods of real-time zero-phase low pass filtering for robust repetitive control, *Proc AIAA/AAS Astrodynam Special Conf Exhibit*, 2002, pp. 4916.1–4916.11.
14. Longman R. W., Iterative learning control and repetitive control for engineering practice, *Int J Contr*, Vol. 73, No. 10, 2000, pp. 930–954.
15. Tomizuka M., Zero phase error tracking algorithm for digital control, *Trans ASME: J Dyn Sys, Meas Contr*, Vol. 109, No. 2, 1987, pp. 65–68.
16. Luo F. L., Jackson R. D. and Hill R. J., Digital closed-loop current controllers using thyristor converters, *IEE-Proc Part B*, Vol. 132, No. 1, January 1985, pp. 46–52.
17. Luo F. L. and Hill R. J., Disturbance response techniques for digital control systems, *IEEE-Trans IE*, Vol. 32, No. 3, August 1985, pp. 245–253.
18. Luo F. L. and Hill R. J., System optimization – self-adaptive controller for digitally controlled thyristor converters, *IEEE-Trans IE*, Vol. 33, No. 3, August 1986, pp. 254–261.

Chapter 12

Applications in Other Branches of Power Electronics

In general, digital control theory is available to be applied in other branches of power electronics such as power factor correction (PFC), static compensation (STATCOM) and flexible AC transmission system (FACTS), reactive power (VAr) compensation, and power quality control (PQC). We will describe digital control theory applying in some branches of power electronics in this chapter.

12.1 INTRODUCTION

Reactive power has no real physical meaning, but is recognized as an essential factor in the design and good operation of power systems. Real and reactive power on a transmission line in an integrated network is governed by the line impedance, voltage magnitudes, the angle difference at the line ends, and the role the line plays in maintaining the network stability under dynamic contingencies. Power transfer in most integrated transmission systems is constrained by transient stability, voltage stability and/or power stability. Reactive power (VAr) compensation or control is an essential part in a power system to minimize power transmission losses, to maximize power transmission capability and to maintain the supply voltage.

It is increasingly becoming one of the most economic and effective solutions to both traditional and new problems in power transmission systems. It is a well-established practice to use reactive at a particular bus bar in any electric power system. In the past, synchronous condensers, mechanically switched capacitors and inductors, and saturated reactors have been applied to control the system voltage in this manner. Since the late 1960s, thyristor controlled reactor (TCR) devices together with fixed capacitors (FCs) or thyristor switched capacitors (TSCs) have been used to inject or absorb reactive power.

Other compensators such as the thyristor controlled series compensator (TCSC) and gate turn-off thyristors (GTO) static VAr compensator (commonly known as advanced static VAr compensator, ASVC) have been applied in power transmission system since 1980s.

12.2 POWER SYSTEMS ANALYSIS

In an ideal electroenergetic system, the voltage and frequency in the various points of power distribution must be constant, presenting only the fundamental component (with no harmonics contents) and a near-unity power factor. In particular, these parameters must be independent of the size and characteristics of the consumer loads; this can be obtained only if these loads are equipped with reactive power compensators to make the network independent from probable changes that appear in the distribution points.

Compensation of the loads is one of the techniques for controlling reactive power; hence to improve the quality of the energy in the AC transmission lines, this technique is generally used for the compensation of individual or a group of loads. This has three essential objectives, namely:

1. power factor correction (PFC);
2. improvement of the voltage regulation;
3. load balancing.

It is noted that PFC and load balancing are desired even when the supply voltage is virtually constant and independent of the load.

12.3 POWER FACTOR CORRECTION

PFC is the capacity of generating or absorbing the reactive power to a load without the use of the supply. The major industrial loads have an inductive power factor (they absorb reactive power); hence the current tends to go beyond the power is usually used for the power conversion, and an excessive load current represents a loss for the consumer, who not only pays for the over-dimensioning of the cable, but also for the excess power loss in the cables. The electric companies do not want to transport the useless reactive power of the alternators toward the loads; these and the distribution network cannot be used at high efficiency, and the voltage regulation in the various points becomes complicated. The principle used by these electric companies almost always penalizes the low-power factor of the clients; hence the great development of systems for power-factor improvement for industrial processes.

PFC technique has the following methods:

- Single-phase active power factor correctors
- Three-phase active power factor correctors

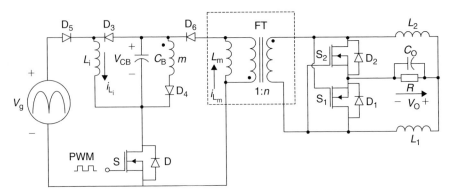

Figure 12.1   Proposed single-stage PFC double-current SR Luo-converter.

- Soft-switching active power factor correctors
- Pulse-width-modulation (PWM) active power factor correctors
- Passive power factor correctors
- Single-stage AC/DC converters.

We take a single-stage high-PFC AC/DC converter as an example for digital control application in this section. The system consists of an AC/DC diode rectifier and a double-current synchronous rectifier (DC-SR) Luo-converter as shown in Figure 7.30 [1]. Although SR DC/DC converters are generally used for low-voltage high-current (LVHC) applications, they are available to perform in normal output voltage level like forward converters.

Figure 12.1 shows the single-stage PFC DC-SR Luo-converter. Suppose that the output inductors L_1 and L_2 are equal to each other, then $L_1 = L_2 = L_O$ is called an output inductor. There are three switches: main switch S, two auxiliary synchronous switches S_1 and S_2. It inherently exhibits high-power factor because the PFC cell operates in continuous conduction mode (CCM). In addition, it is also free to suffer from high-voltage stress across the bulk capacitor at light loads. In order to investigate the dynamical behaviors, the averaging method is used to drive the DC operating point and the small-signal model. Based on the experimental results, the dynamical behavior is verified by the illustration of Bode plots. A proportional-plus-integral-plus-differential (PID) controller is designed to achieve output voltage regulation despite variations in line voltage and load resistance. Finally, a prototype is built and tested to successfully verify the dynamics and performances of the proposed converter.

In power electronic equipments, the PFC circuits are usually added between the bridge rectifier and the loads to eliminate high-harmonics distortion of the line current. In general, they can be divided into two categories: the two-stage approach and the single-stage approach. In the two-stage approach, it includes a PFC stage and a DC/DC regulation stage. It has good PFC and fast output regulations, but the size and cost increase. To overcome the drawbacks, the graft scheme is proposed in Ref. [2]. Many single-stage approaches have been proposed in the literature [3]. It integrates a PFC cell and a DC/DC conversion cell to form a single stage with a common switch. Therefore, the sinusoidal input current waveform and the output voltage regulation can be simultaneously achieved. It thereby meets the requirements of performances and costs.

However, there exists a high-voltage stress across the bulk capacitor C_B at light loads if DC/DC cell operates in discontinuous current mode (DCM). To overcome this drawback, a negative magnetic feedback technique was proposed in literature. However, the dead band exists in the input current and the power factor is thereby degraded. To deal with this problem, the DC/DC cell will operate in DCM. The voltage that crosses the bulk capacitor is independent of loads and the voltage stress is reduced effectively.

In this work, the buck–boost and forward converters are combined to create a single-stage high-PFC converter. The proposed converter inherently exhibits high-power factor because the PFC cell operates in DCM. In addition, it is also free to suffer from high-voltage stress across the bulk capacitor at light loads because the DC/DC cell operates in DCM. The operating principle, steady-state analysis and controller design of the proposed converter are presented in this work. The accuracy of analysis results is verified by experiment and simulation. The output voltage is well regulated despite variations in line voltage and load resistance.

12.3.1 OPERATING PRINCIPLES

Figure 12.1 depicts the proposed forward single-stage high-PFC converter topology. A physical three-winding transformer has turns ratio $1:n:m$. A tertiary transformer winding, in series with diode D_4, is added to the converter for transformer flux resetting. The magnetizing inductance L_m is parallel to the ideal transformer. In the proposed converter, both PFC cell and DC/DC conversion cell are operating in CCM. To simplify the analysis of the circuit, the following assumptions are made:

1. The large-valued bulk capacitor C_B and output capacitor C_O are sufficiently large so that the voltages across the bulk capacitor and output capacitor are approximately constant during one switching period T_S.
2. All switch and diodes of the converter are ideal. The switching time of the switch and the reverse recovery time of the diodes are negligible.
3. The inductors and the capacitors of the converter are considered to be ideal without parasitic components.

Based on the switching-off of the switches and diodes, the proposed converter operating in one switching period T_S can be divided into five linear stages described as follows.

Stage 1 $[0, t_1]$ (S: on, D_1: on, D_2: off, D_3: off, D_4: off, D_5: on, D_6: on): In the first stage, the switch S is turned on. The diodes (D_1, D_5, D_6) are turned on and the diodes

(D_2, D_3, D_4) are turned off. Power is transferred from bulk capacitor C_B to the output via the transformer.

Stage 2 [t_1, t_2] (S: off, D_1: off, D_2: on, D_3: on, D_4: on, D_5: off, D_6: off): The stage begins when the switch S is turned off. The diodes (D_2, D_3, D_4) are turned on and the diodes (D_1, D_5, D_6) are turned off. The current i_{L_i} flows through the diode D_3 and charges the bulk capacitor C_B. The diode D_4 is turned on for transformer flux resetting. In this stage, the output power is provided by the inductor L_O.

Stage 3 [t_2, t_3] (S: off, D_1: off, D_2: on, D_3: off, D_4: on, D_5: off, D_6: off): The stage begins at t_2 when the input current i_{L_i} falls to zero and thus diode D_3 is turned off. The switch S is still off. All diodes, except D_3, maintain their states as shown in the previous stage. During this stage, the voltages $(-v_{C_B}/m)$ and $(-v_O)$ are applied across the inductors L_m and L_O, and thus the inductor currents continue to linearly decrease. The output power is also provided by the output inductor L_O.

Stage 4 [t_3, t_4] (S: off, D_1: off, D_2: off, D_3: off, D_4: on, D_5: off, D_6: off): The stage begins when the current i_{L_O} decreases to zero and thus diode D_2 is turned off. The switch S is still off. The diode D_4 is still turned on and the other diodes (D_1, D_3, D_5, D_6) are still turned off. During this stage, the voltage $(-v_{C_B}/m)$ is applied across inductor L_m. The inductor current continues to linearly decrease. The output power is provided by the output capacitor C_O in this stage.

Stage 5 [t_4, t_5] (S: off, D_1: off, D_2: off, D_3: off, D_4: off, D_5: off, D_6: off): The stage begins when the current i_{L_m} falls to zero and thus diode D_4 becomes off. The switch S is still off and all diodes are also off. The output power is also provided by the output capacitor C_O. The operation of the converter returns back to the first stage when the switch S is turned on again.

According to the analysis of the proposed converter, the key waveforms over one switching period T_S are schematically depicted in Figure 12.2. The slopes of the waveforms $i_{C_O}(t)$ and $i_{C_B}(t)$ are defined as:

$$m_{C_{O1}} = \frac{nv_{C_B} - v_{C_O}}{L_O}, \quad m_{C_{O2}} = -\frac{v_{C_O}}{L_O}, \quad m_{C_{B1}} = -\left[\frac{v_{C_B}}{L_m} + \frac{n(nv_{C_B} - v_{C_O})}{L_O}\right],$$

$$m_{C_{B2}} = -\left(\frac{v_{C_B}}{L_i} + \frac{v_{C_B}}{m^2 L_m}\right), \quad m_{C_{B2}} = -\frac{v_{C_B}}{m^2 L_m} \tag{12.1}$$

12.3.2 MATHEMATICAL MODEL DERIVATION

In this section, the small-signal model of the proposed converter can be derived by the averaging method. The moving average of a variable, voltage or current, over one switching period T_S is defined as the area, encompassed by its waveform and the time axis, divided by T_S.

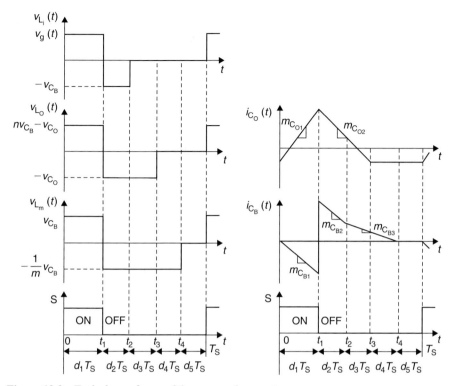

Figure 12.2 Typical waveforms of the proposed converter.

Averaged Model over One Switching Period T_S

There are six storage elements in the proposed converter as shown in Figure 12.1. The state variables of the converter are chosen as the current through the inductor and voltage across the capacitor. Since both PFC cell and DC/DC cells operate in DCM, the initial and final values of the inductor currents vanish in each switching period T_S. From a system point of view, the inductor currents i_{L_i}, i_{L_O} and i_L should not be considered as state variables. Only the bulk capacitor voltage v_{C_B} and the output capacitor voltage v_{C_O} are considered as the state variables of the proposed converter.

For notational brevity, a variable with an upper bar denotes its moving average over one switching period T_S. With the aid of this definition, the averaged state-variable description of the converter is given by

$$C_B \frac{d\bar{v}_{C_B}}{dt} = \bar{i}_{C_B} \tag{12.2}$$

$$C_O \frac{d\bar{v}_{C_O}}{dt} = \bar{i}_{C_O} \tag{12.3}$$

Moreover, in discontinuous conduction, the averaged voltage across each inductor over one switching period is zero. Hence we have three constraints of the form:

$$L_i \frac{d\bar{i}_{L_i}}{dt} = \bar{v}_{L_i} = 0 \tag{12.4}$$

$$L_O \frac{d\bar{i}_{L_O}}{dt} = \bar{v}_{L_O} = 0 \tag{12.5}$$

$$L_m \frac{d\bar{i}_{L_m}}{dt} = \bar{v}_{L_m} = 0 \tag{12.6}$$

The output equation is expressed as:

$$\bar{v}_O = \bar{v}_{C_O} \tag{12.7}$$

Based on the typical waveforms depicted in Figure 12.2, the averaged variables in Equations (12.2) and (12.3) are given by:

$$\bar{i}_{C_B} = \frac{1}{T_S} \sum_{j=1}^{5} \text{area}[i_{C_B(j)}]$$

$$= \frac{1}{T_S} \left\{ d_1^2 T_S^2 m_{C_{B1}} + \frac{1}{2} d_2 T_S^2 [d_2 m_{C_{B2}} + 2(d_3 + d_4) m_{C_{B2}}] + \frac{1}{2} (d_3 + d_4)^2 T_S^2 m_{C_{B3}} \right\}$$

$$\bar{i}_{C_O} = \frac{1}{T_S} \sum_{j=1}^{5} \text{area}[i_{C_O(j)}]$$

$$= \frac{1}{T_S} \left[d_1 T_S^2 (d_1 + d_2 + d_3) \frac{(n\bar{v}_{C_B} - \bar{v}_{C_O})}{2L_O} - T_S \frac{\bar{v}_{C_O}}{R} \right] \tag{12.8}$$

where the notation area $[i_{C_B(j)}]$ denotes the area, encompassed by the waveform $i_{C_B}(t)$ and time axis, during the stage j. Similarly, we have:

$$\bar{v}_{L_i} = \frac{1}{T_S} \sum_{j=1}^{5} \text{area}[v_{L_i(j)}] = \frac{1}{T_S} \left[d_1 T_S \bar{v}_g(t) + d_2 T_S (-\bar{v}_{C_B}) \right]$$

$$\bar{v}_{L_m} = \frac{1}{T_S} \sum_{j=1}^{5} \text{area}[v_{L_m(j)}] = \frac{1}{T_S} \left[d_1 T_S \bar{v}_{C_B} + (d_2 + d_3 + d_4) T_S \left(-\frac{\bar{v}_{C_B}}{m} \right) \right]$$

$$\bar{v}_{L_O} = \frac{1}{T_S} \sum_{j=1}^{5} \text{area}[v_{L_O(j)}] = \frac{1}{T_S} \left[d_1 T_S (n\bar{v}_{C_B} - \bar{v}_{C_O}) + (d_2 + d_3) T_S (-\bar{v}_{C_O}) \right] \tag{12.9}$$

Substituting Equation (12.9) into the constraints (12.4)–(12.6), and performing mathematical manipulations gives:

$$d_2 = \frac{\bar{v}_g(t)}{\bar{v}_{C_B}} d_1 \tag{12.10}$$

$$d_3 = \left(\frac{n\bar{v}_{C_B}}{\bar{v}_{C_O}} - 1 - \frac{\bar{v}_g(t)}{\bar{v}_{C_B}} \right) d_1 \tag{12.11}$$

$$d_4 = \left(m + 1 - \frac{n\bar{v}_{C_B}}{\bar{v}_{C_O}} \right) d_1 \tag{12.12}$$

Now, substituting Equations (12.1) and (12.10)–(12.12) to Equation (12.8), the averaged state equations in Equations (12.2) and (12.3) can be rewritten as:

$$
\begin{aligned}
C_B \frac{d\bar{v}_{C_B}}{dt} &= -d_1^2 T_S \frac{n(n\bar{v}_{C_B} - \bar{v}_{C_O})}{2L_O} + \frac{d_1^2 T_S \bar{v}_g^2(t)}{2L_i \bar{v}_{C_B}} \\
C_O \frac{d\bar{v}_{C_O}}{dt} &= -\frac{\bar{v}_{C_O}}{R} + d_1^2 T_S \frac{n(n\bar{v}_{C_B} - \bar{v}_{C_O})}{2L_O \bar{v}_{C_O}}
\end{aligned}
\tag{12.13}
$$

The averaged rectified line current is given by

$$\bar{i}_g(t) = \frac{1}{T_S}\{\text{area}[i_{L_i(1)}]\} = \frac{1}{T_S} \left[\frac{1}{2}(d_1 T_S)^2 \frac{\bar{v}_g(t)}{L_i} \right] \tag{12.14}$$

It reveals from Equation (12.14) that $\bar{i}_g(t)$ is proportional to $\bar{v}_g(t)$. Thus, the proposed converter is provided with unity power factor.

Averaged Model over One Half Line Period T_L

Based on the derived averaged model described by Equation (12.13) over one switching period T_S, we now proceed to develop the averaged model over one half line period T_L. Since the bulk capacitance and the output capacitance are sufficiently large, both the capacitor voltages can be considered as constants over T_L. Therefore, the state equations of the averaged model over one half line period T_L can be given by:

$$
\begin{aligned}
C_B \frac{d\langle \bar{v}_{C_B} \rangle_{T_L}}{dt} &= \left\langle \frac{d_1^2 T_S}{2} \left[\frac{-n^2 \bar{v}_{C_B} + n\bar{v}_{C_O}}{L_O} + \frac{\bar{v}_g^2(t)}{L_i \bar{v}_{C_B}} \right] \right\rangle_{T_L} \\
&= \frac{1}{\pi} \int_0^\pi \frac{d_1^2 T_S}{2} \left[\frac{-n^2 \bar{v}_{C_B} + n\bar{v}_{C_O}}{L_O} + \frac{v_m^2 \sin^2(\omega t)}{L_i \bar{v}_{C_B}} \right] d(\omega t) \\
&= \frac{d_1^2 T_S}{2} \left[\frac{-n^2 \langle \bar{v}_{C_B} \rangle_{T_L} + n\langle \bar{v}_{C_O} \rangle_{T_L}}{L_O} + \frac{v_m^2}{2L_i \langle \bar{v}_{C_B} \rangle_{T_L}} \right]
\end{aligned}
\tag{12.15}
$$

$$C_O \frac{d\langle \bar{v}_{CO} \rangle_{T_L}}{dt} = \left\langle -\frac{\bar{v}_{CO}}{R} + d_1^2 T_S \frac{(n^2 \bar{v}_{CB}^2 - n\bar{v}_{CB}\bar{v}_{CO})}{2L_O \bar{v}_{CO}} \right\rangle_{T_L}$$

$$= \frac{1}{\pi} \int_0^\pi \left[-\frac{\bar{v}_{CO}}{R} + d_1^2 T_S \frac{(n^2 \bar{v}_{CB}^2 - n\bar{v}_{CB}\bar{v}_{CO})}{L_O \bar{v}_{CO}} \right] d(\omega t)$$

$$= \frac{\langle \bar{v}_{CO} \rangle_{T_L}}{R} + \frac{d_1^2 T_S [-n^2 \langle \bar{v}_{CB} \rangle_{T_L}^2 - n\langle \bar{v}_{CB} \rangle_{T_L} \langle \bar{v}_{CO} \rangle_{T_L}]}{2L_O \langle \bar{v}_{CO} \rangle_{T_L}} \quad (12.16)$$

and the output equation is given by:

$$\langle \bar{v}_O \rangle_{T_L} = \langle \bar{v}_{CO} \rangle_{T_L} \quad (12.17)$$

Notably, Equations (12.15) and (12.16) are non-linear state equations which can be linearized around the DC operating point. The DC operating point can be determined by setting $d\langle \bar{v}_{CB} \rangle_{T_L}/dt = 0$ and $d\langle \bar{v}_{CO} \rangle_{T_L}/dt = 0$ in Equations (12.15) and (12.16). Mathematically, we then successively compute the bulk capacitor voltage V_{CB} and output voltage V_O as:

$$V_{CB} = \frac{1}{2n} \left(\sqrt{\frac{D_1^2 R T_S}{4L_i} + \frac{2L_O}{L_i}} + \sqrt{\frac{D_1^2 R T_S}{4L_i}} \right) \quad (12.18)$$

$$V_O = D_1 \sqrt{\frac{R T_S}{4L_i}} V_m \quad (12.19)$$

The design specifications and component values of the proposed converter are listed in Table 12.1.

According to Table 12.1, it follows directly from Equations (12.18) and (12.19) that $V_{CB} = 146.6$ V and $V_O = 108$ V. Therefore, the proposed converter exhibits low-voltage stress across the bulk capacitor for a voltage of AC (VAC) 110 input voltage.

Table 12.1

Design Specifications and Component Values of the Proposed Converter

Input peak voltage (V_m)	156 V	Duty ratio (D_1)	0.26
Input inductor (L_i)	75 μH	Switching period (T_S)	20 μsec
Magnetizing inductor (L_m)	3.73 mH	Switching frequency (f_s)	50 kHz
Output inductor (L_O)	340 μH	Load resistance (R)	108 Ω
Bulk capacitor (C_B)	330 μF	Turns ratio (1:n:m)	1:2:1
Output capacitor (C_O)	1000 μF	PWM gain (k_{pwm})	1/12 V^{-1}
Bulk capacitor voltage (V_{CB})	146.6 V	Output voltage (V_O)	108 V

After determining the DC operating point, we proceed to derive the small-signal model which is linearized around the operating point. To proceed small perturbations:

$$v_m = V_m + \tilde{v}_m, \quad d_1 = D_1 + \tilde{d}_1, \quad \langle \tilde{v}_{C_B} \rangle_{T_L} = V_{C_B} + \tilde{v}_{C_B}$$
$$\langle \tilde{v}_{C_O} \rangle_{T_L} = V_{C_O} + \tilde{v}_{C_O}, \quad \langle \tilde{v}_O \rangle_{T_L} = V_O + \tilde{v}_O \tag{12.20}$$

with

$$V_m \gg \tilde{v}_m, \quad D_1 \gg \tilde{d}_1, \quad V_{C_B} \gg \tilde{v}_{C_B}, \quad V_{C_O} \gg \tilde{v}_{C_O}, \quad V_O \gg \tilde{v}_O \tag{12.21}$$

are introduced in Equations (12.15) and (12.16), and high-order terms are neglected, yielding the dynamical equations of the form:

$$
\begin{aligned}
C_B \frac{d\tilde{v}_{C_B}}{dt} &= \frac{D_1^2 T_S}{2} \left(-\frac{n^2}{L_O} - \frac{V_m^2}{2 L_i V_{C_B}^2} \right) \tilde{v}_{C_B} + \frac{D_1^2 T_S}{2} \left(\frac{n}{L_O} \right) \tilde{v}_{C_O} + \frac{D_1^2 T_S}{2} \left(\frac{V_m}{L_i V_{C_B}} \right) \tilde{v}_m \\
&\quad + D_1 T_S \left(\frac{-n^2 V_{C_B} + n V_{C_O}}{L_O} + \frac{V_m^2}{2 L_i V_{C_B}} \right) \tilde{d}_1 \\
&= a_{11} \tilde{v}_{C_B} + a_{12} \tilde{v}_{C_O} + b_{11} \tilde{v}_m + b_{12} \tilde{d}_1
\end{aligned}
\tag{12.22}
$$

$$
\begin{aligned}
C_O \frac{d\tilde{v}_{C_O}}{dt} &= \frac{D_1^2 T_S}{2} \left(\frac{2n^2 V_{C_B}}{L_O V_{C_O}} - \frac{n}{L_O} \right) \tilde{v}_{C_B} + \left(-\frac{1}{R} - \frac{D_1^2 T_S}{2} \frac{n^2 V_{C_B}^2}{L_O V_{C_O}^2} \right) \tilde{v}_{C_O} \\
&\quad + 0\tilde{v}_m + D_1 T_S \left(\frac{n^2 V_{C_B}^2}{L_O V_{C_O}} - \frac{n V_{C_B}}{L_O} \right) \tilde{d}_1 \\
&= a_{21} \tilde{v}_{C_B} + a_{22} \tilde{v}_{C_O} + b_{21} \tilde{v}_m + b_{22} \tilde{d}_1
\end{aligned}
\tag{12.23}
$$

The parameters are defined as:

$$a_{11} = \frac{-D_1^2 T_S}{2} \left(\frac{n^2}{L_O} + \frac{V_m^2}{2 L_i V_{C_B}^2} \right), \quad a_{12} = \frac{D_1^2 T_S}{2} \left(\frac{n}{L_O} \right)$$

$$a_{21} = \frac{D_1^2 T_S}{2} \left(\frac{2n^2 V_{C_B}}{L_O V_{C_O}} - \frac{n}{L_O} \right), \quad a_{22} = -\left(\frac{1}{R} + \frac{D_1^2 T_S}{2} \frac{n^2 V_{C_B}^2}{L_O V_{C_O}^2} \right)$$

$$b_{11} = \frac{D_1^2 T_S}{2} \left(\frac{V_m}{L_i V_{C_B}} \right), \quad b_{12} = D_1 T_S \left(\frac{-n^2 V_{C_B} + n V_{C_O}}{L_O} + \frac{V_m^2}{2 L_i V_{C_B}} \right)$$

$$b_{21} = 0, \quad b_{22} = D_1 T_S \left(\frac{n^2 V_{C_B}^2}{L_O V_{C_O}} - \frac{n V_{C_B}}{L_O} \right) \tag{12.24}$$

Mathematically, the dynamical equations (12.22) and (12.23) can be expressed in matrix form as:

$$\begin{bmatrix} \dot{\tilde{v}}_{C_B} \\ \dot{\tilde{v}}_{C_O} \end{bmatrix} = \begin{bmatrix} \frac{a_{11}}{C_B} & \frac{a_{12}}{C_B} \\ \frac{a_{21}}{C_O} & \frac{a_{22}}{C_O} \end{bmatrix} \begin{bmatrix} \tilde{v}_{C_B} \\ \tilde{v}_{C_O} \end{bmatrix} + \begin{bmatrix} \frac{b_{11}}{C_B} & \frac{b_{12}}{C_B} \\ \frac{b_{21}}{C_O} & \frac{b_{22}}{C_O} \end{bmatrix} \begin{bmatrix} \tilde{v}_m \\ \tilde{d}_1 \end{bmatrix} \tag{12.25}$$

$$\tilde{v}_O = \begin{bmatrix} 0 & 1 \end{bmatrix} \begin{bmatrix} \tilde{v}_{C_B} \\ \tilde{v}_{C_O} \end{bmatrix} \tag{12.26}$$

Now taking Laplace transform for the dynamical equation, the resulting transfer functions from line to output and duty ratio to output are given by:

$$\frac{\tilde{v}_O(s)}{\tilde{v}_m(s)} = \frac{\frac{b_{11}a_{21}}{C_B C_O}}{s^2 + \left(-\frac{a_{11}}{C_B} - \frac{a_{22}}{C_O}\right)s + \frac{a_{11}a_{22} - a_{12}a_{21}}{C_B C_O}} \tag{12.27}$$

$$\frac{\tilde{v}_O(s)}{\tilde{d}_1(s)} = \frac{\frac{b_{22}}{C_O}s + \frac{a_{21}b_{12} - a_{11}b_{22}}{C_B C_O}}{s^2 + \left(-\frac{a_{11}}{C_B} - \frac{a_{22}}{C_O}\right)s + \frac{a_{11}a_{22} - a_{12}a_{21}}{C_B C_O}} \tag{12.28}$$

12.3.3 MODEL VALIDATION

One 108 W prototype based on the topology presented in Figure 12.1, with the design specifications and component values listed in Table 12.1, is implemented to verify its operating principle. Substituting the specifications in Table 12.1 into Equation (12.24) gives:

$$\begin{array}{ll}
a_{11} = -1.31 \times 10^{-2} & a_{12} = 3.98 \times 10^{-3} \\
a_{21} = 1.76 \times 10^{-2} & a_{22} = -2.39 \times 10^{-2} \\
b_{11} = 9.59 \times 10^{-3} & b_{12} = 8.98 \times 10^{-2} \\
b_{21} = 0 & b_{22} = 7.69
\end{array} \tag{12.29}$$

From Equations (12.27) and (12.28), the transfer functions from line to output and duty ratio to output are given by:

$$\frac{\tilde{v}_O(s)}{\tilde{v}_m(s)} = \frac{511.95}{s^2 + 63.48s + 733.84} \tag{12.30}$$

$$\frac{\tilde{v}_O(s)}{\tilde{d}_1(s)} = \frac{7689.63(s + 40.19)}{s^2 + 63.48s + 733.84} \tag{12.31}$$

The Bode plots of the transfer function $\tilde{v}_O(s)/\tilde{d}_1(s)$ are presented in Figure 12.3. The curve (1) is the measurement result and curve (2) is the theoretical plot of Equation

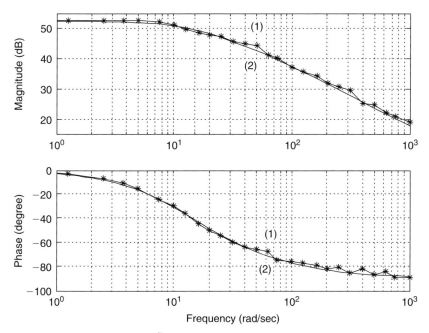

Figure 12.3 Bode plots of $\tilde{v}_O(s)/\tilde{d}_1(s)$.

(12.31) of the implemented converter. It reveals that the theoretical analysis predicts the dynamical behavior of the proposed converter.

12.3.4 SIMULATION RESULTS

The PSpice simulation results presented in Figure 12.4 demonstrate that both PFC and DC/DC cells are operating in DCM. The input inductor current $i_{L_i}(t)$ and output inductor current $i_{L_O}(t)$ both reach zero for the remainder of the switching period. Figure 12.5(a) presents the bulk capacitor voltage $V_{CB} = 149\,\text{V}$ and Figure 12.5 (b) presents the output capacitor voltage $V_{CO} = 110\,\text{V}$. They are close to the theoretical results $V_{CB} = 146.6\,\text{V}$ and $V_{CO} = 1080\,\text{V}$. Figure 12.5 presents the rectified line voltage and current in (a), and the line voltage and current in (b). It reveals that the proposed converter has high-power factor. According to the total harmonic distortion (*THD*) obtained in the simulation results, the power factor is calculated to be $PF = 0.999$.

12.3.5 EXPERIMENTAL RESULTS

One prototype based on the topology depicted in Figure 12.1 is built and tested to verify its operating principle of the proposed converter. The experimental results are depicted in the following figures. Figure 12.6 presents the waveform of the input

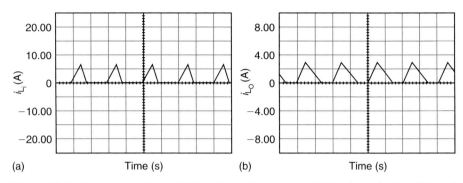

Figure 12.4 Current waveforms: (a) input inductor current $i_{L_i}(t)$ (horizontal: 10 μs/div) and (b) output inductor current $i_{L_O}(t)$ (horizontal: 10 μs/div).

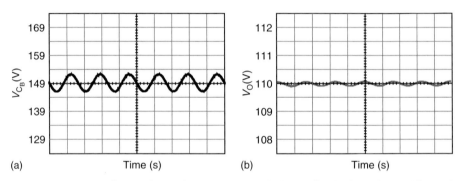

Figure 12.5 (a) Bulk capacitor voltage $V_{C_B}(t)$ (vertical: 5 V/div, horizontal: 5 ms/div) and (b) output capacitor voltage $V_{C_O}(t)$ (vertical: 0.5 V/div, horizontal: 5 ms/div).

voltage and current. Figure 12.7 presents the waveform of the input inductor current $i_{L_i}(t)$ and output inductor current $i_{L_O}(t)$. Figure 12.8 presents the voltage ripples of the bulk capacitor voltage $V_{C_B}(t)$ and output capacitor voltage $V_{C_O}(t)$. Figure 12.9 presents the rectified line voltage and current, and line voltage and current. The proposed converter exhibits low-voltage stress and high-power factor. The measured power factor of the converter is $PF = 0.998$. The efficiency of the proposed converter is about 72%.

12.3.6 CONTROLLER DESIGN

In this section, a PID controller will be employed to regulate output voltage despite variations in the line voltage and load resistance. From a digital control point of view, a DC/DC converter is needed to eliminate the steady-state error. As a result, a PID

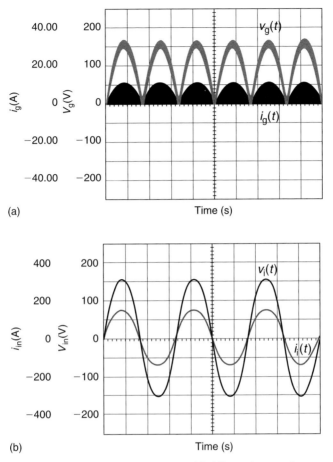

Figure 12.6 The line voltages and currents: (a) rectified line voltage and current (horizontal: 5 ms/div) and (b) input line voltage and current (horizontal: 5 ms/div).

controller is designed with the transfer function:

$$G_{\text{pid}}(s) = p \frac{1 + s\tau_{\text{i}} + s^2 \tau_{\text{i}} \tau_{\text{d}}}{s\tau_{\text{i}}} \qquad (12.32)$$

We select the integral time constant (τ_{i}) and differential time constant (τ_{d}) to eliminate the time constant (τ) and the damping time constant (τ_{d}). We finally obtain the closed-loop transfer function as:

$$G_{\text{C}}(s) = \frac{1}{1 + s\frac{\tau_{\text{i}}}{p}}$$

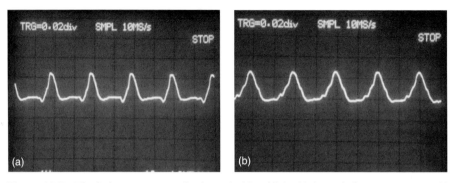

Figure 12.7 The inductor currents (horizontal: 10 μs/div): (a) input inductor currents $i_{L_i}(t)$ (vertical: 5 A/div) and (b) output inductor current $i_{L_O}(t)$ (vertical: 2 A/div).

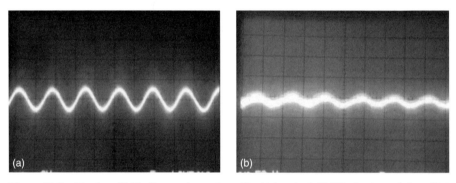

Figure 12.8 Ripples of (a) bulk capacitor voltage $V_{C_B}(t)$ (vertical: 5 V/div, horizontal: 5 ms/div) and (b) output capacitor voltage $V_{C_O}(t)$ (vertical: 0.5 V/div, horizontal: 5 ms/div).

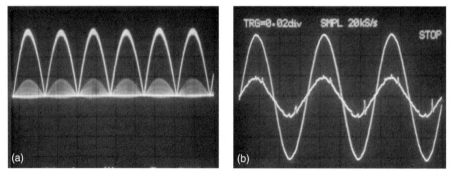

Figure 12.9 The line voltages and currents (horizontal: 5 ms/div): (a) rectified line voltage and current (vertical: 50 V/div, 10 A/div) and (b) input line voltage and current (vertical: 50 V/div, 2 A/div).

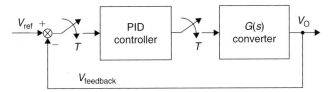

Figure 12.10 The block diagram of the control system.

Figure 12.10 presents the block diagram of the overall system with reference input voltage $V_{ref} = 9.8$ V and regulated output voltage $V_O = 108$ V, as shown in Table 12.1.

The load R is initially set to 218 Ω and the line peak voltage V_m is set to 156 V at $t = 0$ s, then changed to a heavy load $R = 162$ Ω at $t = 0.2$ s and $R = 108$ Ω at $t = 0.3$ s, and then changed to a lower-line peak voltage $V_m = 142$ V at $t = 0.4$ s. From an energy conservation point of view, the average output power $\langle P_O \rangle_{T_S} = V_O^2/R$ is equal to the average input power $\langle P_i \rangle_{T_S} = V_i I_i$. If the output voltage V_O is fixed, the average output power $\langle P_O \rangle_{T_S}$ increases as the load R decreases. The average input I_i thereby increases for a fixed input voltage V_i. Hence the duty ratio D must increase, and thus the control voltage v_{ctrl} increases necessarily. On the other hand, if V_O and R are fixed and V_i decreases, then average input current I_i will increase to maintain the constant average power. As mentioned before, the control voltage v_{ctrl} thereby increases.

In this work, the buck–boost and forward converters are combined to create a single-stage high-PFC converter. The proposed converter exhibits high-power factor, low-voltage stress and output voltage regulation. The operating principle, the operating point and the small-signal model of the proposed converter are also studied in this work. Based on the classical control theory, a PID controller is designed to achieve fast output voltage regulation. An AC/DC prototype with output power 108 W is built and tested to verify its operating principle of the proposed converter. It reveals from the simulation and experimental results that the designed controller can be used, without degrading the power factor of the converter, to achieve output voltage regulation despite variations in line voltage and load resistance.

12.4 STATIC COMPENSATION (STATCOM)

We investigate the topic of STATCOM, titled Trinary Hybrid Multilevel Inverter Used in STATCOM with Unbalanced Voltages, in this section. A trinary hybrid multilevel inverter is applied Synchronous STATCOM with unbalanced voltages. Benefiting from trinary hybrid topology of the inverter, the cost of STATCOM is reduced because of not only fewer switching components but also of reduced cost of DC capacitors. The combination of vector control based on synchronous frame and staircase modulation is used in presented STATCOM system to regulate reactive power or balance bus voltages under balanced or unbalanced conditions. To achieve this control aim, a new method based on the comparison of reference amplitudes and reference signals is presented in this chapter. The performance of the proposed control strategy is confirmed by simulation and experiment.

Synchronous STATCOM is a FACTS device, which is connected as a shunt to the network, for generating or absorbing reactive power. STATCOM can be utilized to regulate voltage, control power factor and stabilize power flow [12]. Many inherent benefits of multilevel inverter have led to their increased interest in STATCOM. In Refs [13,14], cascade multilevel inverters have been used in STATCOM. Furthermore, the application of binary hybrid multilevel inverters in STATCOM has also been investigated because the binary hybrid multilevel inverter can generate more voltage levels than the cascade multilevel inverter with the same number of switches [15–17]. Recently, trinary hybrid multilevel inverter have been presented and attracted more interest since it is said that it can generate most voltage levels among existing multilevel inverters [18,19].

The application of trinary hybrid multilevel inverter in STATCOM is investigated in this section. In this topology, not only fewer switches are required, but also the cost of DC capacitors is decreased. Moreover, the problem of regenerative power in trinary hybrid multilevel inverters mentioned in Ref. [20] is avoided because the STATCOM mainly generates or absorbs reactive power.

Voltage imbalance is a problem that STATCOM must deal with in the distribution system. Steady-state voltage imbalance can arise from unequal loading on each phase or from unbalanced faults on the power system, which cause single-phase voltage sags. These sags are detrimental since they cause heating in motors and affect sensitive single-phase loads. The presented STATCOM can balance bus voltages under unbalanced conditions.

The staircase PWM is widely used in STATCOM [23,24] since GTOs with lower-switching frequency are employed as switches in such applications of high power and high voltage [25–27]. The vector control based on synchronous frame transform has been used successfully in STATCOM to regulate reactive power [13,21] and reduce negative sequence component of the bus voltage [22]. In this chapter, the vector control and staircase modulation are combined to reach the control aims. The challenge here is that the conventional method based on the comparison of switching angles and phase angles [13,23,24] to generate the switching signals cannot work well in such control system. A new method based on the comparison of reference amplitudes and reference signals is proposed in this chapter. Furthermore, dead-zone control is used to improve the performance of the inverter. The performance of the proposed control system is confirmed by simulation and experiment.

12.4.1 SYSTEM CONFIGURATION

System configuration is described in this section.

Configuration of STATCOM System

Figure 12.11 shows one line diagram of a distribution system with STATCOM and Figure 12.12 shows a simplified model of Figure 12.11. The STATCOM that is based on a three-phase 9-level trinary hybrid multilevel inverter is connected to bus B through the interface impedance Z_I. Z_S is the equivalent impedance of the source and Z_L is the

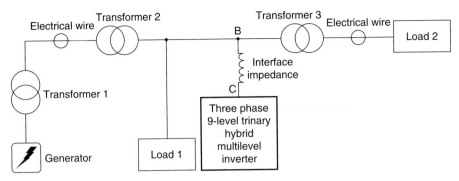

Figure 12.11 Distribution system with the STATCOM.

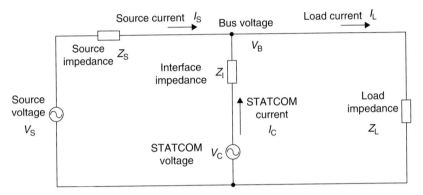

Figure 12.12 Simplified model of the distribution system with the STATCOM.

equivalent impedance of the load. In steady states and balanced conditions, voltages and currents can be expressed as phasors. In Figure 12.12, V_S is the source voltage, V_B is the bus voltage, V_C is the generated voltage of the STATCOM, I_C is the current generated by the STATCOM, I_S is the source current and I_L is the load current. Parameters of the compensator and the distribution system are shown in Table 12.2. The STATCOM in steady states will generate a leading reactive current when the amplitude of V_C is larger than that of V_B, and vice verse, it will draw a lagging current from the source.

Three-Phase 9-Level Trinary Hybrid Multilevel Inverter

Figure 12.13 shows a three-phase Y-configured 9-level trinary hybrid multilevel inverter used in the STATCOM. The inverter has separate DC capacitors for each H-bridge unit of each phase. To get maximum output voltage levels of the inverter, the ratio of DC capacitor voltages is arranged as 1:3, so that the inverter can output 9 V levels for each phase. With two H-bridges per phase, however, a cascade multilevel inverter only can output 5 V levels for each phase and a binary hybrid multilevel inverter only can output 7 V levels for each phase. The more the output voltage levels a multilevel inverter

Table 12.2

Parameters of the Compensator and the Distribution System for Simulations and Experiments

	Simulations	Experiments
Operating frequency	50 Hz	50 Hz
Rating of source voltage (line-to-line rms value)	13.5 kV	196 V
Rating of reactive power	10 MVAr	2000 VAr
Rating of STATCOM current (phase rms value)	428 A	6 A
Interface impedance per phase	$\omega L_I = 3.64\,\Omega$, $R_I = 0.3\,\Omega$	$\omega L_I = 3.64\,\Omega$, $R_I = 0.3\,\Omega$
Source impedance per phase	$\omega L_S = 2.2\,\Omega$, $R_S = 0.3\,\Omega$	$\omega L_S = 2.2\,\Omega$, $R_S = 0.3\,\Omega$
Unit voltage of DC capacitors (U_D)	3.3 kV	48 V
DC capacitor in the H-bridge with DC voltage U_D	2933 μF	3300 μF
DC capacitor in the H-bridge with DC voltage $3U_D$	1114 μF	1100 μF

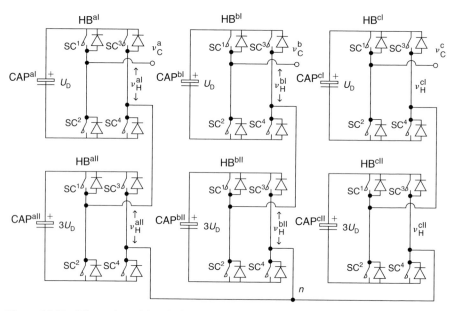

Figure 12.13 Three-phase 9-level trinary hybrid multilevel inverter.

has, the more the similar sinusoid waveform can be synthesized. Thereby, more lower-order harmonics can be eliminated and total harmonic distortion (*THD*) can be reduced greatly. Since the ratio of DC capacitor voltages of the inverter is trinary, this kind of inverter can be called trinary hybrid multilevel inverter.

Three phases of the inverter are controlled separately and the operation principle of each phase is identical. In the following, the A-phase of the inverter is analyzed. HB^{ak}

Table 12.3

Values of Switching Functions for Different Values of v_C^a

v_C^a	$4U_D$	$3U_D$	$2U_D$	U_D	0	$-U_D$	$-2U_D$	$-3U_D$	$-4U_D$
SF^{aI}	1	0	−1	1	0	−1	1	0	−1
SF^{aII}	1	1	1	0	0	0	−1	−1	−1

represents the kth H-bridge in the A-phase leg of the inverter, where superscript "a" means A-phase and k can be I or II. v_H^{ak} and v_D^{ak} represent the output voltage and the DC capacitor voltage of the HB^{ak}, respectively. A switching function, SF^{ak}, is used to relate v_H^{ak} and v_D^{ak} as:

$$v_H^{ak} = SF^{ak} \cdot v_D^{ak} \quad k = I, II \qquad (12.33)$$

The value of SF^{ak} can be either 1, or −1 or 0. For the value 1, switching components SC^1 and SC^4 need to be turned on. For the value −1, switches SC^2 and SC^3 need to be turned on. For the value 0, switches SC^1 and SC^3, or SC^2 and SC^4 need to be turned on. The A-phase voltage of the inverter, v_C^a, is represented as:

$$v_C^a = \sum_{k=I}^{II} (SF^{ak} \cdot v_D^{ak}) \qquad (12.34)$$

The unit voltage of DC capacitors is U_D, i.e. v_D^{aI} and v_D^{aII} are U_D and $3U_D$, respectively. So v_C^a has nine levels totally. Table 12.3 shows the values of switching functions for different values of v_C^a.

Two types of modulation for multilevel inverters have been presented in power applications: carried-based PWM strategies and optimal programmed PWM strategies [0]. With much lower-switching frequency, the optimal programmed PWM strategies can eliminate the same number of lower-order harmonics as the carried-based PWM strategies. In this distribution system, the proposed STATCOM works under high voltages, so GTOs are selected as the switching components that cannot switch at high frequency. The staircase PWM, the most popular one among optimal programmed PWM strategies, is used in the proposed STATCOM.

Figure 12.14 shows the A-phase waveforms of the inverter (only waveforms with solid lines are considered). By applying the Fourier transform to v_C^a, the amplitude of any odd nth harmonic of v_C^a can be expressed as Equation (12.35), whereas the amplitudes of all even harmonics are zero:

$$\left| v_C^a \right|_n = \frac{4U_D}{n\pi} \sum_{j=1}^{4} \cos(n\phi_j) \quad n = 1, 3, 5, \ldots \qquad (12.35)$$

The switching angles, ϕ_1, ϕ_2, ϕ_3 and ϕ_4, are chosen so as to cancel predominant lower-frequency harmonics. For the 9-level case in Figure 12.14, the 5th, 7th and 11th harmonics can be eliminated with the appropriate choice of the switching angles. One

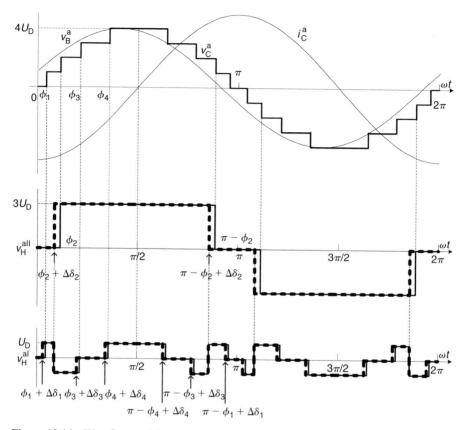

Figure 12.14 Waveforms of the output voltage of A-phase inverter, the A-phase current and the output voltages of H-bridges of A-phase inverter.

degree of freedom is used so that the magnitude of the output waveform corresponds to the modulation index of A-phase, M^a, which is expressed as:

$$M^a = \frac{\pi \cdot |v^a_C|_1}{16 U_D} \tag{12.36}$$

where $|v^a_C|_1$ is the amplitude of the fundamental component of v^a_C. Let the equations from Equation (12.35) be as follows:

$$\begin{aligned}
\cos(\phi_1) + \cos(\phi_2) + \cos(\phi_3) + \cos(\phi_4) &= 4M^a \\
\cos(5\phi_1) + \cos(5\phi_2) + \cos(5\phi_3) + \cos(5\phi_4) &= 0 \\
\cos(7\phi_1) + \cos(7\phi_2) + \cos(7\phi_3) + \cos(7\phi_4) &= 0 \\
\cos(11\phi_1) + \cos(11\phi_2) + \cos(11\phi_3) + \cos(11\phi_4) &= 0
\end{aligned} \tag{12.37}$$

Table 12.4

Table of Theoretical Switching Angles for Different Modulation Index

Modulation index (M^a)	ϕ_1(rad)	ϕ_2(rad)	ϕ_3(rad)	ϕ_4(rad)
0.56	0.59438	0.85477	1.0383	1.3207
⋮	⋮	⋮	⋮	⋮
0.78	0.17641	0.39873	0.726	1.086
0.79	0.17371	0.37637	0.6986	1.0709
0.8	0.17175	0.35575	0.6703	1.0545
⋮	⋮	⋮	⋮	⋮
0.85	0.078667	0.35911	0.48161	0.95097

Table 12.5

Comparison of GTO Counts

	Trinary hybrid 9-level multilevel inverter	Cascade 9-level multilevel inverter
GTOs in series per valve in the H-bridge with DC voltage U_D	3 (1 redundant)	3 (1 redundant)
GTOs in series per valve in the H-bridge with DC voltage $3U_D$	6 (1 redundant)	
Total number of GTOs	108	144

Table 12.4 shows the off-line calculated switching angles, which are stored in a look-up table.

Counts of GTOs

GTOs are selected as switching components in the STATCOM. A readily available GTO (MITSUBISHI GTO FG1000BV-90DA) has a typical repetitive peak off-state voltage of 4.5 kV and repetitive controlled on-state current of 1 kA [29]. Normally, the GTO repetitive peak off-state voltage and repetitive controlled on-state current are chosen to be 2–3 times of the system nominal ratings. In each H-bridge module, GTOs are connected in series to make up rated DC source voltage to satisfy the redundancy requirement. The redundancy requirement is that if any single GTO fails (such as short circuit) in one inverter arm, the remaining functional GTOs can sustain continuous operation until the next planned maintenance outage. The number of GTOs required in trinary hybrid 9-level inverter and common cascade 9-level inverter are given in Table 12.5, where unit voltage of DC capacitors, U_D, is 3.3 kV. The comparison shows the trinary hybrid multilevel inverter which uses fewer GTOs since the fewer redundant switches are needed in trinary hybrid topology.

Series Connection of GTOs

One of the advantages of a multilevel inverter is to achieve high voltage without having to connect switching devices in series directly. But with the increase of power and voltage of the applications, the series connection of switching devices is inevitable. The total DC link voltages in each phase is 13.2 kV in this chapter and in Ref. [30], 16 kV in Ref. [31] and 38.4 kV in Ref. [32]. If the cascade multilevel inverters are used in these systems and each valve contains only one GTO specified in the chapter, each phase of inverter will contain 8 H-bridges in this chapter and Ref. [30], 10 H-bridge in Ref. [31] and 23 H-bridges in Ref. [32]. Too many H-bridges result in very complicated power circuits, too many control signals and bulky system. Therefore, in very high-power and high-voltage application, the series connection of power semiconductors is necessary.

Early, a large snubber is used to limit the dv/dt of the switching component during the turn-off period [33], and a variable inductor is to be placed between each gate drive circuit and the corresponding switching component in order to control the rising/falling time and to adjust the transient voltage [30,35]. These modifications will increase the losses and reduce the dynamic performance, so a system with fewer GTOs connected in series provides better voltage sharing, and improves efficiency and dynamic performance [15]. Recently, however, the technology that allows the robust, reliable and cost-efficient series connection of GTOs is industrially mature [30]. Especially, in Ref. [32], with a digital control circuit for extremely accurate adjustment of gate turn-off timing in units of 0.1 μs, up to 16 GTOs connected in series shared the voltage uniformly in the turn-off period. This technology of adjustment can be available even if the number of GTOs is further increased for a higher-voltage converter in the future. And with such technology, a system with more GTOs connected in series has almost the same voltage sharing, dynamic performance and efficiency as a system with fewer GTOs connected in series. Thanks to the precise gate turn-off timing adjustment, presented in Ref. [32], the trinary inverter in which up to six GTOs connected in series are not only feasible, but also have almost the same performances as the cascade multilevel inverter in which three GTOs are connected in series.

Device Power Loss and the Cost of Cooling Systems

Figures 12.15 and 12.16 show waveforms of the currents flowing through arms of the HB^{al} and the HB^{all}, respectively. i_A^{aij} ($i =$ I, II, $j = 1, \ldots, 4$) is the current flowing through the jth arm of the HB^{ai}. The positive value of i_A^{aij} means that the current flows through GTOs, while the negative value of i_A^{aij} implies that the current flows through antiparallel diodes. As mentioned previously, if the output voltage of a H-bridge is zero, the switches SC^1 and SC^3 will be turned on or the switches SC^2 and SC^4 will be turned on. For the purpose of balancing the current stresses and power losses of switches, both of these two switching states for the zero output voltage of a H-bridge are used and each switching state is used in an alternative cycle. In Figure 12.5, if v_H^{all} is 0, the SC^1 and SC^3 of the HB^{all} are turned on when ωt is from ϕ_2 to $\phi_2 + 2\pi$, but the SC^2 and SC^4 of the HB^{all} are turned on when ωt is from 0 to ϕ_2 or from $\phi_2 + 2\pi$

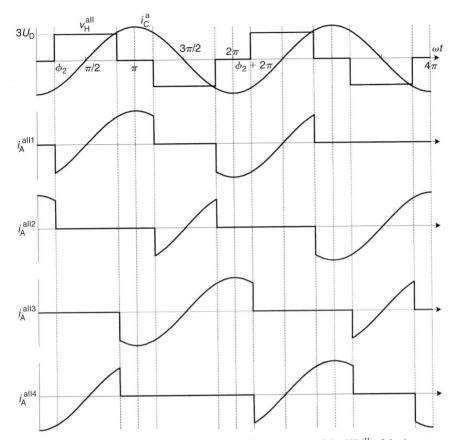

Figure 12.15   Waveforms of the currents flowing through arms of the HB^{all} of the inverter.

to 2π. In Figure 12.6, if v_H^{al} is 0, the SC^1 and SC^3 of the HB^{al} are turned on when ωt is from ϕ_1 to $\phi_1 + 2\pi$, but the SC^2 and SC^4 of the HB^{al} are turned on when ωt is from 0 to ϕ_1 or from $\phi_1 + 2\pi$ to 2π.

From Figures 12.15 and 12.16, we can find out that the combination of waveforms of the non-zero current that flows through an arm during two periods is just the waveform of i_C^a in a complete period. So the on-state power losses of GTO, $P_{G,ON}$, and the on-state power losses of antiparallel diode, $P_{D,ON}$, can be expressed as:

$$P_{G,ON} = \frac{V_G}{2T} \int_{\pi/2}^{3\pi/2} i_C^a \, \mathrm{d}(\omega t) \tag{12.38}$$

$$P_{D,ON} = \frac{V_D}{2T} \int_{\pi/2}^{3\pi/2} i_C^a \, \mathrm{d}(\omega t) \tag{12.39}$$

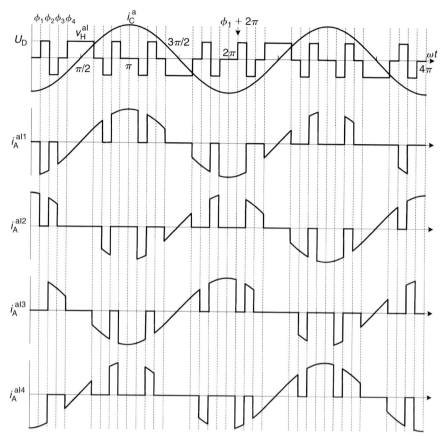

Figure 12.16 Waveforms of the currents flowing through arms of the HB^{al} of the inverter.

where V_G and V_D are the voltage drops of a GTO and a diode, respectively, if they are in on-state. T is the period of i_C^a. V_G is 2.8 V approximately according to the data sheet of the GTO and V_D is 1.3 V approximately. So, from Equations (12.38), (12.39) and Table 12.2, we can get $P_{G,ON}$ as 269 W and $P_{D,ON}$ as 125 W for the worst cases.

As shown in Figure 12.15, the current that flows through a GTO in the HB^{all} is $\left| i_C^a \right|_1 \sin \phi_2$ before the GTO is turned off, where $\left| i_C^a \right|_1$ is the amplitude of the A-phase STATCOM current. For the worst case, the current is 454 A. Figure 12.17 shows the data sheet of turn-on and turn-off switching energy of the GTO (MITSUBISHI GTO FG1000BV-90DA) when the DC off-state voltage is 2250 V. Based on Figure 12.17, the turn-off switching energy is 1.6 J when DC off-state voltage is 2250 V. As shown in Table 12.5, there are five GTOs (not including redundant one) connected in series in an arm of the HB^{all}, so the GTOs endure the off-state voltage, $3U_D/5$ (1980 V), after they are turned off. The turn-off switching energy of the GTO in the HB^{all} can be calculated as $1.6 \times 1980/2250 = 1.4$ J. From Figure 12.15, one can find out the switching losses of the GTO in the HB^{all} are due to a switching-off process of the GTO in a period. So,

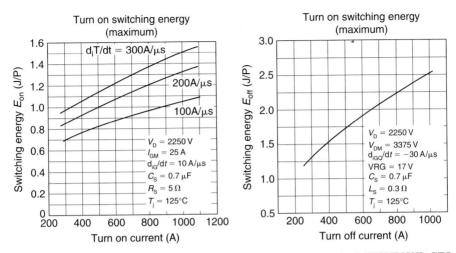

Figure 12.17 Turn-on and turn-off switching energy of the GTO (MITSUBISHI GTO FG1000BV-90DA).

Table 12.6

Comparison of Device Power Losses and Cost of Cooling Systems

		Worst cases considered (redundant GTOs excluded)			Cost of cooling system for a GTO and an antiparallel diode ($)	Total cost of cooling systems in three-phase inverter ($)
	H-bridge (DC voltage)	GTO on-state power losses (W)	GTO switching losses (W)	Antiparallel diode on-state power losses (W)		
Trinary hybrid multilevel inverter	HB (U_D)	269	194	125	$588k_{\text{cooling}}$	$54,576k_{\text{cooling}}$
	HB ($3U_D$)	269	70	125	$464k_{\text{cooling}}$	
Cascade multilevel inverter	HB (U_D)	269	51	125	$445k_{\text{cooling}}$	$65,644k_{\text{cooling}}$
	HB (U_D)	269	62	125	$456k_{\text{cooling}}$	
	HB (U_D)	269	66	125	$460k_{\text{cooling}}$	
	HB (U_D)	269	69	125	$463k_{\text{cooling}}$	

this switching losses can be calculated as $P^{\text{all}}_{\text{G,SW}} = 1.4 \times 50 = 70$ W for the worst case. The switching losses of the GTO in the HB^{al} are due to three switching-off processes and two switching-on processes in a period. By the same method, the switching losses of the GTO in the HB^{al} are calculated as $P^{\text{al}}_{\text{G,SW}} = 194$ W for the worst case.

If a 9-level cascade multilevel inverter is used in such a STATCOM system, the on-state losses and switching losses of the GTOs for the worst cases can be calculated by the same method above and the calculating results are shown in Table 12.6.

Suppose the cost of cooling system is proportional to the power losses, i.e.:

$$C_{cooling} = k_{cooling} P_{loss} \tag{12.40}$$

where $C_{cooling}$ is the cost of cooling system, P_{loss} is power losses and $k_{cooling}$ is the coefficient whose unit is \$/W. As shown in Table 12.6, the cost of the cooling system for a GTO and an antiparallel diode in the H-bridge with $3U_D$ DC voltage in the trinary hybrid multilevel inverter is almost the same as that in the cascade multilevel inverter. Comparatively, the cost of the cooling system for a GTO and an antiparallel diode in the H-bridge with U_D DC voltage in the trinary hybrid multilevel inverter are a little higher, since the GTO switches are at higher frequency. However, the total cost of cooling systems for the trinary hybrid multilevel inverter are lower than that for the cascade multilevel inverter as shown in Table 12.6 since fewer GTOs are used in trinary hybrid multilevel inverters.

Cost of DC Capacitors

The trinary hybrid inverter not only need fewer GTOs than the cascade multilevel inverter, but also has less cost of DC capacitors. Firstly the required capacitances of DC capacitors of the trinary hybrid multilevel inverter are analyzed. Figure 12.14 shows A-phase waveforms of the inverter (only waveforms with solid lines are considered). The STATCOM supplies reactive power, so the output voltage and output current of the inverter is orthogonal. Suppose the current generated by the STACOM is sinusoidal and $\left| i_C^a \right|_1$ is the amplitude of A-phase current. The A-phase current of the STATCOM in Figure 12.14 can be expressed as:

$$i_C^a = \left| i_C^a \right|_1 \sin(\omega t - \pi/2) \tag{12.41}$$

where ω is radian frequency. Assuming initial voltages of A-phase DC capacitors are U_D and $3U_D$, respectively.

Table 12.7 shows capacitor voltages during first half cycle in Figure 12.14, where

$$m_1 = \frac{\left| i_C^a \right|_1}{\omega C^{al}}, \qquad m_2 = \frac{\left| i_C^a \right|_1}{\omega C^{all}} \tag{12.42}$$

The ε, DC voltage regulation factor, is selected as 5%, which means the DC capacitor voltages fluctuate within 0.95–1.05 times of the normal value. To keep the DC capacitor voltages within this range, the capacitances of DC capacitors are expressed as:

$$C^{al} = \frac{\left| i_C^a \right|_1 \text{MAX}[\text{ABS}(\sin \phi_1 - \sin \phi_2), \text{ABS}(\sin \phi_1 - 2\sin \phi_2 + \sin \phi_3),}{\text{ABS}(\sin \phi_1 - 2\sin \phi_2 + \sin \phi_3 + \sin \phi_4 - 1)]}{2\omega\varepsilon U_D} \tag{12.43}$$

$$C^{all} = \frac{\left| i_C^a \right|_1 (1 - \sin \phi_2)}{6\omega\varepsilon U_D} \tag{12.44}$$

Table 12.7

A-Phase DC Capacitor Voltages During Half Cycle

Instant	$v_{\mathrm{D}}^{\mathrm{aI}}$	$v_{\mathrm{D}}^{\mathrm{aII}}$
0	U_{D}	$3U_{\mathrm{D}}$
$\phi_1/(\omega t)$	U_{D}	$3U_{\mathrm{D}}$
$\phi_2/(\omega t)$	$U_{\mathrm{D}} - m_1(\sin\phi_1 - \sin\phi_2)$	$3U_{\mathrm{D}}$
$\phi_3/(\omega t)$	$U_{\mathrm{D}} - m_1(\sin\phi_1 - 2\sin\phi_2 + \sin\phi_3)$	$3U_{\mathrm{D}} - m_2(\sin\phi_2 - \sin\phi_3)$
$\phi_4/(\omega t)$	$U_{\mathrm{D}} - m_1(\sin\phi_1 - 2\sin\phi_2 + \sin\phi_3)$	$3U_{\mathrm{D}} - m_2(\sin\phi_2 - \sin\phi_4)$
$\pi/(2\omega t)$	$U_{\mathrm{D}} - m_1(\sin\phi_1 - 2\sin\phi_2 + \sin\phi_3 + \sin\phi_4 - 1)$	$3U_{\mathrm{D}} - m_2(\sin\phi_2 - 1)$
$(\pi - \phi_4)/(\omega t)$	$U_{\mathrm{D}} - m_1(\sin\phi_1 - 2\sin\phi_2 + \sin\phi_3)$	$3U_{\mathrm{D}} - m_2(\sin\phi_2 - \sin\phi_4)$
$(\pi - \phi_3)/(\omega t)$	$U_{\mathrm{D}} - m_1(\sin\phi_1 - 2\sin\phi_2 + \sin\phi_3)$	$3U_{\mathrm{D}} - m_2(\sin\phi_2 - \sin\phi_3)$
$(\pi - \phi_2)/(\omega t)$	$U_{\mathrm{D}} - m_1(\sin\phi_1 - \sin\phi_2)$	$3U_{\mathrm{D}}$
$(\pi - \phi_1)/(\omega t)$	U_{D}	$3U_{\mathrm{D}}$
$\pi/(\omega t)$	U_{D}	$3U_{\mathrm{D}}$

Table 12.8

Prices of Capacitors with High Voltage and High Capacitance

Capacitors	Series	Prices (Europe $)
A	GMKPg 3.6 kV/1114 µF	1873
B	GMKPg 2.6 kV/4400 µF	2565
C	GMKPg 1.9 kV/4000 µF	1248
D	GMKPg 1 kV/9000 µF	943
E	GMKPg 0.9 kV/12,000 µF	1598

where ABS is the function of absolute value, MAX is the function of selecting one with the maximum value. Based on the consideration of the worst case from Table 12.4, the required capacitances C^{aI} and C^{aII} are 2580 µF and 1050 µF, respectively.

If a 9-level cascade multilevel inverter is used in such STATCOM system, the required capacitances are calculated by Equation (12.45) below for comparison. From Equation (12.45) and Table 12.4, for the worst cases, the required capacitances are 5380, 3960, 3130 and 1060 µF:

$$C^{\mathrm{a}k} = \frac{|i_{\mathrm{C}}^{\mathrm{a}}|_1(1 - \sin\phi_k)}{2\omega\varepsilon U_{\mathrm{D}}} \tag{12.45}$$

Table 12.8 shows the prices of high-voltage high-capacitance capacitors [36], and Table 12.9 shows the counts and cost of capacitors. Capacitors are connected to form an array of capacitors to satisfy required capacitance and rating voltage. The array has n rows in parallel and each row includes m capacitors connected in series. Moreover, an additional row is used to satisfy the redundancy requirement. Normally, the peak voltage and current rating of the capacitor array are chosen to be 2–3 times of the system nominal voltage. The comparison results show the cost of DC capacitors in

Table 12.9

Comparison of Counts and Costs of Capacitors

	H-bridge (DC voltage) per phase	Capacitance (mF)/rating voltage (kV) of DC capacitor	Capacitor	m series	$n+1$ parallel	Total price (Europe $)
Trinary hybrid multilevel inverter	HB (U_D)	2.58/3.3	B	3	2+1	305,253
	HB ($3U_D$)	1.05/9.9	A	6	6+1	
Cascade multilevel inverter (without balancing stresses)	HB (U_D)	5.38/3.3	B	3	4+1	333,816
	HB (U_D)	3.96/3.3	B	3	3+1	
	HB (U_D)	3.13/3.3	B	3	3+1	
	HB (U_D)	1.06/3.3	A	2	2+1	
Cascade multilevel inverter (with balancing stresses)	Four HBs (U_D)	5.38/3.3	B	3	4+1	461,700

trinary hybrid inverters which is less than that in cascade multilevel inverters in which the current and voltage stresses of switches are balanced [23] or not balanced [13]. There are two reasons why the trinary hybrid inverter has less cost of DC capacitors. Firstly, it needs less redundancy capacitors. Secondly, the capacitor CAP^{al} was both charged and discharged in quarter cycles ($0 \sim \pi/2$, $\pi/2 \sim \pi$, $\pi \sim 3\pi/2$ or $3\pi/2 \sim 2\pi$) as shown in Figure 12.14, so the required C^{al} is smaller.

12.4.2 CONTROL SYSTEM OF THE STATCOM

Figure 12.18 shows the control system of STATCOM. The power control module not only controls the reactive power but also the active power which compensates the power losses of the inverter and interface impedance. The input signals of the power control module are positive sequence components of bus voltages and inverter output currents. The function of unbalanced voltage control module is to eliminate the negative sequence components of the bus voltages, so that the bus voltages can be balanced. The reference output voltage of inverter v_C^{abc*} is the addition of the output of power control module v_P^{abc*} and the output of unbalanced voltage control module v_U^{abc*}. Each phase of the inverter is controlled separately by the inverter control module A, B or C. Inverter control modules not only control the output voltage waveform of the inverter, v_C^{abc}, but also are responsible for the balancing of each DC capacitor voltage. In the following, we first introduce vector representation and transformation of instantaneous three-phase quantities, and then specify these modules one by one.

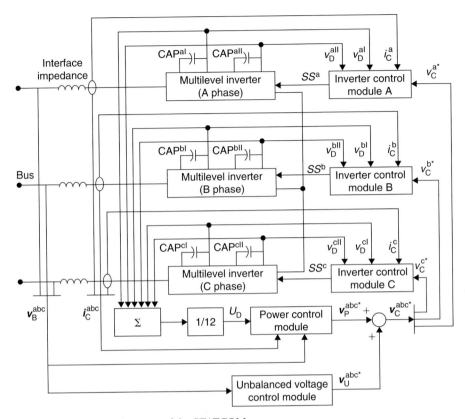

Figure 12.18 Control system of the STATCOM.

Vector Representation and Transformation of Instantaneous Three-Phase Quantities

A set of three instantaneous phase variables γ^a, γ^b and γ^c that sum to be zero can be uniquely represented in the $\alpha\beta$-phase frame through the abc $\rightarrow \alpha\beta$ transformation $[\mathbf{T}^{abc\rightarrow\alpha\beta}]$ as follows:

$$[\mathbf{T}^{abc\rightarrow\alpha\beta}] = \frac{2}{3}\begin{bmatrix} 1 & -\frac{1}{2} & -\frac{1}{2} \\ 0 & \frac{\sqrt{3}}{2} & -\frac{\sqrt{3}}{2} \\ \frac{1}{\sqrt{2}} & \frac{1}{\sqrt{2}} & \frac{1}{\sqrt{2}} \end{bmatrix} \tag{12.46}$$

The $\alpha\beta \rightarrow$ abc transformation $[\mathbf{T}^{\alpha\beta\rightarrow abc}]$ is the inverse of $[\mathbf{T}^{abc\rightarrow\alpha\beta}]$, which is defined as follows:

$$[\mathbf{T}^{\alpha\beta\rightarrow abc}] = [\mathbf{T}^{abc\rightarrow\alpha\beta}]^{-1} = \frac{3}{2}[\mathbf{T}^{abc\rightarrow\alpha\beta}]^{T} \tag{12.47}$$

Thus,

$$\begin{bmatrix} \gamma^\alpha \\ \gamma^\beta \\ 0 \end{bmatrix} = \begin{bmatrix} \mathbf{T}^{\alpha\beta \to abc} \end{bmatrix} \gamma^{abc}, \qquad \gamma^{abc} = \begin{bmatrix} \mathbf{T}^{abc \to \alpha\beta} \end{bmatrix} \begin{bmatrix} \gamma^\alpha \\ \gamma^\beta \\ 0 \end{bmatrix} \tag{12.48}$$

where

$$\gamma^{abc} = \begin{bmatrix} \gamma^a \\ \gamma^b \\ \gamma^c \end{bmatrix} \tag{12.49}$$

Furthermore, one can get $dq+$ or $dq-$ co-ordinate expressions by using the positive or negative sequence synchronous reference frame transformations $\begin{bmatrix} \mathbf{T}^{\alpha\beta \to dq+} \end{bmatrix}$ or $\begin{bmatrix} \mathbf{T}^{\alpha\beta \to dq-} \end{bmatrix}$, respectively:

$$\gamma^{dq+} = \begin{bmatrix} \mathbf{T}^{\alpha\beta \to dq+} \end{bmatrix} \gamma^{\alpha\beta}, \qquad \gamma^{\alpha\beta} = \begin{bmatrix} \mathbf{T}^{dq+ \to \alpha\beta} \end{bmatrix} \gamma^{dq+} \tag{12.50}$$

$$\gamma^{dq-} = \begin{bmatrix} \mathbf{T}^{\alpha\beta \to dq-} \end{bmatrix} \gamma^{\alpha\beta}, \qquad \gamma^{\alpha\beta} = \begin{bmatrix} \mathbf{T}^{dq- \to \alpha\beta} \end{bmatrix} \gamma^{dq-} \tag{12.51}$$

where

$$\gamma^{\alpha\beta} = \begin{bmatrix} \gamma^\alpha \\ \gamma^\beta \end{bmatrix}, \quad \gamma^{dq+} = \begin{bmatrix} \gamma^{d+} \\ \gamma^{q+} \end{bmatrix}, \quad \gamma^{dq-} = \begin{bmatrix} \gamma^{d-} \\ \gamma^{q-} \end{bmatrix} \tag{12.52}$$

$$\begin{bmatrix} \mathbf{T}^{\alpha\beta \to dq+} \end{bmatrix} = \begin{bmatrix} \cos\theta & \sin\theta \\ -\sin\theta & \cos\theta \end{bmatrix}, \qquad \begin{bmatrix} \mathbf{T}^{dq+ \to \alpha\beta} \end{bmatrix} = \begin{bmatrix} \mathbf{T}^{\alpha\beta \to dq+} \end{bmatrix}^T \tag{12.53}$$

$$\begin{bmatrix} \mathbf{T}^{\alpha\beta \to dq-} \end{bmatrix} = \begin{bmatrix} \cos\theta & -\sin\theta \\ -\sin\theta & -\cos\theta \end{bmatrix}, \qquad \begin{bmatrix} \mathbf{T}^{dq- \to \alpha\beta} \end{bmatrix} = \begin{bmatrix} \mathbf{T}^{\alpha\beta \to dq-} \end{bmatrix}^T \tag{12.54}$$

$$\theta = \int \omega t + \theta_0 \tag{12.55}$$

where θ_0 is determined by the definition of the $dq+$ co-ordinate frame.

Power Control Module

The power control module regulates the positive sequence reactive power and active power injected into the bus. Three-phase bus voltages v_B^{abc} and three-phase inverter output currents i_C^{abc} can be transformed into $v_B^{\alpha\beta}$ and $i_C^{\alpha\beta}$ in $\alpha\beta$-phase frame by Equation (12.48). The reactive power and active power can be shown as:

$$P = \frac{3}{2}(v_B^\alpha i_C^\alpha + v_B^\beta i_C^\beta), \qquad Q = \frac{3}{2}(v_B^\alpha i_C^\beta - v_B^\beta i_C^\alpha) \tag{12.56}$$

The $\mathbf{v}_B^{\alpha\beta}$ and $\mathbf{i}_C^{\alpha\beta}$ can be transformed into $\mathbf{v}_B^{dq+}$ and $\mathbf{i}_C^{dq+}$ in $dq+$ frame by Equation (12.50). The $dq+$ co-ordinate frame is defined where $d+$ axis is always coincident with the instantaneous voltage vector and the $q+$ axis is in quadrature with it, i.e.:

$$\theta = a\tan\left(\frac{v_B^\beta}{v_B^\alpha}\right) \tag{12.57}$$

Under balanced steady-state conditions:

$$\mathbf{v}_B^{dq+} = \begin{bmatrix} |\mathbf{v}_B^{abc}| \\ 0 \end{bmatrix} \tag{12.58}$$

where $|\mathbf{v}_B^{abc}|$ is the amplitude of phase voltage of the bus.

Therefore, the reactive power and active power can be expressed as:

$$P = \frac{3}{2}|\mathbf{v}_B^{abc}|i_{Cd}^+, \qquad Q = \frac{3}{2}|\mathbf{v}_B^{abc}|i_{Cq}^+ \tag{12.59}$$

In Figure 12.12, the resistance and inductance of interface impedance are expressed as R_I and L_I. From Figure 12.12, we have:

$$L_I\frac{d\mathbf{i}_C^{abc}}{dt} + R_I\mathbf{i}_C^{abc} = \mathbf{v}_C^{abc} - \mathbf{v}_B^{abc} \tag{12.60}$$

From Equations (12.48) and (12.60), we have:

$$L_I\frac{d\mathbf{i}_C^{\alpha\beta}}{dt} + R_I\mathbf{i}_C^{\alpha\beta} = \mathbf{v}_C^{\alpha\beta} - \mathbf{v}_B^{\alpha\beta} \tag{12.61}$$

From Equations (12.50), (12.57) and (12.61), we have:

$$L_I\frac{d}{dt}\begin{bmatrix} i_C^{d+} \\ i_C^{q+} \end{bmatrix} + \omega L_I\begin{bmatrix} -i_C^{q+} \\ i_C^{d+} \end{bmatrix} + R_I\begin{bmatrix} i_C^{d+} \\ i_C^{q+} \end{bmatrix} = \begin{bmatrix} v_C^{d+} - v_B^{d+} \\ v_C^{q+} - v_B^{q+} \end{bmatrix} \tag{12.62}$$

Thus, under balanced conditions, the plant of the STATCOM system can be expressed as Equation (12.63) below in the s-domain, as shown in Figure 12.19:

$$\begin{bmatrix} v_C^{d+} - v_B^{d+} + \omega L_I i_C^{q+} - R_I i_C^{d+} \\ v_C^{q+} - v_B^{q+} - \omega L_I i_C^{d+} - R_I i_C^{q+} \end{bmatrix} = \begin{bmatrix} sL_I i_C^{d+} \\ sL_I i_C^{q+} \end{bmatrix} \tag{12.63}$$

A PI controller is used for both active and reactive current control loop as shown in Figure 12.20. Under balanced conditions, the inverter can be regarded as a unit function. v_C^{d+} and v_C^{q+} in Figure 12.19 are equal to v_C^{d+*} and v_C^{q+*} in Figure 12.20, i.e.,

$$\begin{aligned} v_C^{d+} &= v_C^{d+*} \\ v_C^{q+} &= v_C^{q+*} \end{aligned} \tag{12.64}$$

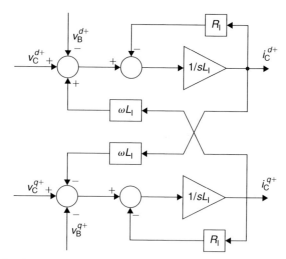

Figure 12.19   The plant of the STATCOM system in the *s*-domain.

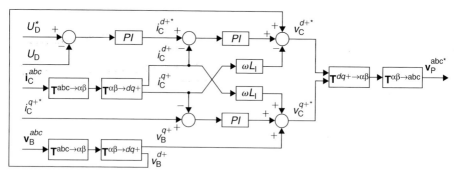

Figure 12.20   Power control module.

Figure 12.21 shows the equivalent control diagrams for i_C^{d+} and i_C^{q+}, which is derived from Figures 12.19 and 12.20, and Equation (12.64). The controlled system is reduced to a first-order transfer function [24].

Active power flowing into STATCOM will regulate DC capacitor voltages of the inverter. U_D^* is the reference value of unit voltage of DC capacitors. U_D is the unit voltage of DC capacitors and can be calculated as follows:

$$U_D = \frac{1}{12}\left(v_D^{aI} + v_D^{aII} + v_D^{bI} + v_D^{bII} + v_D^{cI} + v_D^{cII}\right) \tag{12.65}$$

The active current reference, i_C^{d+*}, is generated from a PI controller, which controls U_D. The reactive current reference, i_C^{q+*}, is given according to different compensation

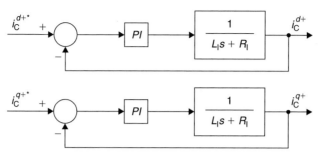

Figure 12.21 The equivalent control diagram for i_C^{d+} and i_C^{q+}.

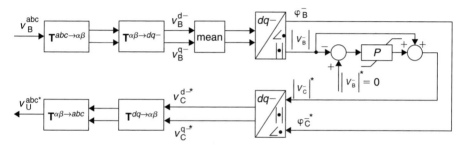

Figure 12.22 Unbalanced voltage control module.

aims. For example, for an STATCOM to compensate the reactive power of a load, it will be the load reactive current. Under balanced conditions, the $dq+$ component of the bus voltages is shown in Equation (12.58). The output of this module, v_P^{abc*}, is obtained from v_C^{d+*} and v_C^{q+*} through $\mathbf{T}^{dq+\to\alpha\beta}$ and $\mathbf{T}^{\alpha\beta\to abc}$, respectively, as previously mentioned.

Unbalanced Voltage Control Module

Assuming that the sequence components are not coupled, Figure 12.12 can be thought as separately representing either the positive or negative sequence. Considering the case of the negative sequence components and using the phasors, V_C^- represents the negative sequence component of compensator voltage generated by the STATCOM and V_B^- means the negative sequence component of bus voltage. Setting V_C^- equal to k times of V_B^-, it can be expressed as:

$$V_B^- = V_S^- \frac{Z_I Z_L}{Z_S Z_I + Z_I Z_L + (1-k)Z_S Z_L} \tag{12.66}$$

A P controller in the synchronous reference $dq-$ frame is used to produce the amplitude of $\mathbf{v}_C^-$ from the amplitude of $\mathbf{v}_B^-$ as shown in Figure 12.22. A large P can reduce the negative sequence component of the bus voltage greatly, which is derived from Equation (12.66). The output of transform $T^{\alpha\beta \to dq-}$ contains second harmonic components with frequency 100 Hz in addition to DC components. A mean function that generates the average value of input during last 0.01 s is used to eliminate the second harmonic components. Thus, in the $dq-$ frame, the regulated quantities appear as DC. When the STATCOM is used to balance the bus voltages, there is a problem that the inverter current may be over rating. Under unbalanced conditions, the output of the P controller in Figure 12.22 is a signal corresponding to the voltage drop across the STATCOM interface impedance. By limiting the value of this voltage drop, the inverter current is limited.

When the STATCOM balances the bus voltages, the negative sequence power that the inverter sends can be expressed as:

$$P^- = \frac{V_C^-(V_C^- - V_B^-)R_1}{R_1^2 + (\omega L_1)^2} \tag{12.67}$$

Since $R_1 \ll \omega L_1$, the P^- is quite small. This small deviation of DC capacitor voltages that is caused by the P^- can be balanced by ejecting or absorbing additional positive sequence power.

Inverter Control Modules

Figure 12.23 shows the inverter control module A. The operation and principle of the inverter control modules B and C are the same as that of the inverter control module A. The inverter control module A can be divided as parts A and B as shown in Figure 12.23.

The part A of Figure 12.23 addresses the issue of balancing individual capacitor voltages v_D^{aI} and v_D^{aII}. Without additional control for balancing the individual capacitor voltages, the capacitor voltages will become unequal under unbalanced conditions or during transient process. Additionally, each DC capacitor voltage may not exactly be balanced even under steady balanced conditions since inverter devices are not ideal and have different tolerance errors. Figure 12.14 shows the waveforms when the STATCOM supplied reactive power to the system.

Firstly, the second H-bridge of A-phase, HB^{aII}, is analyzed. When the output voltage of HB^{aII}, v_H^{aII}, has the same direction as i_C^a, the capacitor CAP^{aII} is discharged and vice versa. If the v_H^{aII} is shown as real line in Figure 12.14, the average charge into the capacitor CAP^{aII} over each half cycle is zero. However, if v_H^{aII} is shifted to $\Delta\delta_2$ by the dark dashed line, the charge over each half cycle can be expressed as:

$$Q^{aII} = \int_{\phi_2 + \Delta\delta_2}^{\pi - \phi_2 + \Delta\delta_2} 3U_D |i_C^a|_1 \cos\theta \, d\theta = -6U_D |i_C^a|_1 \cos\phi_2 \sin\Delta\delta_2 \tag{12.68}$$

where i_C^a is sinusoidal and $|i_C^a|_1$ is amplitude of i_C^a. Q^{aII} is proportional to $\Delta\delta_2$ when $\Delta\delta_2$ is small. So approximately Q^{aII} can be written as:

$$Q^{aII} = -6U_D |i_C^a|_1 \cos\phi_2 \cdot \Delta\delta_2 \tag{12.69}$$

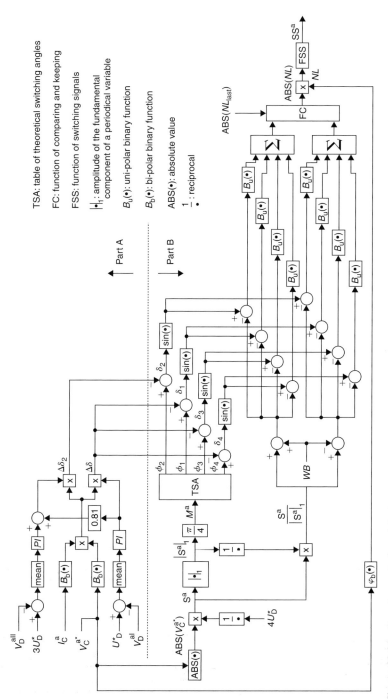

Figure 12.23 Inverter control module A.

Therefore, the capacitor voltage v_D^{all} can be controlled by slightly shifting the switching pattern. For high-power high-voltage applications, the total power loss of the inverter is less than 1%, and hence $\Delta\delta_2 \ll 0.1$ rad [13]. The shifted switching angles about HB^{all} during 0 to 2π are $\phi_2 + \Delta\delta_2$, $\pi - (\phi_2 - \Delta\delta_2)$, $\pi + (\phi_2 + \Delta\delta_2)$ and $2\pi - (\phi_2 - \Delta\delta_2)$. Suppose

$$\delta_2 = \phi_2 - B_b(v_C^a i_C^a)\Delta\delta_2 \tag{12.70}$$

where $B_b(\cdot)$ is bi-polar binary function and can be expressed as:

$$B_b(\tau) = \begin{cases} 1 & \tau > 0 \\ 0 & \tau = 0 \\ -1 & \tau < 0 \end{cases} \tag{12.71}$$

Therefore, the shifted switching angles about HB^{all} during 0 to 2π are δ_2, $\pi - \delta_2$, $\pi + \delta_2$ and $2\pi - \delta_2$.

The average charge current for CAP^{all} can be expressed as:

$$i_D^{all} = 100Q^{all} \tag{12.72}$$

The relationship between the current of CAP^{all}, i_D^{all}, and the voltage of CAP^{all}, v_D^{all}, can be expressed as:

$$i_D^{all} = C^{all}\frac{dv_D^{all}}{dt} \tag{12.73}$$

From Equations (12.69), (12.72) and (12.73), the transfer function from $\Delta\delta_2$ to v_D^{all} in s-domain can be written as:

$$\frac{v_D^{all}}{\Delta\delta_2} = \frac{k_1}{s} \quad \left(k_1 = \frac{-600U_D|i_C^a|_1\cos\phi_2}{C^{all}}\right) \tag{12.74}$$

Once the switching angles of HB^{all} is decided, the switching angles of HB^{al} will be regulated for controlling the DC capacitor voltage of HB^{al}, v_D^{al}. As shown in Figure 12.14, the switching angles of HB^{all} over the first quarter cycle is $\phi_2 + \Delta\delta_2$, so the second switching angles of HB^{al} over the first quarter cycle must be $\phi_2 + \Delta\delta_2$, otherwise the inverter will generate voltage spikes. If the switching angles of HB^{al} over the first half cycle are ϕ_1, $\phi_2 + \Delta\delta_2$, ϕ_3, ϕ_4, $\pi - \phi_4$, $\pi - \phi_3$, $\pi - \phi_2 + \Delta\delta_2$ and $\pi - \phi_1$, the charge to CAP^{al} during the first half cycle can be expressed as:

$$Q^{al'} = 4U_D|i_C^a|_1 \cos\phi_2 \sin\Delta\delta_2 \tag{12.75}$$

For balancing the CAP^{al}, other switching angles will shift slightly as shown in Figure 12.14. Then the charge to the CAP^{al} during half cycle can be expressed as:

$$Q^{al} = -2U_D|i_C^a|_1(-2\cos\phi_2\sin\Delta\delta_2 + \cos\phi_1\sin\Delta\delta_1 + \cos\phi_3\sin\Delta\delta_3$$
$$+ \cos\phi_4\sin\Delta\delta_4) \tag{12.76}$$

To shift switching angles in average, $\Delta\delta_1$, $\Delta\delta_3$ and $\Delta\delta_4$ are set equal to $\Delta\delta$. So Equation (12.76) can be rewritten as:

$$Q^{al} = -2U_D|i_C^a|_1[-2\cos\phi_2\sin\Delta\delta_2 + (\cos\phi_1 + \cos\phi_3 + \cos\phi_4)\sin\Delta\delta] \tag{12.77}$$

In s-domain, the v_D^{al} can be expressed as:

$$v_D^{al} = \frac{k_2}{s}\Delta\delta_2 + \frac{k_3}{s}\Delta\delta \tag{12.78}$$

where

$$k_2 = \frac{200U_D|i_C^a|_1\cos\phi_2}{C^{al}}, \quad k_3 = \frac{-200U_D|i_C^a|_1(\cos\phi_1 + \cos\phi_3 + \cos\phi_4)}{C^{al}} \tag{12.79}$$

PI controllers are used to regulate the DC capacitor voltages as shown in Figure 12.24. In the control loop, additional feed forward path (bold part) can enhance dynamic response. The relationship between $-k_2/k_3$ and modulation index is shown in Figure 12.25. In general, the inverter runs at the modulation index higher than 0.7. So $-k_2/k_3$ is selected as 0.81 approximately.

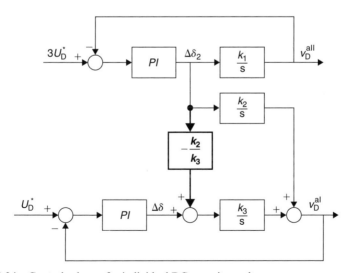

Figure 12.24 Control scheme for individual DC capacitor voltages.

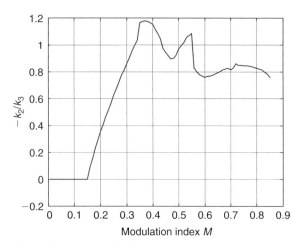

Figure 12.25   Relationship between $-k_2/k_3$ and modulation index M.

The part B of Figure 12.23 shows the main control scheme to generate desired switching signals from the reference voltages of the inverter. In this chapter and Refs [13,24], the staircase PWM is used. In Refs [13,24], only the balanced condition is considered and the control aim is just to regulate the reactive power, so the amplitude of inverter voltage can be controlled by control loop for reactive power and the phase angle of inverter voltage can be controlled by the control loop for active power. The above method cannot be applied in the STATCOM control system presented in this chapter since, in addition to regulation of reactive power, balance of bus voltages during unbalanced conditions is involved in the control aims. To achieve these aims, the reference voltages of the inverters, v_C^{abc*}, are the addition of the resulting signals of the power control module v_P^{abc*} and v_U^{abc*} as shown in Figure 12.18. Based on the reference voltages of the inverter, the switching signals are produced to control the inverter.

Under balanced conditions, the reference voltages of the inverter are quite close to pure sinusoidal waveforms since they only contain higher-order harmonic components whose amplitudes are very low. The 5th-, 7th- and 11th-order harmonics of the output voltage of the A-phase (B-phase or C-phase) inverter are nearly eliminated by the staircase modulation, so these harmonic components of the STATCOM currents and bus voltages are very small. The output voltage of the A-phase (B-phase or C-phase) inverter contains triple-order harmonic components. Under balanced conditions, the amplitudes of triple-order harmonic components of the output voltage of the A-phase inverter are the same as those of the B-phase inverter and the C-phase inverter, so triple-order harmonic components of the STATCOM currents and bus voltages do not exist with the proper connection of the STATCOM system. Amplitudes of other higher-order harmonic components of the STATCOM currents and bus voltages are very low. As stated above, the reference voltages of the inverter are the addition of resulting signals of the power control module and the unbalanced voltage control module that are fed by the STATCOM currents and the bus voltages as shown in Figure 12.18. As shown in Figure 12.22, the unbalanced voltage control module contains the mean functions that eliminate the effect of harmonic components of the STATCOM currents and the bus voltages. But the power control module does not contain them to keep high-dynamic performance. So the reference voltages of the inverter contain higher-order harmonic components whose amplitudes are very low.

Under unbalanced conditions, the reference voltages of the inverters are far from pure sinusoidal waveforms since they contain lower-order harmonic components whose amplitudes are high. Under unbalanced conditions, the amplitudes of output voltages of A-phase inverter, B-phase inverter and C-phase inverter are not identical, so the amplitudes of triple-harmonic components of these output voltages are not identical. It causes that the STATCOM currents and the bus voltages contain high triple-order harmonic components, especially the third-order harmonic components. The reference voltages of the inverter also contain high triple-order harmonic components that passed from the STATCOM currents and the bus voltages through the power control module. So the reference voltages of the inverter are far from pure sinusoidal waveforms. Without good sinusoidal reference voltages of the inverter, the 5th-, 7th- and 11th-order harmonic components of the output voltages of the A-phase, B-phase and C-phase inverters cannot be eliminated effectively by the staircase modulation. Thus, the STATCOM currents and the bus voltages under unbalanced conditions contain higher 5th-, 7th- and 11th-order harmonic components than those under balanced conditions. Therefore, the reference voltages of the inverter are far from pure sinusoidal waveforms because of not only high triple-order harmonic components but also 5th-, 7th- and 11th- order harmonic components. If mean functions or filters are added into the power control module to eliminate the lower-order harmonic components of the reference voltages of the inverter, the dynamic performance of the reactive and active power control will be worse. It is undesirable since the main purpose of STATCOM is to regulate reactive power quickly and the STATCOM works under balanced conditions at most of the time.

Therefore, a robust control method is needed in this STATCOM system, which must satisfy the following two items. Firstly, under balanced conditions, reactive power can be regulated rapidly and v_C^a do not contain 5th-, 7th- and 11th-harmonics. Secondly, under unbalanced conditions, the STATCOM can work steadily and the bus voltages can be balanced. To achieve the above aims, a new control method is proposed in this chapter. By this method, v_C^a is synthesized and satisfies the following two items. Firstly, under balanced conditions, the amplitude and phase angle of fundamental component of v_C^a are the same as those of v_C^{a*}. Moreover, v_C^a do not contain 5th, 7th and 11th harmonics. Secondly, under unbalanced conditions, v_C^a can follow the track of v_C^{a*}.

The method by which v_C^a is synthesized is shown in the part B of Figure 12.23. ABS(v_C^{a*}) is the absolute value of v_C^{a*}. The reference signal S^a equal to ABS(v_C^{a*})/($4U_D^*$). $|S^a|_1$ is the amplitude of fundamental component of S^a. From Equation (12.36), one can get modulation index M^a that is just $|S^a|_1\pi/4$. From the table of switching angles (TSA) shown in Table 12.3, the theoretical switching angles, ϕ_1 to ϕ_4, can be gained.

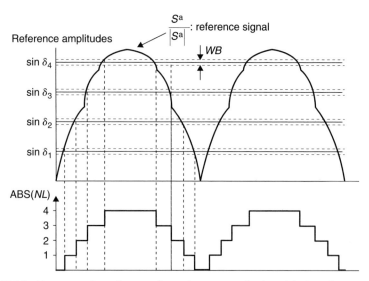

Figure 12.26   Demonstration of comparing reference amplitudes with the reference signal in the inverter control model A.

The theoretical switching angles are shifted slightly to balance individual capacitor voltages by the control loop as shown in the part A of Figure 12.23. Thus, the final switching angles are δ_1 to δ_4.

The following is the key part of the new method. The conventional method used in Refs [13,23,24] is to compare the phase angle ωt with switching angles to determine the switching states. The new method is to compare the reference signal ($S^a/|S^a|_1$) with a series of reference amplitudes (sin δ_1 to sin δ_4) as shown in Figure 12.26.

Firstly, the case in which v_C^{a*} is a perfect sinusoidal waveform is considered. In the first quarter cycle, S^a is a perfect sinusoidal waveform and can be expressed as

$$S^a = |S^a|_1 \sin(\omega t) \tag{12.80}$$

In the first quarter cycle,

$$\frac{S_a}{|S^a|_1} > \sin \delta_i \quad \Leftrightarrow \quad \omega t > \delta_i \quad (i = 1, 2, 3, 4) \tag{12.81}$$

The Equation (12.81) shows, as v_C^{a*} is perfect sinusoidal, that the new method has the same function as the conventional method by which the lower-order harmonics can be eliminated. In practice, under balanced conditions, v_C^{a*} is quite close to a sinusoidal waveform. So, under balanced conditions, the new method can also eliminate lower-order harmonics just like the conventional method.

Secondly, under unbalanced conditions, v_C^{a*} is far from a sinusoidal waveform because of lower-order harmonics. With the new method as shown in Figure 12.26, one can get v_C^a whose waveform is quite similar to that of v_C^{a*}. Moreover, the new method is more robust than the method in which the switching angles are compared with ωt that is gotten by Phase Lock Loop (PLL). Under unbalanced condition and during transient process, the increasing rate of the value of ωt gotten by PLL is not very stable, so a small deviation of this rate will result in a large deviation of comparison result. Therefore, the new method based on the comparison of amplitudes is more robust than the conventional method based on the comparison of angles under unbalanced conditions and during transient processes.

Thus, with the new method, the two aims mentioned previously are achieved. Moreover, the dead zone control is used to avoid high-frequency switching-off of switches in a short interval. In Figure 12.26, *WB* is the width of dead zone. The control system of dead zone is shown in Figure 12.23. The values of $(S^a/|S^a|_1 + WB)$ and $(S^a/|S^a|_1 - WB)$ are compared with $\sin \delta_1$ and $\sin \delta_4$, respectively. The comparison results are the inputs of $B_u(\cdot)$, which is a uni-polar binary function shown as follows:

$$B_u(\tau) = \begin{cases} 1 & \tau \geq 0 \\ 0 & \tau < 0 \end{cases} \qquad (12.82)$$

The addition results of $B_u(\cdot)$ are compared in the function of comparing and keeping, FC, as shown in Figure 12.27. Suppose *NL* is the expected level number of v_C^a and ABS(*NL*) is the absolute value of *NL*. If the two addition results are different, the FC outputs the ABS(*NL*$_{\text{last}}$) (the last values of ABS(*NL*)). If the two addition results are identical, the FC outputs this addition result. The *NL* is gotten from ABS(*NL*) and the polar of v_c^{a*}. Finally, based on Table 12.2 and definition of the switching function, the switching signals for the A-phase inverter can be gotten from *NL*. From Figure 12.26, one can see that the waveform of ABS(*NL*) will slightly shift to the right because of the dead zone, which will result in additional charge or discharge of DC capacitors. But, with the control loop shown in the part A of Figure 12.23, the DC capacitor voltages can be balanced.

12.4.3 SIMULATION RESULTS

The performance of the STATCOM system presented above has been verified under balanced and unbalance conditions by simulation. The simulation investigations were performed with MATLAB Simulink. The parameters of the distribution system and STATCOM are shown in Table 12.2. In Table 12.2, the capacitance of DC capacitors used in simulation is calculated based on Table 12.8 and Table 12.9 (redundancy capacitors are not considered). The parameters of the GTOs are shown in Table 12.10.

Figure 12.27 shows simulated waveforms under balanced conditions. The step changes of the reference signal of the reactive power, Q^*, is from 0 to 7 MVAr at

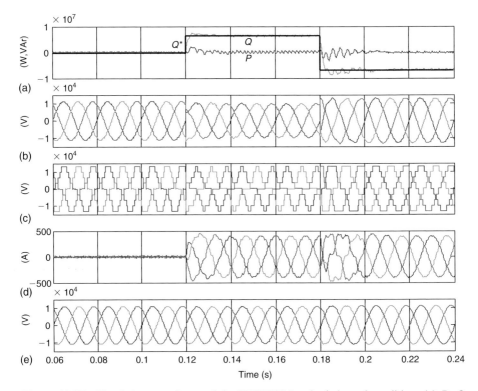

Figure 12.27 Simulation waveforms of the STATCOM under balanced condition: (a) P, Q and Q^*: active power, reactive power and reference value of reactive power; (b) v_C^{a*}, v_C^{b*} and v_C^{c*}: reference voltages of the inverter; (c) v_C^a, v_C^b and v_C^c: the output voltage of A-phase, B-phase and C-phase inverter; (d) phase current of the STATCOM and (e) line neutral voltages of the bus.

0.12 s and from 7 to -7 MVAr at 0.18 s. It is seen that the reactive power Q rapidly tracks the step-changing reference while the active power maintains zero. Complete decoupled control is achieved. As maximum rms value of output voltage of one phase of the inverter is bounded at $2\sqrt{2}\pi(4U_D^*)$, the respond speed of this system is only constrained by a practical DC voltage. Figure 12.28 shows simulated frequency spectrums of the A-phase STATCOM current, the A-phase line-neutral voltage of the bus and the reference voltage of the A-phase inverter when Q^* is -7 MVAr. In them, the triple-order harmonic components are nearly zero and the 5th-, 7th- and 11th-order harmonic components are low. The reference voltages of the inverter still contain 13th- and higher-order harmonics which may result in additional switching-off of the switches. Two methods are used to avoid the additional switching. Firstly, smaller P in PI controller as shown in Figure 12.29 will reduce the amplitude of harmonics of v_C^{abc*}. P and I are adjusted to 9 and 500, respectively, so that the inverter has enough dynamic response and the amplitude of harmonics of v_C^{abc*} is limited in appropriate

Table 12.10

Parameter of the GTO (MITSUBISHI GTO
FG1000BV-90DA)

Forward voltage	2.3 V
Turn-on resistance	0.002 Ω
Turn-on inductance	10 μH
Current falling time	10 μs
Current tail time	20 μs
Diode forward voltage	1.2 V
Diode turn-on resistance	0.0005 Ω
Snubber resistance	10 Ω
Snubber capacitor	0.7 μF

range. Furthermore, the dead zone control as shown in the part B of Figure 12.23 eliminates the effect of the harmonics. The width of dead zone, *WB*, is selected as 0.008. From Figure 12.27, one can see that voltage levels of the inverter is identical, which means the DC capacitor voltages are balanced by the control loops as shown in the part A of Figure 12.23 in which *P* and *I* are selected as 0.001 and 0.01, respectively.

Figure 12.29 shows simulated waveforms of the STATCOM during unbalanced conditions. Before 0.2 s and after 0.4 s, the source voltage is balanced. From 0.2 to 0.3 s, the source voltages are unbalanced with 0.25 per-unit negative sequence voltage components. Figure 12.29(a) and (b) shows the bus voltages without compensation and with compensation. With compensation, the bus voltage is balanced. By limiting the value of P controller in Figure 12.26, the current sent by the STATCOM is limited within the rating values as shown Figure 12.29(d).

When compensator is active, the negative sequence component of the bus voltage is reduced. The amount of compensation is subjected to the limitation of the inverter current. In the unbalanced voltage control module as shown in Figure 12.26, the *P* is selected as 10 and limitation of P controller is from −7000 to 7000. Figure 12.30 shows simulated frequency spectrums of the STATCOM currents, the bus voltages and the reference voltages of the inverter during the unbalanced conditions with compensation. The inverter is controlled well to compensate the bus voltages in spite of high lower-order harmonic components in the reference voltages of the inverter, which proves that the new method as shown in the part B of Figure 12.23 is effective. The dominant lower-order harmonic components in the STATCOM currents are the third-order harmonics, whose amplitudes are lower than 15% of the rating value of the STATCOM currents. In the worst case, the voltage ripple of a DC capacitor caused by the third-order harmonic component of the STATCOM current is less than 0.5% of the normal voltage of a DC capacitor. The effect of other harmonic components on the voltage ripple of a DC capacitor is much lower than that of the third-order harmonic. And the durations of unbalanced conditions are generally short. So the effect of harmonic components of the STATCOM currents on the voltage ripples of DC capacitors are small and transitory. The determination of

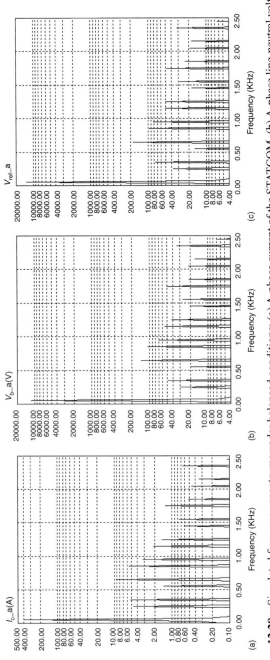

Figure 12.28 Simulated frequency spectrums under balanced conditions: (a) A-phase current of the STATCOM; (b) A-phase line-neutral voltage of the bus and (c) reference voltage of the A-phase inverter.

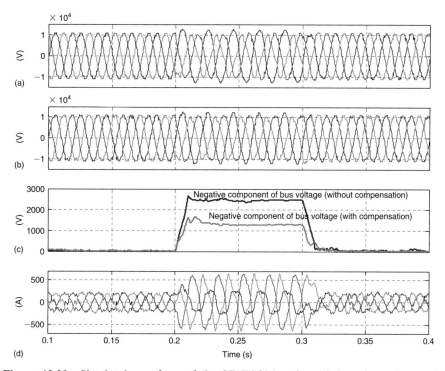

Figure 12.29 Simulated waveform of the STATCOM under unbalanced conditions: (a) line-neutral voltages of the bus without compensation, (b) line-neutral voltages of the bus with compensation, (c) negative component of the bus voltages with and without compensation and (d) phase currents of the STATCOM with compensation.

DC capacitance can still be based on the assumption of a sinusoidal current from the STATCOM.

12.4.4 EXPERIMENTAL RESULTS

To verify the performance of the proposed compensator experimentally, a hardware prototype has been built in the laboratory using the scaled system parameters as shown in Table 12.2. For the experimental system, a programmed ac source is used to represent the voltage source of the system. The STATCOM consists of a three-phase 9-level MOSFET inverter which is controlled using a TMS320F240 controlled card, and three inductances.

Figure 12.31 shows the output voltages of A-phase, B-phase and C-phase inverters and the phase currents. From 0 to 20 ms, the reference value of reactive power that the STATCOM sends is set as zero. At 20 ms, there is a step change of the reference value of reactive power from 0 to 850 VAr. At 60 ms, there is a step change of the

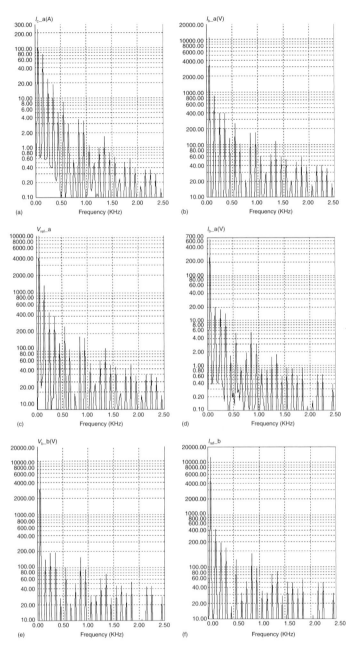

Figure 12.30 Simulated frequency spectrums under unbalanced conditions: (a) A-phase current of the STATCOM, (b) A-phase line-neutral voltage of the bus, (c) reference voltage of the A-phase inverter, (d) B-phase current of the STATCOM, (e) B-phase line-neutral voltage of the bus, (f) reference voltage of the B-phase inverter.

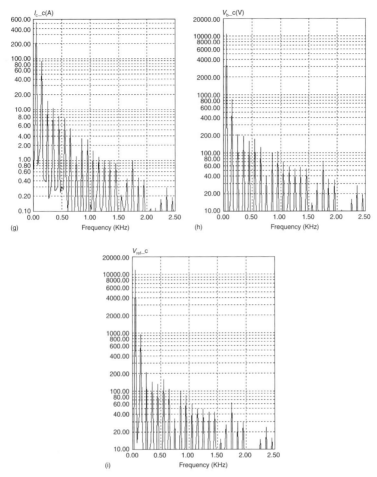

Figure 12.30 (Continued) (g) C-phase current of the STATCOM, (h) C-phase line-neutral voltage of the bus and (i) reference voltage of the C-phase inverter.

reference value from 850 to −850 VAr. The results show excellent dynamic response to the step changes.

Figure 12.32 shows the line-to-line bus voltages without compensating unbalanced voltages. Figure 12.33 shows the line-to-line bus voltages and phase currents of the STATCOM with compensation of unbalanced voltages. From 0 to 40 ms and from 120 to 200 ms, the source voltages are balanced. From 40 to 120 ms, the source voltages are unbalanced with 0.25 per-unit negative sequence voltage components added. With compensation, the bus voltages are balanced partially. The extent of compensation is constrained by the current of the STATCOM which is limited within the normal value as shown in Figure 12.33.

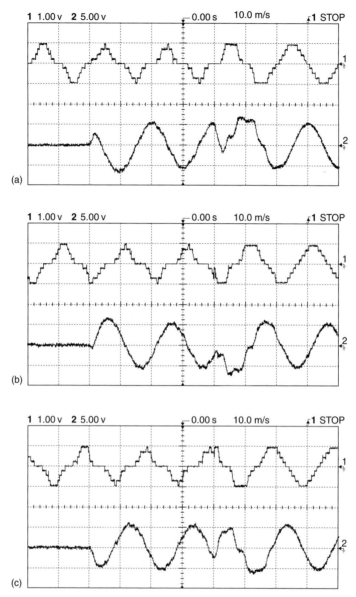

Figure 12.31 Experimental waveforms of the STATCOM under balanced conditions: from 0 to 20 ms, the reference value of reactive power that the STATCOM sends is zero; from 20 to 60 ms, the reference value is 850 VAr; from 60 to 100 ms, the reference value is −850 VAr. (a) CH1: output voltage of the A-phase inverter (200 V/div); CH2: A-phase current of the STATCOM (1 A/div). (b) CH1: output voltage of the B-phase inverter (200 V/div); CH2: B-phase current of the STATCOM (1 A/div). (c) CH1: output voltage of the C-phase inverter (200 V/div); CH2: C-phase current of the STATCOM (1 A/div).

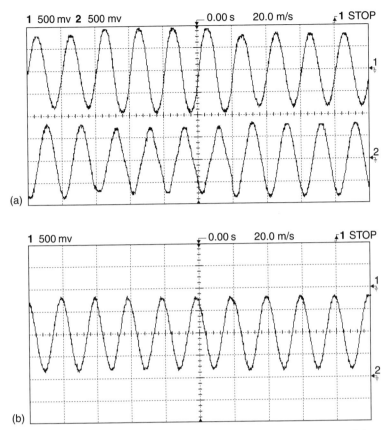

Figure 12.32   Experimental waveforms of the STATCOM under unbalanced conditions without compensation: from 0 to 40 ms and from 120 to 200 ms, the source voltages are balanced; from 40 to 120 ms, the source voltages are unbalanced. (a) CH1: AB line-to-line voltage of the bus (200 V/div); CH2: BC line-to-line voltage of the bus (200 V/div). (b) CH1: CA line-to-line voltage of the bus (200 V/div).

12.4.5 SUMMARY

This chapter investigates the application of trinary hybrid multilevel inverter in STATCOM with unbalanced voltages, which is cost-effective because of reduced cost of switching components, cooling systems and DC capacitors. The staircase modulation permits the inverter run at lower frequency. Vector control based on synchronous frame transform lead to high-dynamic performance of STATCOM. Moreover, the bus voltages are rebalanced during the unbalanced conditions and the compensation current is limited within normal values. The new method by which the switching signals are generated from the reference inverter voltages are based on the comparison of

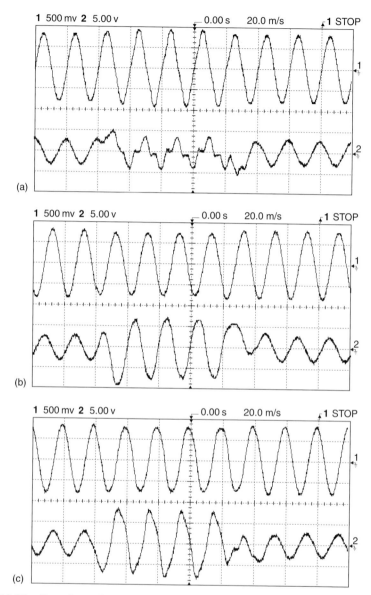

Figure 12.33 Experimental waveforms of the STATCOM under unbalanced conditions with compensation: from 0 to 40 ms and from 120 to 200 ms, the source voltages are balanced; from 40 to 120 ms, the source voltages are unbalanced (a) CH1: AB line-to-line voltage of the bus (200 V/div); CH2: A-phase current of the STATCOM (1 A/div). (b) CH1: BC line-to-line voltage of the bus (200 V/div); CH2: B-phase current of the STATCOM (1 A/div). (c) CH1: CA line-to-line voltage of the bus (200 V/div); CH2: C-phase current of the STATCOM (1 A/div).

amplitudes instead of angles. By this method, the output voltage of the inverter does not contain lower-order harmonics under stable balanced conditions and the inverter can keep high-dynamic performance under unbalanced conditions or transient processes.

FURTHER READING

1. Luo F. L. and Ye H., *Advanced DC/DC Converters*, CRC Press LLC, Boca Raton, Florida, USA, 2004. **ISBN: 0-8493-1956-0**.
2. Wu T. F. and Chen Y. K., A systematic and unified approach to modeling PWM dc/dc converters based on the graft scheme, *IEEE Trans Ind Electron*, Vol. 45, No. 1, 1998, pp. 88–98.
3. Kheraluwala M. H., Steigerwald R. L. and Gurumoorthy R. A., Fast-response high power factor converter with a single power stage, *Proc IEEE Power Electron Spec Conf*, 1991, pp. 769–779.
4. Lee Y. S. and Siu K. W., Single-switch fast-response switching regulators with unity power factor, *Proc IEEE Appl Power Electron Conf*, 1996, pp. 791–796.
5. Redl R., Balogh L. and Sokal N. O., A new family of single-stage isolated power-factor corrector with fast regulation of the output voltage, *Proc IEEE Power Electron Spec Conf*, 1994, pp. 1137–1144.
6. Luo F. L. and Ye H., *Advanced Multi-Quadrant Operation DC/DC Converters*, Taylor & Francis Group, LLC, Boca Raton, Florida, USA, 2005. **ISBN: 0-8493-7239-9**.
7. Zhao Q., Lee F. C. and Tsai F. S., Voltage and current stress reduction in single-stage power factor correction ac/dc converters with bulk capacitor voltage feedback, *IEEE Trans Power Electron*, Vol. 17, No. 4, 2002, pp. 477–484.
8. Shen M. and Qian Z., A novel high-efficiency single-stage PFC converter with reduced voltage stress, *IEEE Trans Ind Appl*, Vol. 38, No. 2, 2002, 507–513.
9. Redl R. and Balogh L., Design consideration for single-stage isolated power-factor-corrected power supplies with fast regulation of the output voltage, *Proc IEEE Power Electron Spec Conf*, 1995, pp. 454–458.
10. Rodriguez E., Canales F., Najera P. and Arau J., A novel isolated high quality rectifier with fast dynamic output response, *Proc IEEE Power Electron Spec Conf*, 1997, pp. 550–555.
11. Qiu M., Moschopoulos G., Pinheiro H. and Jain P., Analysis and design of a single stage power factor corrected full-bridge converter, *Proc IEEE Appl Power Electron Conf*, 1999, pp. 119–125.
12. Gyugyi L., Dynamic compensation of AC transmission line by solid-state synchronous voltage source, *IEEE Trans Power Deliv*, Vol. 9, No. 2, 1994, pp. 904–911.
13. Peng F. Z., Lai J. S., McKeever J. W. and VanCoevering J., A multilevel voltage-source inverter with separate DC sources for static VAr generation, *IEEE Trans Ind Appl*, Vol. 32, No. 5, 1996, pp. 1130–1138.
14. Ainsworth J. D., Davies M., Fitz P. J., Owen K.E. and Trainer D.R., Static var compensator (STATCOM) based on single-phase chain circuit converters, *IEE Proc Gener Trans Distrib*, Vol. 145, No. 4, 1998, pp. 381–386.
15. Lee C. K., Leung J. S. K., Hui S. Y. R. and Chung H. S. H., Circuit-level comparison of STATCOM technologies, *IEEE Trans Power Electron*, Vol. 18, No. 4, 2003, pp. 1084–1092.
16. Tan P. C., Loh P. C. and Holmes D. G., A robust multilevel hybrid compensation system for a 25-kV electrified railway applications, *IEEE Trans Power Electron*, Vol. 19, No. 4, 2004, pp. 1043–1052.

17. Patil K. V., Mathur R. M., Jiang J. and Hosseini S. H., Distribution system compensation using a new binary multilevel voltage source inverter, *IEEE Trans Power Deliv*, Vol. 14, No. 2, 1999, pp. 459–464.
18. Lai Y. S. and Shyu F. S., Topology for hybrid multilevel inverter, *IEE Proc Electron Power Appl*, Vol. 149, No. 6, 2002, pp. 449–458.
19. Luo F.L. and Liu Y., A new asymmetric hybrid multilevel inverter, *IEE Conf Int Power Eng Conf*, Vol. 1, 2003, pp. 311–316.
20. Rech C., Grundling H. A., Hey H. L., Pinheiro H. and Pinheiro J. R., A generalized design methodology for hybrid multilevel inverters, *IEEE Conf Ind Electron Soc Annu Conf*, Vol. 1, 2002, pp. 834–839.
21. Schauder C. and Mehta H., Vector analysis and control of advanced static VAR compensator, *IEE Proc-C Gener Trans Distrib*, Vol. 140, No. 4, 1993, pp. 299–306
22. Hochgraf C. and Lasseter R. H., Statcom controls for operation with unbalanced voltages, *IEEE Trans Power Deliv*, Vol. 13, No. 11, 1998, pp. 538–544.
23. Tolbert L. A., Peng F. Z., Cunnyngham T. and Chiasson J. N., Charge balance control schemes for cascade multilevel converter in hybrid electric vehicles, *IEEE Trans Ind Electron*, Vol. 49, No. 5, 2002, pp. 1058–1064.
24. Peng F. Z. and Lai J. S., Dynamic performance and control of a static VAr generator using cascade multilevel inverter, *IEEE Trans Ind Appl*, Vol. 33, No. 3, 1997, pp. 748–755.
25. Edwards C. W., Mattern K. E., Stacey E. J., Nannery P. R. and Gubernick J., Advanced static Var-generator employing GTO thyristors, *IEEE Trans Power Deliv*, Vol. 3, No. 2, 1998, pp. 1622–1627.
26. Cho G. C., Jung G. H., Choi N. S. and Cho G. H., Analysis and controller design of static Var compensator using three-level GTO inverter, *IEEE Trans Power Electron*, Vol. 11, No. 1, 1996, pp. 57–65.
27. Trainer D. R., Tennakoon S. B. and Morrison R. E., Analysis of GTO-based static VAr compensator, *IEE Proc Electron Power Appl*, Vol. 141, No. 6, 1994, pp. 293–302.
28. Li L., Czarkowski D. and Dzieza J., Optimal surplus harmonic energy distribution, *IEEE Ind Electron Soc Annu Conf*, Vol. 2, 1998, pp. 786–791.
29. www.mitsubishichips.com
30. Steimer P. K., Gruning H. E., Werninger J. and Schroder D., State-of-the-art verification of the hard-driven GTO inverter development for a 100-MVA intertie, *IEEE Trans Power Electron*, Vol. 13, No. 6, 1998, pp. 1182–1190.
31. Nakajima T., Suzuki K.-I., Yajima M., Kawakami N., Tanomura K.-I. and Irokawa S., A new control method preventing transformer DC magnetization for voltage source self-commutated converters, *IEEE Trans Power Deliv*, Vol. 11, No. 3, 1996, pp. 1522–1528.
32. Suzuki H., Nakajima T., Izumi K., Sugimoto S., Mino Y. and Abe H., Development and testing of prototype models for a high-performance 300 MW self-commutated AC/DC converter, *IEEE Trans Power Deliv*, Vol. 12, No. 4, 1997, pp. 1589–1601.
33. Seki N. and Uchino H., Which is better at a high power reactive power compensation system, high PWM frequency or multiple connection, *IEEE Ind Appl Soc Annu Meet*, Vol. 2, 1994, pp. 946–953.
34. Ichikawa F., Suzuki K., Nakajima T., Irokawa S. and Kitahara T., Development of self-commutated SVC for power system, *Power Convers Conf*, Yokohama, 1993, pp. 609–614.
35. Mori S., Matsuno K., Hasegawa T., Ohnishi S., Takeda M., Seto M., Murakami S. and Ishiguro F., Development of a large static VAr generator using self-commutated inverters for improving power system stability, *IEEE Trans Power Sys*, Vol. 8, No. 1, 1993, pp. 371–377.
36. www.vishay.com

Index